AMERICAN FARMING CULTURE AND THE HISTORY OF TECHNOLOGY

Presenting a history of agriculture in the American Corn Belt, this book argues that modernization occurred not only for economic reasons but also because of how farmers use technology as a part of their identity and culture.

Histories of agriculture often fail to give agency to farmers in bringing about change and ignore how people embed technology with social meaning. This book, however, shows how farmers use technology to express their identities in unspoken ways and provides a framework for bridging the current rural-urban divide by presenting a fresh perspective on rural cultural practices. Focusing on German and Jeffersonian farmers in the 18th century and Corn Belt producers in the 1920s, the Cold War, and the recent period of globalization, this book traces how farmers formed their own versions of rural modernity. Rural people use technology to contest urban modernity and debunk yokel stereotypes and women specifically employed technology to resist urban gender conceptions. This book shows how this performance of rural identity through technological use impacts a variety of current policy issues and business interests surrounding contemporary agriculture from the controversy over genetically modified organisms and hog confinement facilities to the growth of wind energy and precision technologies. Inspired by the author's own experience on his family's farm, this book provides a novel and important approach to understanding how farmers' culture has changed over time, and why machinery is such a potent part of their identity.

This book will be of great interest to students and scholars of agricultural history, technology and policy, rural studies, the history of science and technology, and the history of farming culture in the USA.

Joshua T. Brinkman, Ph.D., J.D., is Adjunct Professor of History at Elon University and Science, Technology, and Society at Virginia Tech, USA. His work examines how technology and identity shape one another as well as energy, environmental, and agricultural law and policy. He is a contributing author in the *Routledge Handbook of Energy Transitions*.

Routledge Studies in Food, Society and the Environment

For more information about this series, please visit: www.routledge.com/Routledge-Studies-in-Food-Society-and-the-Environment/book-series/RSFSE

AMERICAN FARMING CULTURE AND THE HISTORY OF TECHNOLOGY

Joshua T. Brinkman

LONDON AND NEW YORK

from Routledge

Designed cover image: © Bill Chizek, Getty

First published 2024
by Routledge
4 Park Square, Milton Park, Abingdon, Oxon OX14 4RN

and by Routledge
605 Third Avenue, New York, NY 10158

Routledge is an imprint of the Taylor & Francis Group, an informa business

British Library Cataloguing-in-Publication Data
A catalogue record for this book is available from the British Library

ISBN: 978-1-032-63794-5 (hbk)
ISBN: 978-1-032-63790-7 (pbk)
ISBN: 978-1-032-63795-2 (ebk)

DOI: 10.4324/9781032637952

Typeset in Times New Roman
by codeMantra

*To Bethany and Meg
and to my parents and Glen, Virginia, Dennis,
and Jan Brinkman*

CONTENTS

ACKNOWLEDGMENTS

In writing this book, I am indebted to Bethany Brinkman for her love and support. I could not have changed careers or had the time to write this book without her. I want to thank Meg for distracting me when I needed to be reminded of what is important in life. I also thank Susan and Doug Brinkman. Our many trips back to the Midwest as a child inspired this study and allowed me to learn about rural culture firsthand. The insights I brought to this book come, in part, from visiting my Iowa farm family, and I want to thank John L. and Alvina Howe, Glen and Virginia Brinkman, and Dennis, Ann, and Jan Brinkman for including me in their world.

I would like to thank Richard F. Hirsh for aiding me in research and writing as well as Saul Halfon, Daniel Breslau, Matthew Goodrum Ashley Shew, Barbara Allen, Sonja Schmid, and Matthew Wisnioski for helping me with academic insight that shaped my thinking and aided me in forming my arguments for this book. Kathleen Araújo has aided me in embracing and valuing my perspectives from the history of technology and science and technology studies as well as introducing my work to this press. I want to thank Routledge-Tylor & Francis Group for their guidance and encouragement as well as the journals *Technology and Culture* and *Energy Research & Social Science* for embracing my work, including the encouragement I have received from Benjamin K. Sovacool. Virginia Tech and Elon University have supported my teaching and academic endeavors, and I appreciate Daniel Breslau and Rod Clare for the teaching opportunities. Finally, I want to thank Greg D. Horstmeier from DTN/*Progressive Farmer*, Roselynn Lemm from Dotdash Media Inc./*Successful Farming*, Matthew Bormann with Bormann Ag., Joe and Casey Everett of Everett Farms, Inc., Cathi Utley with the *Herman Advertiser-Courier*, Dave Barnhouse of Dave Barnhouse Fine Art, Marianne Swan, and Frederick Courtright with the Permission Company, LLC for granting me permission to use their work and materials to discuss American farming culture.

INTRODUCTION

Posing with Metal

> The most romantic act of my Father's life, courtship aside, about which I know
> next to nothing, was to join his brother farming.
>
> David Hamilton, *Deep River*, 2001[1]

My inspiration for exploring the history of technology use in rural America began by asking one simple question: why do farmers take pictures with technology? While many people have family photos of vacations or holiday meals, farm families in the American Midwest have boxes with photos of various family members posing with machinery. My grandparents and uncle, whose ancestors have farmed for generations in Germany and northeastern Iowa, certainly had such boxes of photos, as well as framed pictures on the walls of various family members with farm equipment.[2] In fact, the first order of business any time I visited the family farm in Iowa included taking photos with different pieces of farm machinery much like families vacationing in Paris, France pose with the Eiffel Tower. Today, I still have childhood pictures of myself standing next to the newest tractor, the largest combine, and the semi-truck with the name "Brinkman Farms" painted on the side. When my wife first visited the farm, her initiation into the family included sitting on the early 1950s Farmall M my father drove as a teenager. After all, in my family, one cannot be "a true Brinkman" without sitting on the old Farmall. While the family had long ago abandoned the antiquated tractor for newer equipment, my uncle and grandfather kept it in mint condition so that various relatives could sit on it and take photos. The pictures of family members with machinery, whether of grandparents in the 1940s or uncles in the 2000s, tend to take the form of two common poses. The first type of photo, found in Figure 0.1a and 0.1b, shows the farmer posing with the latest unit of production.

DOI: 10.4324/9781032637952-1

(a)   (b)

FIGURE 0.1   A farmer in Sheridan County, Kansas, poses with his new John Deere tractor in 1939 and a farmer in Greene, Iowa posing with his newest Case International combine in Iowa in the mid-2000s. "Mr. Germeroth, Farm Security Administration client sitting on tractor which was bought by means of FSA loan, Sheridan County, Kansas," *Library of Congress*, https://www.loc.gov/pictures/item/2017740694/ (accessed 8/9/23); "Dennis Brinkman Posing with Case International Combine," Author's personal records. Greene, IA, 2000.

The second type of photo, shown in Figure 0.2a and 0.2b, involves the rest of the family standing on or adjacent to the new machinery, sometimes with the farmer himself taking the picture and, at times, with the whole family.

By posing with the latest technologies for family photos, farmers reveal their unique relationship with technology.[3] These families use tractors and combines not just to plant or harvest corn more efficiently, but to showcase their identities as modern people, to reinforce agrarian ideas about the farmers' high moral status, and to display their wealth in ways considered proper within an agrarian producer ethic.[4] Technology, when combined with the camera's gaze, constructs the agriculturalist as a modern producer and reinforces rural discourses and identities. By posing with the latest technologies, farmers combat urban "hick" stereotypes.[5] A photograph captures embodied rural ideas about moral ways to acquire and display wealth and success.[6] The Corn Belt farmer shows status by having the newest combine or planting straight corn rows, not through a "frivolous" display of wealth evidenced by his house or his clothing.[7] In practicing his identity as a modern businessperson, the farmer must do so through a culturally determined practice of signaling success through utilitarian objects of production. The tractor is a material embodiment of the self, and the photograph serves as a theatrical performance and realization of this embodiment.

(a)

(b)

FIGURE 0.2 A farm family in Greene, Iowa posing with a threshing machine in 1940 and a farm family posing with a tractor in LuVerne, Iowa in 2014. "Farmer and Family Posing with Threshing Machine." Author's personal records. Greene, IA, 1940; MNB Farms, Ltd. Webpage, http://bormannag.com/history (accessed 12/1/14).

Farm equipment also incorporates and produces cultural memory. The use of old equipment embodies past labor activities, which defined family members and relationships. The old Farmall M tractor that my grandfather kept in mint condition served the sole function of preserving memory and reinforcing identities. Photographing farm machinery both preserves memory and gives future generations a way to see how far they have come since the "olden days."

For a human construct such as identity to become real, it must be constantly performed and reinforced. Farmers perform their sense of a modern, productive, independent, and family-oriented self simply by *using* technology. Every time a farmer starts an expensive combine and engages the positioning devices and monitoring units on the console, he or she reinforces this identity and combats urban yokel stereotypes, both for himself and others. This identity reinforcement occurs through an embodied and unspoken process of using technologies. The technology represents productivity but, more importantly, modern and independent productivity. I personally recall an example of this sense of productive modernity: as a child, I remember standing next to my uncle as hundreds of bushels of corn shot into the back of a semi-truck and with him proudly exclaiming "This corn is going all over the world!"

Clearly, farmers and their families attach deep social and cultural meanings to the machinery they use. Hence, any analysis of Corn Belt agrarians and their relationship to technology must account for more than simply economics or the scientific rationalism demanded by globalized industrial agriculture. Many economists, historians, and journalists have uncritically attributed the use of new technologies in American agriculture solely to the competitiveness of farming.[8] These observers of change either neglect to explain the origins of this cutthroat economic landscape or assume that it inevitably results from material conditions. Such economic or competitive circumstances do not necessarily dictate how and why farmers use technology. Indeed, the way people use material objects and the economic conditions both have cultural dimensions. While a new combine may increase corn production and reduce fuel costs by measurable amounts, the farmer does not adopt the new technology for this reason alone. Such an economic-centered analysis overlooks important social and cultural views about technology that are deeply and historically embedded in rural farming communities.

To understand how and why people act in the world, one must understand how they use technology *to view themselves*.[9] As such, this book is from the farmer's perspective. Importantly, this sense of self is often embodied and unspoken. This book argues that people use technology not just for rational or articulable reasons or for economic gain, but to form and perform their identities in unspoken ways, a process I call *performative use*. Such performative use of technology also has real consequences in the world in terms of how people behave in a variety of domains, including debates over policy. In addition, how groups of people use technologies to form identities, and what that sense of self looks like, arises not just from material conditions, but historical and cultural influences as well. I use the Midwest farmer as a case study to demonstrate this notion of the embodied use of technology to perform identity, but all people undertake the same process on a daily basis. In the case of the American Corn Belt, attitudes towards technology among farmers and non-farming residents may be driven not only by simple economic factors, but also by a strong cultural tendency to imbue artifacts with values and ideologies, such as modernity, used for productive purposes.[10]

Users implant their identities within the technologies they use. Economists and other scholars have emphasized material conditions leading to technological change, but have sometimes missed significant cultural dimensions.[11] In the case of agriculture, several historians have provided detailed accounts of the economic and political structures in the U.S. that led to 20th-century mechanization.[12] Rather than rehash these thorough and convincing structural accounts, I seek to add to them by highlighting unexpressed identities as important factors in driving technological change.[13] This book seeks to bring insight from historians of technology and science and technology scholars exploring the social construction of technology. Social constructivism recognizes that social, cultural, political, environmental, and ideological factors shape the design and use of technology.[14] In other words, nothing predetermines the uses and developments of technologies. More specifically, my analysis builds on work by David Edgerton, Rosalind Williams, and Ruth Schwartz Cowan, who focus on the everyday use of material objects rather than on macro-level studies of large systems or on celebrated inventors.[15] But I expand upon Edgerton and Cowan's theoretical frameworks to explore how people use material objects in their daily work lives for performative purposes to establish identities and notions of modernity.[16] As Eric Schatzberg demonstrated, ideologies and symbolic meanings often guide technological choices by designers.[17] Indeed, these ideologies become fixed within the artifacts themselves. I add to this insight by suggesting that users also implant their identities within the technologies they use. Instead of portraying a world in which people choose between an old fashioned, non-mechanized existence and an industrialized and mechanized way of life, I propose a more subtle theory of modernity and use of technology that incorporates ideology and identity as essential elements.[18]

On the one hand, farmers have faced strong material pressures to adopt new technologies leveed by economic and political interests as past scholars have demonstrated.[19] Nevertheless, it is not enough to explain the adoption of new technologies through the platitude "farming is competitive," because the competition itself has resulted from the way people have formed identities in response to cultural conditions. The specific ways in which economic action is taken and technologies are used are not inevitable according to a set of material conditions, but also determined by how identity and notions of modernity are strategically formed through a social and historical process. Users of material objects do not face a simple economic decision of whether to adopt one technology or not, but many choices presenting themselves all at once. Thus, the use of technology is not rationally based on purely economic conditions, but instead, is also part of the performance of the user's identity.

People use these material objects in embodied ways that are not fully articulated, but that are still strategic. Further, while identity may form strategically, it is not necessarily the result of immediate economic conditions. For example, when interviewed during the financial crisis in the 1980s that exposed Midwest farm families to prolonged hunger, farmers in Kansas gave these reasons for not abandoning farming: "Farming is the only thing I know," and "I want to farm. I want to raise my kids on a farm."[20] Another Iowan explained, "You farm the soil yourself. You work hard and it gives you a wonderful feeling.... It's not just my job that's threatened. It's my way of life. Farming is who I am."[21] All of these farmers resisted the rational economic choice of abandoning farming for higher paying work because of the way they conceived of their identities. Surely when farmers, and other people for that matter, use technologies they do so in part for the noneconomic purpose of forming and reinforcing this sense of self.

Focusing on Technological Use and Granting Farmers Agency

In this book, I also aim to highlight the perspective of men and women on Midwest farms and give them agency in bringing about technological change. Focusing on the history of technology from a user's perspective allows scholars to deconstruct the social power relationships embedded in large technological systems.[22] Highlighting use allows historians to abandon the concept of technology for the concept of materiality, thereby broadening the scope of scholarship. For David Edgerton, the Western progress narrative constructed by 20th-century corporations as a means of selling products has created a cultural bias that causes people to focus only on the newest and most complex technologies rather than the way people use technology in their lives. By associating technology with only great inventors, Edgerton argues, history becomes a series of epochs such as "the computer age" that fails to capture how most people live. The spotlight on great inventors for Edgerton also makes technological development appear linear because it draws attention away from older material objects.[23] In short, focusing on use gets the progress narrative out of the historian's way allowing for a more honest view of how people use technology.

A focus on use also provides a window into what a Society for the History of Technology (SHOT) panel in 2015 called "diversity as method" because it allows the scholar to emphasize material use by people excluded by dominant social institutions without "naturalizing" social categories.[24] The focus on use allows historians of technology to enact Donna Haraway's appeal to abandon social dichotomies because it requires an examination of material use from the perspective of the "outsider." Use reveals how those people excluded from the dominant institutions that pursue technological development utilize objects to form identities and discourses either as forms of resistance to social power or as a way of carving out cultural "spaces."[25]

Several historians have already focused on use and incorporating diversity as method. Nina Lerman showed that exhibits of "material arts" included basket weaving by black women alongside machinery as late as the 1870s. For Lerman, the fact that these baskets later became labeled "craft" rather than "technology," suggests that technology is not even an artifact, but a social category of exclusion. The category "technology" became mapped onto existing social hierarchies such that objects resulting from the activities of those already with social power became "technology." A focus on only "great inventors" would ignore the function of technology as reinforcing social hierarchies.[26]

Other historians focusing on use have revealed a multiplicity of ideologies, identities, and discourses. Rayvon Fouché, for example, views material objects not as only suppressing racial minorities, but also as a means towards "black technological agency."[27] Fouché asserts that black people have not only been oppressed by technology, such as slave ships, but have also used material objects as "black vernacular technology" to preserve and promote their identities organized around a unique black aesthetic.[28] In focusing on use, Fouché can see technology as a site of contestation in which people co-construct material objects and racial identities, such as a DJ redesigning a conventional turntable. A study simply focusing on the invention of the turntable would view it as neutral artifact that serves one design function rather than a technology for determining the meaning of "blackness" in America and as means for social protest.

An analysis of use also promises to provide fresh insights into how farmers have constructed their identities as modern producers. Ronald Kline has studied how farmers in the Midwest in the early 20th century modified technology for their own purposes. For example, Kline shows how rural people transformed the automobile from a form of transportation to a general source of power.[29] I seek to combine Kline's focus on use with Fouché's concept of technology as a site of resistance and identity formation. Rather than expressing a vernacular, farmers use technology to form a non-dominant identity of rural modernity that incorporates traditional rural values such as independence and the importance of the nuclear family. Eric Schatzberg has shown how military engineers favored metal airplane construction in the early 20th century not for purely technical reasons, but because they associated metal with modernity.[30] In addition, David Nye has argued that electricity constituted a powerful symbol of modernity between 1885 and 1915 in the form of world's fairs, theatres, public events and electric advertising signs.[31] I suggest that using technology in ways described by Kline allows farmers to continuously perform their identities as modern and to combat urban "yokel" stereotypes.[32]

This book seeks to explain how people use technologies to develop coexisting *"discourse-identity bundles"* to construct moral selves. My view, that people use technology to perform what they regard as a moral identity, concurs with Edward Jones-Imhotep's notion of the "technological self." In his history of mid-20th century pianist Glenn Gould's obsession with recording technology, Jones-Imhotep argues that Gould worked with material objects not just to achieve a certain aesthetic but as a "moral project" to create "the kinds of people we ought to be."[33] My theory of performative use to form and reinforce bundles of discourses and identities projects Jones-Imhotep's concept of the technological self beyond the "lone wolf" individual represented by Gould, to broad groups of social actors. My theory views the technological self not as a conscious conceptualization by a single individual, but as an unarticulated identity influenced by historical and social contexts.[34]

The Rural-Urban Divide and the *Pattern of Audience*

Throughout U.S. history, a *pattern of audience* developed involving performance by agrarians for the benefit of "outsiders." According to this pattern, the audience most often regards the farmer as a less urbane "other" despite his display of wealth and productive capacity. The ritual of social elites observing an agrarian culture performing an ambiguous mix of sophistication and backwards traditionalism would become reframed in the early 20th century as a rural-urban conflict with updated bundles of identities and discourses. This pattern of audience continues today across domains from technological use to politics. Contemporary political commentators as well as recent scholarship have noted the prominence of a rural-urban conflict driving the current American political divide, but they have often treated it as a recent phenomenon related to one factor such as current right-wing populism or the presence of social media.[35] This book suggests that the current rural-urban conflict is only the most recent manifestation of a long historical pattern of rural resentment related to a much broader set of historical circumstances related to many factors, including technology and production. This book will show that this pattern of audience has existed in the U.S. since at least the 18th century.

Similarly, many traditional narratives of agricultural change portray the American farmer as clinging to Jeffersonian yeoman ideologies and resisting industrialization until New Deal policies ushered in a modern era.[36] A more careful analysis of cultural trends of technological use contained in farm journal advertisements, farmers' quotes, family farm photos, and action by farmers using artifacts reveals that many farmers had formed their own version of rural modernity much earlier. Beginning in the 1920s, using technology allowed farmers to continuously perform their identities as modern to contest negative views of rural life by urban dwellers. In doing so, some farmers could resist efforts by interests outside the farm to define *rural modernity* as *urban industrialization* and debunk "rube" stereotypes. As Joseph Frazier Wall notes in his bicentennial history of Iowa, farming especially presents a need for identity-reinforcing devices in an American cultural context:

> The Iowa farmer had always been glibly and easily characterized as to personality and attitudes by historians and political commentators as he has been in dress by cartoonists. He has been portrayed as being basically conservative, isolationist, the last exponent of true laissez-fair in economics, and intolerant of alien peoples and new ideas. Much of this characterization is false, but not easy to refute perhaps because *the farmer has had more difficulty than anyone else in our society of establishing his own identity*. He can identify with both management and labor in his aspirations and economic objectives. He is both producer and at the same time a heavy consumer of basic capital goods.
> *[emphasis added]*[37]

Wall's description presents a repeated American cultural practice of Iowa farmers using technology performatively as one means of "establishing his own identity."

Definition of "Modernity" and "Multiple Modernities"

Before describing each type of discourse-identity bundle, I must define "modernity." I adopt sociologist Anthony Giddens' theory that across cultures, modernity constituted a state of mind characterized by constant reflexivity with an emphasis on the new. This reflexivity involved basing actions on the repeated testing of knowledge rather than wholly on custom. Hence, modernity at its core sought to enact a set of shared discourses and sensibilities in the project of perfectibility of man and society to reach a kind of European utopia of "newness." This modern sensibility tends to view reason and technology as leading to personal and social progress.[38] In addition modernity associates truth with quantification.[39]

Conceptualizing modernity as a sensibility rather than a strict set of rules or definitions allows the historian to understand why a group such as Midwest farmers could develop a discourse of modernity containing traditional elements that seemed to contradict the dominant modern conception. As Avi Rubin argued in his study of the Ottoman court system, people have constructed multiple modern identities that have combined a general sensibility favoring the new, emanating from Europe, while still retaining traditional norms and values from their own cultures.[40] Farmers, for example, could both work with progressive reformers and extension systems in the early 20th century and resent these same outside reformers. Rural denizens took progressive ideas about progress and technology and re-packaged them with other rural values to form a distinctly rural form of modernity. The rural discourse that developed in the Midwest Corn Belt embraced universally modern ideas about the faith of technology, rationalism, modern business practices, and capitalist competition in bringing about progress while injecting traditional notions about morality, prosperity, and ethical ways of living with origins in Jeffersonian and German agrarianism. Farmers in the Midwest Corn Belt helped to develop a discourse based on rural identity containing seemingly opposing views and positions that differed from purely urban discourses of industrialism. This book recognizes those multiple modernities.

Discourse and Unarticulated Identities

This book uses discourse as a means of identifying and describing unarticulated cultural values. I have approached this problem by taking an expansive view of discourse that seeks to reveal broad trends in the ways people speak, act, and see the world. In doing so, one must appreciate the difference between rhetoric and discourse. An examination of rhetoric traces the use of specific words, while discourse analysis employs language in a contextual framework to more fully understand the motivations of historical actors.[41] In contrast to rhetoric, Gilbert Weiss and Ruth Wodak characterize discourse analysis as a mediation of social and linguistic theory that views language as a social practice. This approach critically analyzes speech and writing qualitatively with the view that language is both socially constructing and conditioned.[42] Put differently, discourse determines the boundaries within which people think about the world around them and, in doing so, constructs social relations and identities. Alternatively, the cultural, historical, and political contexts in which words exist also shape discourse. This reciprocal nature of discourse means that it serves an important function in producing power relations and constructing boundaries between different social groups.[43]

In focusing on discourse, the scholar aims to reveal meanings and assumptions hidden or masked by traditional interviewing focusing on articulated language.[44] Importantly, actions, such as decisions about the use of various technologies, can also constitute discourse while rhetoric limits its focus on language. One would not expect farmers to use the term "modernity" to describe why they bought a tractor any more than settlers in the 1840s would use the term "Manifest Destiny" to explain why they left their eastern homes for western territories. Commentators and historians invent these terms to understand the ideologies that drove social actors. A purely rhetorical search of settlers' journals for the term "Manifest Destiny" would likely yield little. But this lack of word use at the time does not mean that we can no longer think of those pioneers as motivated by a belief in a divine directive to expand the country.[45] Similarly, farmers in the 1920s (or even in the 2000s) did not often use the term "modernity" to describe their adoption of what we now identify as a modern state of mind. These farmers often employed the term "modern," but also terms such as "efficiency" and "progress" to express an up-to-date and future-oriented sensibility, or they simply *acted* in ways consistent with such notions.

The Six Identities Relating to Farming and Book Outline

I attempt to see the world through the *farmer's* cultural lens. My decision to focus on how the world looked to rural Americans by no means implies that others saw the world more "rationally." I have no intent of taking such a pejorative stance towards the farmer. All social actors see the world through cultural filters in which identity plays a critical role.[46] I only single out the identities and discourses motivating farmers, rather than scientists, for example, in using material objects because of my own interest in agriculture. This book seeks to describe embodied practices that occur when farmers use technologies on a daily basis rather than make claims about how "things really are" in the world.[47]

Namely, I present the history of American Midwest agriculture as experiencing *six discourse identity bundles*, each of which speak to the relationship between farmers and technology: *traditional German agrarianism, traditional Jeffersonian or English agrarianism, urban industrialism (or urban industrial modernity), rural capitalistic modernity, rural globalized ultramodernity, and organic reformist discourse*. Except for Jeffersonian agrarianism, which I borrow from other scholars, I have created these identities to conceptualize American farming culture. Social actors have developed these discourse-identity bundles historically and they exist up in the social space, not within people themselves. Once created due to historical factors, these bundles of discourse and identities exist in the social ether of rural America. Any individual can adopt these discourses and identities, but they commit the technological user to different forms of performative use.

Rural discourses and identities are not only unstable and constantly negotiated, but often conflict. These discourse-identity bundles develop within social and historical contexts include the scientific, political, technological and economic pressures exerted on farmers that previous historians have outlined.[48] As a result, rural Americans do not form one monolithic identity through technological use over time, but multiple combinations of rural discourses and self-images that form, break apart, or interact in a space of negotiation and conflict over the morality of rural ways of life.

Much as a child blows bubbles that float in the air, join, and break apart, these *discourse identity bundles* form and exist in the social space of rural America in which farmers help create and make use of them (Figures 0.3 and 0.4). Discourse identity bundles exist in the social space allowing people to answer basic questions like "how should the land look?," "what is the moral way to produce food?," and

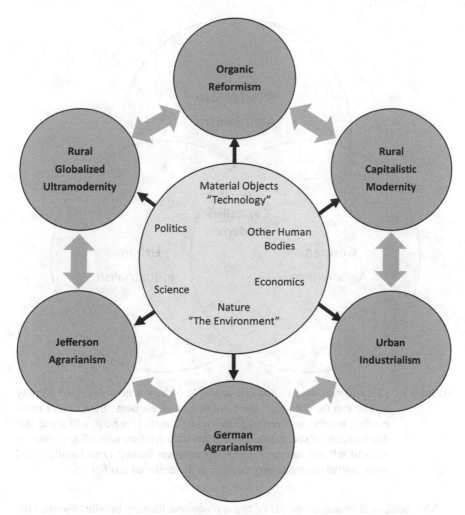

Organic
Reformism

Rural
Globalized
Ultramodernity

Rural
Capitalistic
Modernity

Material Objects
"Technology"

Politics

Other Human
Bodies

Economics

Science

Nature
"The Environment"

Jefferson
Agrarianism

Urban
Industrialism

German
Agrarianism

FIGURE 0.3 Theory of discourse identity bundling in visual form. The outside circles
contain the six discourse identity bundles encountered by farmers. The
arrows between each bundle represent sites of conflict. The large circle in
the middle represents society showing factors that shape each discourse-
identity bundle. Technological users then repeatedly perform each bundle
of discourses and identities through technological use.

"what should women and men do on a farm?" Farmers interact with technologies
as a way of forming and performing these bundles of discourses and identities.
These discourse identity bundles may hold firm, repel one another, or combine to
form new moral selves as a result of farmers' strategic responses to new realities
in their surroundings (which can include structural factors such as economics or
politics or environmental ones).

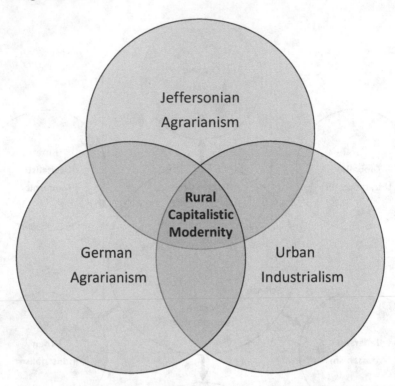

FIGURE 0.4 This diagram depicts just one example of how a new discourse identity bundle can form through the combination of elements from two or more existing bundles as a response to social factors. This book will argue that this blending of some features of urban industrialism with a few characteristics of Jeffersonian and German agrarianism formed a new bundle called rural capitalistic modernity (shown in the center of this figure).

This book will examine several of these discourse identity bundles through the lens of performative use in response to "new" realities challenging rural Americans from the 18th century to the present. These six rural discourses and identities form a rough chronological outline of this book.

Chapter 1 outlines two 18th- and 19th-century bundles of discourses and identities informing later identities in the Midwest: traditional German agrarianism and Jeffersonian agrarianism. *Traditional German agrarianism* preferred a strong production ethos that tended to view material objects, personal property, and an ordered/productive nature as moral, and it understood farming methods in terms of religion and tradition. In addition, German agrarianism contained somewhat amorphous gender roles because work and land ownership constituted values transcending sex. A virtuous marriage involved both men and women bringing land and a work ethic to a shared production process. Importantly, German agrarianism also included a strong practice of performativity through material objects. While perhaps not standing for "gender equality" in a contemporary sense, both genders performed their identities through technologies used for productive purposes and a clear gendering of house and farm labor often did not exist. While not necessarily displaying "modernity" in either its 18th- or 20th-century forms, German agrarianism at least performs success and morality through the use of artifacts. Under this discourse-identity bundle, the family and successful production practices also imbue the German farmer's way of life with morality.

Jeffersonian agrarianism arose out of traditional English notions that saw the small yeoman farmer as a frontier hero braving obstacles to fulfill a divine mission. Associated most directly with Thomas Jefferson, this discourse-identity bundle sees the farmer as occupying a mythical pastoral space, which supports and upholds both American democracy and urban industry and grants him the highest moral standing. As a result, the farmer must preserve the pastoral not in a contemporary environmentalist sense, but to prevent urban industrial encroachment. As a small yeoman preserving democracy against external danger, Jeffersonian agrarianism encourages the farmer to think and act communally with other agrarians but take an "independent" or "individualist" posture when threatened by an "other." While Jefferson himself embraced a democratized version of technology, observers often regard Jeffersonian agrarianism's commitment to the small yeoman farmer as anti-technological. Both Jeffersonian and German agrarianism view personal land ownership and family-based production practices as virtues, although Jeffersonian notions maintain stricter gender roles in which male dominance over land and family is more important than work and production. Unlike traditional German agrarians, land and capacity for production for the English Jeffersonian still passes through the oldest son. German agrarianism imbues work as a virtue whereas English farmers dreamed of ultimately avoiding manual labor.

Chapter 2 discusses how the rural-urban conflict beginning in the early 20th century included rural resentment of the growing economic, political, and cultural power of cities as well as a "rube" stereotype about farmers. While Midwest agrarians worked with progressive reformers, they also resisted a type of discourse-identity bundle I call *urban industrialism*, which they perceived among some urban voices from outside the farm. Urban industrialism sees production and technology as exclusively male domains and consumption and family as female spheres. Importantly, urban industrialism views rationality, reductionism, and efficiency as virtues that trump any of the moral values maintained by rural identities. As a result of the rural-urban conflict in the 1920s and the continuing rural-to-urban demographic shift in American society, urban industrialism sees the farmer as a backward "other" stubbornly resisting change and modernity. Rather than a hero of Jeffersonian agrarianism, the farmer becomes the "rube," "bumpkin," "hick," or "yokel." As such, modern urban actors such as academics, bankers, government agencies, and corporations must modernize agriculture by making every farm a factory. Urban industrialism promoted a view that the farm should resemble a factory characterized by centralized management, heavy mechanization, and a lack of control over work processes on the part of the farmer. Urban industrialism views the farmer as simply an uneducated input in a large, well-managed machine and technology as scientific reductionism and order.[49]

Midwest farmers both embraced elements of urban industrialism and opposed it based on a perception of urban encroachment on rural life as part of the rural-urban conflict. From rural perspectives, urban notions of modernity can violate important values of Jeffersonian and German agrarianism such as independence, ownership of personal property, and the farmer's control of a family-based production process. Rural capitalistic modernity often views urban industrialism as threatening the farmer's control over property and work. I argue that the perceived threat of urban industrialism motivated many farmers to use technology to perform an alternative rural modernity that borrowed progressive values while preserving elements of German and Jeffersonian agrarianism important to rural Midwesterners.

Chapter 3 shows how farmers used technology to develop *rural capitalistic modernity* in the rural Midwest in the early 20th century and how they both borrowed from and resisted urban industrialism. The chapter describes how the rural-urban conflict of the 1920s combined with the rise of progressive reform motivated farmers to use technology to form a uniquely rural modernity. Farm journals often served as discursive devices for developing and re-enforcing a new rural identity.[50] Rural capitalistic modernity builds on German and Jeffersonian agrarianism with some important differences. First, this identity embraced a modern sensibility including a faith in science and technology, an emphasis on the new as inherently "good," an acceptance of a reductionist approach to nature not found in notions of the "pastoral," and a desire to think rationally and economically about business practices.[51] Second, rural capitalistic modernity changed the farmer from seeing himself or herself as a communal preserver of democracy to an individual competitor who maintained true free enterprise capitalism. Farmers went from traditional community-based production practices to competing with their neighbors as individual producers resulting, in the long run, with farmers failing, thinner profit margins, and larger land and capital investment.

Third, rural capitalistic modernity sought to draw a thin and precarious line between a modern sensibility and complete urban industrialization of the pastoral and the independent, family-based, German/Jeffersonian agrarian. The farmer must maintain his or her role as resisting encroachment by an urban "other" by contesting urban versions of modernity and promoting rural ones in a world where the line between these two modern identities sometimes blurs. Modern farmers guard their control over work and technology.

Importantly, I reject the model in which a group of farmers resisted all technological change while another group embraced technology. To be sure Amish or other farmers have rejected technological change, but for a majority of farmers such a view of technology use is overly simplistic.[52] Instead, I show that starting in the 1920s, most farmers encountered many possible technologies, and they embraced some and rejected others through the lens of a dominant identity of rural modernity that arose due to historical factors in the Midwest.

Chapter 4 discusses the perspective of rural women performatively using farm technology in the American Midwest during the early 20th century and how the rural-urban conflict, for these women, had a gender component. As with German agrarianism, which sees both genders as occupying the realm of production and family, rural capitalistic modernity applies to both men and women with certain technologies becoming gendered in the Jeffersonian tradition. Men and women view themselves as modern producers with certain technologies viewed as masculine (although not necessarily an exclusively male domain). Rather than posing the all-to-common abstract black-and-white question of whether men and women were "equal" or whether a male patriarchy existed, I explore what rural women thought of themselves when using technology. From the perspective of women on farms, they too have used technology to perform a rural modern identity as producers. Again, *my interest is how users themselves perform their identities from their point of view*, including rural women. Rural capitalistic modernity concerns itself not only with combating urban stereotypes of the farmer as an anti-modern "hick," but also the view of rural women as equivalent to their urban sisters occupying a consumer role in the home. Rural women have used technology to reinforce their identities as modern producers and combat this urban conception of them as unhappy home-bound consumers.

Chapter 5 demonstrates how rural capitalistic modernity became the dominant discourse-identity bundle in the rural Midwest during the Cold War and took a nationalistic form that intertwined technology, capitalism, and patriotism.

Chapter 6 shows how farmers since the 1970s have used technology to perform a discourse-identity bundle I call ***rural globalized ultramodernity***. Rural capitalistic modernity builds on prior rural identities. It incorporates elements of rural capitalistic modernity arising out of the conflict between rural and urban conceptions of "modern" with a few significant differences. First, rural globalized ultramodernity sees the farmer as not simply equal to urban actors, but as more modern, or as the ultimate technical expert. The farmer not only uses technology that urban denizens do not use, but also designs and employs artifacts practically to produce an abundance of food. This feature of ultramodernity represents an extreme version of the embattled Jeffersonian frontier hero because the farmer sees himself as victorious in meeting the challenge of feeding a growing world population through use of the latest technologies. It is also an elevated brand of German agrarianism because it uses complex technologies and systems to demonstrate success.

Second, the farmer still suspects that urban dwellers see him or her as a backward yokel even though his or her use of technology far exceeds that of his unappreciative urban cousins. Third, rural ultramodernity abandons the nationalistic notions of rural modernity and embraces the farmer's role in a global food network. The global nature of the farmer's business accentuates his view of himself as ultramodern. His challenges no longer come exclusively from the untamed frontier of Jeffersonian myth but from a menacing global food shortage. Fourth, the farmer frames her identity in biological rather than religious terms. Namely, rather than seeing herself as an agent of God, the farmer sees herself as an "inborn innovator" who inherited modernity from her modern grandmother or grandfather. Fifth, both genders still see themselves as ultramodern and technology and ordered nature as benign means of preserving "the family farm." As such, rural globalized ultramodernity retains German agrarian conceptions of production, nature, and work as non-gendered. Corn Belt denizens resent both urban yokel and gender stereotypes.

Chapter 7 shows how urban "yokel" stereotypes have continued in multiple cultural domains despite the performative use of technology by rural Americans to express a rural modern identity. This continued stereotyping creates a sense of frustration and inferiority among farmers further driving performative use.

Chapter 8 describes how my notion of performative use and description of rural modern identities can serve as a useful tool for understanding why farmers embrace or reject new technologies. Again, the idea that rural America is divided into anti-modern farmers that reject technology and modern farmers that embrace technology is too simplistic. Rather, users accept some technologies and reject others. What matters is *why* the user accepts or rejects the technology and what that technology means to her, not just whether a particular technology was adopted or not. Seeing technological use as a performance of a particular rural identity offers a way to understand, for example, why many farmers celebrate self-driving tractors but reject large hog lots or welcome wind turbines but eschew combine software.

Chapters 9 and 10 describe the rise of *organic reformist identity* in the rural Midwest since the 1970s arguing that this bundle of discourses and identities tends to alienate many farmers. To be sure, farmers have adopted some organic and sustainable agricultural practices, but they have done so through a rural ultramodern identity. Organic reformist identity views the Jeffersonian pastoral as an environment untouched by technology and regards this version of nature and work as more virtuous than technological production. Organic reformist identity often upholds the pre-technological Jeffersonian yeoman as an idealized hero and reject rural identities that see complex technologies as preserving the family. As such, many organic reformists still use technology performatively, but they do so by employing less advanced objects that they view as more moral. Rural organic identity frequently regards complex technologies as male and immoral rather than as non-gendered and modern. In addition, it often views nature as female.

Most importantly, organic reformist discourses tend to draw a clear dichotomy between industrial and family farms in ways that other rural identities and discourses do not. For many Corn Belt denizens, organic reformist discourses update urban industrialism's view of the rural farmer as backward, but rather than viewing the farmer as anti-modern, it sees him or her as anti-contemporary and unable to grasp environmental or health concerns. For agrarians who have experienced a pattern of audience involving rube stereotypes, they assume the organic reformer sees the farmer as both too ignorant to understand the impact of industrial agriculture and too wedded to an outdated modernity characterized by a blind faith in science and technology. From the ultramodern farmer's perspective, under both organic and industrial identities, an outsider threatens his control over work. According to many Midwesterners, both of these urban discourses constitute threats from an "other" who fails to truly appreciate the life or the virtues of the "real" American farmer. Additionally, farmers in the Corn Belt often see reformers as advocating an organic reformist identity containing elements that conflict with rural versions of modernity. Rural identities and performative use among farmers influence controversies involving networks of scientific knowledge and material objects (so called "technoscience") such as the debate over genetically modified organisms (GMOs).

In summary, rural identities change over time. According to Jon Gjerde, the Midwest did not develop one monolithic identity but "composite nationality" incorporating both American and old-world European values.[53] Gjerde in his extensive social history of the Midwest in the 19th century described it as a region of "composite nationality" characterized by conflict in which immigrant traditions became reframed rather than wholly assimilated with American "Yankee" culture.[54] In keeping with Gjerde's hybrid conception, the six rural identities borrow some elements from one another while rejecting others. Further, the rural identities do not vanish but remain after spawning new versions of themselves. I acknowledge, for example, that some organic discourses presented to farmers retain the Jeffersonian pastoral ideal while rural capitalistic modernity rejects Jefferson's pastoral for a more German or urban modern tendency to order nature. These identities and discourses exist in the social milieu of rural America at the same time and farmers can perform any of them while using technology.

One can see how farmers use technology as a means of performing their identities in the memoir of English professor David Hamilton who recalled growing up on a Missouri farm in the late 1930s and early 1940s:

> It was my small theater of possibility as one winter afternoon after another, after walking home from school, I went downstairs, climbed into the tractor's seat, and steered toward that imaginary horizon. The only window was high, small, behind me to the wintery north. A gray wash of shadow surrounded my play. But for me, that gloomy wall opened to a sunswept expanse of grain. I worked long rolling fields, under sky blue to the horizon, with wheat spreading on all sides like my expectation of summer. I thought it heroic to bring all the grain to harvest. On that tractor, I farmed more acreage than my family would ever know, and night after night, my mother had to call and call again to tell me to turn out the light and come upstairs for supper.[55]

As a young boy, Hamilton had already inherited an embodied cultural practice of performance by rural Americans before he could even drive the tractor he sat on. Later in his memoir, he states poetically, "To farm is to hold onto something, and a farm is land to grasp." That "something" noted by Hamilton is not, in fact, a tangible object such as money, land, or machinery, but the farmer's sense of self. Farming in the Corn Belt involves, most importantly, grasping farmers' identities, not just their land, and technology serves as an important means of "holding on." Hamilton recognized this identity-forming function of his uncle's tractor when he wrote, "Standing on his tractor, Uncle Henry was grasping more than the wheel. He had grown up farming. He had studied agriculture at the university."[56] Hamilton then described the pride his uncle and father had at "'being his own boss,' able to make his own decisions and work through his mistakes." Hamilton's father saw the tractor as a tool to perform his identity as an independent, land-owning farmer through repeated use of the artifact itself.

This relationship between Corn Belt denizens and artifacts was not inevitable. Several cultural and historical forces in the century before Hamilton farmed his imaginary acreage shaped this performative use of technology. I now turn to my narrative of the development of performative use among farmers in the American Midwest, beginning with German agrarianism.

Notes

1 David Hamilton, *Deep River: A Memoir of a Missouri Farm* (Columbia, MO: University of Missouri Press, 2001), 150.

2 The first use of the word "Corn Belt" among farmers that I found appears in 1920 when Estelline Bennett discussed the farming methods by native Americans in an article for *Better Farming*, in which she stated, "On the Fort Bethold reservation in North Dakota many miles north of what white men consider the corn belt, the Indians still are raising the Mandan corn with the same sureness of crop that is a matter of history." Estelline Bennett, "Mandarin Corn, the Little Mother of the Great Crop," *Better Farming* 43, no. 8 (March 1920): 7, 13. For other uses of the term "Corn Belt" in the 1920s see also A.L. Haecker, "'Do I Need a Silo?' Half a Million Farmers in the U.S. Have Answered 'Yes,'" *Better Farming* 43, no. 6 (June 1920): 6.

3 I consider the Corn Belt to comprise an expansive area consisting of Ohio, Illinois, Indiana, Missouri, Iowa, Indiana, southern Minnesota and Wisconsin and eastern North Dakota, South Dakota, Nebraska, and Kansas, and northern Kentucky. J.E. Spencer and Ronald J. Horvath, "How Does an Agricultural Region Originate?" *Annals of the Association of American Geographers* 53, no. 1 (1963): 80–81.

4 See also Jonas Larsen, "Families Seen Sightseeing: Performativity of Tourist Photography," *Space and Culture* 8 (2005): 416–434.

5 Stereotypes play a crucial role in forming the dominant images that comprise cultural understandings of groups of people. For a discussion of the notion of culture as dominant images that confront people on a daily basis see Gary Downey, "What Is Engineering Studies For? Dominant Practices and Scalable Scholarship," *Engineering Studies* 1, no. 1 (2009): 55–76.

6 Phillip Auslander regards photography "as an access point to the reality of the performance." Philip Auslander, "The Performativity of Performance Documentation," *PAJ: A Journal of Performance and Art* 28, no. 3 (2006): 1–10.

7 In many ways, Corn Belt agrarians take the opposite approach from the leisure class discussed by Thorstein Veblen. Veblen notes a sense of taste developed by the upper classes for items that mark the ability to abstain from manual labor. Through a process of "conspicuous consumption," the leisure classes reinforce this sense of taste and send

signals to others of their status. Pierre Bourdieu added depth to this theory of taste by coining the term "habitus" to convey the process by which people growing up in a certain culture or class attain those habits, tastes, and ways of thinking and behaving such that they become embodied. Thorstein Veblen, *The Theory of the Leisure Class* (Mentor Books: New York 1953, originally published in 1899); Pierre Bourdieu, *Distinction: A Social Critique of the Judgment of Taste* (Cambridge, MA: Routledge, 1984), 169.

8 See for example Alan L. Olmstead, "The Mechanization of Reaping and Mowing in American Agriculture, 1833–1870," *The Journal of Economic History* 35, no. 2 (June 1975): 327–352; Paul W. Rhode, *Arresting Contagion: Science, Policy, and Conflicts over Animal Disease Control* (Cambridge, MA: Harvard University Press, 2015); Wayne D. Rasmussen, "The Impact of Technological Change on American Agriculture," *Journal of Economic History* 22 (1992): 578–599; Deborah Fitzgerald, *Every Farm a Factory: The Industrial Era in American Agriculture* (New Haven, CT: Yale University Press, 2003), 1–9; David B. Danbom, *Born in the Country: A History of Rural America* (Baltimore, MD: The Johns Hopkins University Press, 1995), 234–248; Shane Hamilton, "Agribusiness, the Family Farm, and Politics of Technological Determinism in the Post-World War II United States," *Technology and Culture* 55, no. 3 (2014): 560–590; Robert H. Wiebe, *The Search for Order, 1877–1920* (New York: Hill and Wang, 1967), viii; J.L. Anderson, *Industrializing the Corn Belt* (DeKalb: Northern Illinois University Press, 2009); Clifford B. Anderson, "The Metamorphosis of American Agrarian Idealism in the 1920s and 1930s," *Agricultural History* 35, no. 4 (1961): 182–188, 182; Danbom, "Romantic Agrarianism in Twentieth-Century America," *Agricultural History* 65, no. 4 (1991): 1–12; Tarla Rai Peterson, "Jefferson's Yeoman Farmer as Frontier Hero: A Self Defeating Mythic Structure," *Agriculture and Human Values* 7, no. 1 (1999): 9–19; Paul K. Conklin, *A Revolution Down on the Farm: The Transformation of American Agriculture Since 1929* (Lexington: University Press of Kentucky, 2008): 51–91; Carolyn Dimitri, Anne Effland and Neilson Conklin, "Economic Research Services/USDA," *The 20th Century Transformation of U.S. Agriculture/EIB3* (Washington, DC: US Government Publishing Office, 2005), 9–12; Giovanni Federico, "Not Guilty? Agriculture in the 1920s and the Great Depression," *The Journal of Economic History* 65, no. 4 (December 2005): 949–976.

9 David Edgerton, *Shock of the Old* (Oxford: Oxford University Press, 2007), ix–xviii; and Ruth Schwartz Cowan, *More Work for Mother: The Ironies of Household Technology from the Open Hearth to the Microwave* (New York: Basic Books, 1983); Mats Fridlund, "Buckets, Bollards and Bombs: Towards Subjective Histories of Technologies and Terrors," *History and Technology* 27, no. 4 (2011): 391–416; Nelly Oudshoorn and Trevor Pinch, eds., *How Users Matter: The Co-Construction of Users and Technology* (Cambridge, MA: MIT Press, 2003).

10 Many other scholars have noted the fragile nature of Corn Belt identity centered on a farming mentality" organized around a perception of shared technologies and production processes that do not actually reflect reality. See Spencer and Horvath, "How Does an Agricultural Region Originate? 81–21," William Warntz, "An Historical Consideration of the Terms 'Corn' and 'Corn Belt' in the United States," *Agricultural History* 31, no. 1 (1957): 40–45; Derwent Whittlesey, "Major Agricultural Regions of the Earth," *Annals of the Association of American Geographers* 26, no. 4 (1936): 211–212; James R. Shortridge, "The Emergence of the 'Middle West' as an American Regional Label," *Annals of the Association of American Geographers* 74, no. 2 (1984): 209–220. This conception differs from some historians who have incorrectly assumed that the region acquired the term "Corn Belt" because the corn-hog economy has in fact always dominated the area. See for example, Mark Essig, *Lesser Beasts: The Snout to Tail History of the Humble Pig* (New York: Basic Books, 2015), 153–155. Geographers as early as the 1930s have noted that the Corn Belt has implied "distinctions which have little or no geographic validity when checked on the ground." Whittlesey, "Major Agricultural Regions of the Earth." See also Douglas MacLeod, "The Corn Belt: An Exercise to Define the Limits of a Region," *Journal of Geography* 110, no. 1 (2011): 32–46.

11 The anthropologist Jane Adams, for example implies that farmers in the Corn Belt clung to traditionalism until forced to adopt new hardware as a result of economic pressures. As a result Adams' account demonstrates the reliance on exclusively economic explanations for increased mechanization, a view commonly shared view in American discourse in a wide variety of cultural, political, and economic domains. Jane Adams, *The Transformation of Rural Life* (Chapel Hill: University of North Carolina Press, 1994), 81–83.

12 For examples of these histories highlighting structural influences on technological change in American agriculture, see, Wiebe, *The Search for Order*, viii; Hamilton, "Agribusiness, the Family Farm, and Politics of Technological Determinism," 560–590; Anderson, "The Metamorphosis of American Agrarian Idealism in the 1920s and 1930s," 182; Danbom, "Romantic Agrarianism in Twentieth-Century America," 1–12; Danbom, *Born in the Country*; Peterson, "Jefferson's Yeoman Farmer as Frontier Hero," 9–19; Conklin; Dimitri, Effland, and Conklin, "Economic Research Services/ USDA," 9–12; Dennis S. Nordin and Roy V. Scott, *From Prairie Farmer to Entrepreneur: The Transformation of Midwestern Agriculture* (Bloomington: Indiana University Press, 2005), 147–178; Arnold Pacey, *Meaning in Technology* (Cambridge, MA: MIT Press, 1999).

13 The Historian Ronald Kline particularly has drawn attention to technological use by the farmers themselves in the fist half of the 20th century and has debunked the notion of the agrarian who passively accepted industrial attitudes drawn from the cities. His work, along with Debra Fitzgerald's attention to rural ideologies, opens up new possibilities to take a more nuanced look at social and cultural attitudes in the rural Midwest. Ronald R. Kline, *Consumers in the Country: Technology and Social Change in Rural America* (Baltimore, MD: Johns Hopkins University Press, 2000); Fitzgerald, *Every Farm a Factory*; Debra Fitzgerald, "Blinded by Technology: American Agriculture in the Soviet Union, 1928–1932," *Agricultural History* 70, no. 3 (1996): 459–486.

14 Wiebe E. Bijker, Thomas P. Hughes and Trevor J. Pinch, *The Social Construction of Technological Systems: New Directions in the Sociology and History of Technology* (Cambridge, MA: MIT Press, 1987); Langdon Winner, "Do Artifacts have Politics?" In *The Whale and the Reactor: A Search for Limits in an Age of High Technology* (Chicago, IL: University of Chicago Press, 1986), 19–39; Noble, David F. "Social Choice in Machine Design: The Case of Automatically Controlled Machine Tools," in *The Social Shaping of Technology*, ed. Donald MacKenzie and Judy Wajcman (Buckingham: Open University Press, 1999).

15 Edgerton, *Shock of the Old*, ix–xviii; and Cowan, *More Work for Mother*.

16 See also Rosalind Williams, *The Triumph of Human Empire* (Chicago, IL: University of Chicago Press, 2013); part of this paragraph is published in Brinkman, Joshua T. and Richard F. Hirsh, "Welcoming Wind Turbines and the PIMBY ('Please in My Backyard') Phenomenon: The Culture of the Machine in the Rural American Midwest," *Technology and Culture* 58, no. 2 (2017): 335–367.

17 Eric Schatzberg, "Ideology and Technical Choice: The Decline of the Wooden Airplane in the United States, 1920–1945," *Technology and Culture* 35 (1994): 34–69.

18 See also Joshua T. Brinkman and Richard F. Hirsh, "The Effect of Unarticulated Identities and Values on Energy Policy," in *The Handbook of Energy Transitions*, ed. Katherine Araújo (London: Routledge, 2022); Brinkman and Hirsh, "Welcoming Wind Turbines and the PIMBY ('Please in My Backyard') Phenomenon."

19 See for example, Fitzgerald, *Every Farm a Factory.*

20 Keith Schneider, "New Product on Farms in Midwest: Hunger," *New York Times* (September 29, 1987), A1, B24.

21 Andrew H. Malcolm, "Problems on Farms Take Toll on Family Life," *New York Times* (November 20, 1984), A1, A17.

22 For a discussion of how scholars should view technology as part of large technological systems, see Thomas P. Hughes, "Technological Momentum," in *Does Technology*

Drive History? The Dilemma of Technological Determinism, ed. Merritt Roe Smith and Leo Marx (Cambridge, MA: MIT Press, 1994), 101–113; Wiebe E. Bijker, Thomas P. Hughes and Trevor J. Pinch, *The Social Construction of Technological Systems*.

23 Edgerton, *Shock of the Old*, Introduction.

24 Ruth Schwartz Cowan and Francesca Bray, Organizers, "Presidential Roundtable: Diversity as Method in the History of Technology" (presentation, Society for the History of Technology Annual Meeting, Albuquerque, NM, October 9, 2015).

25 Donna Haraway, "A Cyborg Manifesto: Science, Technology, and Social-Feminism in the Late Twentieth Century," in *Simians, Cyborgs and Women: The Reinvention of Nature*, ed. Donna Haraway (New York: Routledge, 1991), Chapter 8.

26 Nina E. Lerman, "Categories of Difference, Categories of Power Bringing Gender and Race to the History of Technology," *Technology and Culture* 51, no. 4 (October 2010): 893–918.

27 Fouché uses as an example of the "black vernacular" the "blues muse," a term coined by the notable black studies scholar Guthrie Lewis, Jr. to denote an aesthetic found in early blues and up to black music in the present as a means of coping with while subtly protesting social hierarchies by creating an "othered" cultural space. Guthrie Ramsey, Jr., *Race Music: Black Cultures from Bebop to Hip Hop* (Berkeley: University of California Press, 2003).

28 Rayvon Fouché, "Say it Loud, I'm Black and I'm Proud: African Americans, American Artifactual Culture, and Black Vernacular Technological Creativity," *American Quarterly* 58 (2006): 639–661.

29 Kline, *Consumers in the Country*.

30 Eric Schatzberg, "Ideology and Technical Choice," 34–69.

31 David E. Nye, *Electrifying America: Social Meaning of a New Technology* (Cambridge, MA: MIT Press, 1992).

32 See also Brinkman and Hirsh, "Welcoming Wind Turbines and the PIMBY ('Please in My Backyard') Phenomenon."

33 Edward Jones-Imhotep, "Malleability and Machines: Glenn Gould and the Technological Self," *Technology and Culture* 57, no. 2 (2016): 287–321.

34 See Carl Elliott, *Better than Well: American Medicine Meets the American Dream* (W.W. Norton & Co.: New York, 2003), 100–127.

35 Katherine J. Cramer, *The Politics of Resentment: Rural Consciousness in Wisconsin and the Rise of Scott Walker* (Chicago, IL: University of Chicago Press, 2016); Kai A. Shchafft, "Rurality and Crises of Democracy: What Can Rural Sociology Offer the Present Moment?" *Rural Sociology* 86, no. 3 (2021): 393–418; Al Cross, "'Stop Overlooking Us!:' Missed Intersections of Trump, Media, and Rural America," in *The Trump Presidency, Journalism, and Democracy* (New York: Routledge, 2018), 231–256.

36 See, for example, see Dimitri, Effland and Conklin, "Economic Research Services/ USDA," 9–12; Wiebe, *The Search for Order*, viii; Hamilton, "Agribusiness, the Family Farm, and Politics of Technological Determinism," 560–590; Adams, *The Transformation of Rural Life*, 81–83; Anderson, "The Metamorphosis of American Agrarian Idealism in the 1920s and 1930s;" Danbom, "Romantic Agrarianism in Twentieth-Century America," 1–12; Peterson, "Jefferson's Yeoman Farmer as Frontier Hero," 9–19; Dimitri, Effland and Conklin, "Economic Research Services/USDA." This sentence is published in Brinkman and Hirsh, "Welcoming Wind Turbines and the PIMBY ('Please in My Backyard') Phenomenon."

37 Joseph Frazier Wall, *Iowa: A Bicentennial History* (New York: W.W. Norton & Company, 1978), 127–128.

38 Bruno Latour, *Science in Action* (Cambridge, MA: Harvard University Press, 1987); Ulrich Beck, *Risk Society: Towards a New Modernity* (London: Sage Publications, 1986); Scott Lash, Beck and Anthony Giddens, *Reflexive Modernization. Politics, Tradition and Aesthetics in Modern Social Order* (Cambridge: Polity Press, 1994).

39 Lorraine Daston and Peter Galison, *Objectivity* (New York: Zone Books, 2010); Theodore M. Porter, *Trust in Numbers* (Princeton, NJ: Princeton University Press, 1995).

40 Avi Rubin, *Ottoman Nizamiye Courts: Law and Modernity* (New York: Palgrave Macmillan, 2011).

41 Marianne Jørgensen and Louis Phillips, *Discourse Analysis as Theory and Method* (London: Sage, 2002), 1–2; Rodney H. Jones, "Creativity and Discourse," *World Englishes* 29, no. 4 (2010): 471–474.

42 Gilbert Weiss and Ruth Wodak, "Introduction: Theory, Interdisciplinarity and Critical Discourse Analysis," in *Critical Discourse Analysis: Theory and Interdiscilinarity*, ed. Gilbert Weiss and Ruth Wodak (New York: Palgrave Macmillan, 2003), 13; see also Stephen Bix, *Discourse and Genre* (New York: Palgrave Macmillan, 2011), 20–35.

43 Jørgensen and Phillips, *Discourse Analysis as Theory and Method*, 1–2; see also Michel Foucault, *Discipline and Punish: The Birth of the Prison* (New York: Vintage Books, 1979); Foucault, Security, *Territory, and Population: Lectures at the Collége de France 1977–1978* (New York: Palgrave Macmillan, 2007).

44 My use of discourse analysis draws on the work of Ernesto Laclau and Chantal Mouffe. See Jørgensen and Phillips, *Discourse Analysis as Theory and Method*, 7–13. Also see Jones, "Creativity and Discourse," 467–480.

45 The term, "Manifest Destiny" first appeared in articles published by John O'Sullivan in 1845 to argue in support of the annexation of Texas and U.S. claims to the Oregon territory. See Albert K. Weinberg, *Manifest Destiny: A Study of Nationalist Expansionism in American History* (Gloucester, MA: Peter Smith, 1958), 144; Richard De Zoysa, "America's Foreign Policy: Manifest Destiny or Great Satan?" *Contemporary Politics* 11, no. 2–3 (2005): 133–156.

46 For scholars theorizing identity in a variety of domains, see Elliott; John A. Livingston, "Other Selves," in *Rooted in the Land: Essays on Community and Place*, ed. William Vitek and Wes Jackson (New Haven, CT: Yale University Press, 1996), 134–139; Ramsey, Jr., *Race Music: Black Cultures from Bebop to Hip Hop*, 35–39; Benjamin B. Ringer and Elinor R. Lawless, *Race-Ethnicity and Society* (New York: Routledge, 1989); Stuart Hall, "Ethnicity: Identity and Difference," *Radical America* 23, no. 4 (1987): 9–20; George Lipsitz, *Time Passages: Collective Memory and American Popular Culture* (Minneapolis,: University of Minnesota Press, 1990); Erving Goffman, *The Presentation of Self in Everyday Life* (New York: Anchor, 1959); Amy Gutman, ed., *Multiculturalism: Examining the Politics of Identity* (Princeton, NJ: Princeton University Press, 1994); Arnold Ludwig, *How Do We Know Who We Are? A Biography of the Self* (Oxford: Oxford University Press, 1997); Charles Taylor, *The Ethics of Authenticity* (Cambridge, MA: Harvard University Press, 1991); Ann Swidler, "Culture in Action: Symbols and Strategies," *American Sociological Review* 51 (1986): 273–286; Jean L. Cohen, "Strategy or Identity: New Theoretical Paradigms and Contemporary Social Movements," *Social Research* 52 (1985): 663–667; Karen A. Cerulo, "Identity Construction: New Issues, New Directions," *Annual Review of Sociology* 23 (1997): 385–409; Hank Johnson, Enrique Larana and Joseph R. Gusfield, "Identities, Grievances and New Social Movements," in *New Social Movements: From Ideology to Identity*, ed. Larana, Johnston and Gusfield (Philadelphia, PA: Temple University Press, 1994), 3–35; Sandra Harding, *Whose Science? Whose Knowledge?: Thinking from Women's Lives* (Ithaca, NY: Cornell University Press, 1991), 272–277.

47 As Eric Schatzberg demonstrated, ideologies and symbolic meanings often guide technological choices. Indeed, these ideologies become fixed within the artifacts themselves. Schatzberg, "Ideology and Technical Choice," 34–69; see also Nye, *Electrifying America*.

48 Perhaps the most comprehensive history of these structural pressures on farmers to mechanize in the 20th century can be found in Deborah Fitzgerald, *Every Farm a Factory*.

49 As described thoroughly in Fitzgerald, *Every Farm a Factory*.
50 See John J Fry, "'Good Farming-Clear Thinking-Right Living:' Midwest Farm News-papers, Social Reform, and Rural Readers in the Early Twentieth Century," *Agricultural History* 7, no. 1 (2004): 34–49.
51 Anthony Giddens, *The Consequences of Modernity* (Stanford, CA: Stanford University Press, 1990).
52 See for example Kline, "Resisting Development, Reinventing Modernity: Rural Elec-trification in the United States before World War II," *Environmental Values* 11, no. 3 (August 2002): 327–344.
53 Jon Gjerde, *The Minds of the West: Ethnocultural Evolution in the Rural Midwest, 1830–1917* (Chapel Hill: University of North Carolina Press, 1997), 3–20.
54 Ibid., 3–20.
55 Hamilton, *Deep River*, 5–6.
56 Ibid.,155–156.

Bibliography

Adams, Jane. *The Transformation of Rural Life*. Chapel Hill: University of North Carolina Press, 1994.
Anderson, Clifford B. "The Metamorphosis of American Agrarian Idealism in the 1920s and 1930s." *Agricultural History* 35, no. 4 (1961): 182–188.
Anderson, J.L. *Industrializing the Corn Belt*. DeKalb: Northern Illinois University Press, 2009.
Auslander, Philip. "The Performativity of Performance Documentation." *PAJ: A Journal of Performance and Art* 28, no. 3 (2006): 1–10.
Bennett, Estelline. "Mandarin Corn, the Little Mother of the Great Crop." *Better Farming* 43, no. 8 (March 1920): 7, 13.
Bijker, Wiebe E., Thomas P. Hughes and Trevor J. Pinch. *The Social Construction of Tech-nological Systems: New Directions in the Sociology and History of Technology*. Cam-bridge, MA: MIT Press, 1987.
Bourdieu, Pierre. *Distinction: A Social Critique of the Judgment of Taste*. Cambridge, MA: Routledge, 1984.
Brinkman, Joshua T. and Richard F. Hirsh. "Welcoming Wind Turbines and the PIMBY ('Please in My Backyard') Phenomenon: The Culture of the Machine in the Rural Ameri-can Midwest." *Technology and Culture* 58, no. 2 (2017): 335–367.
———. "The Effect of Unarticulated Identities and Values on Energy Policy." In *The Hand-book of Energy Transitions*, edited by Katherine Araújo, 71–85. London: Routledge, 2022.
Cerulo, Karen A. "Identity Construction: New Issues, New Directions." *Annual Review of Sociology* 23 (1997): 385–409.
Cohen, Jean L. "Strategy or Identity: New Theoretical Paradigms and Contemporary Social Movements." *Social Research* 52 (1985): 663–667.
Conklin, Paul K. *A Revolution Down on the Farm: The Transformation of American Agri-culture Since 1929*. Lexington: University Press of Kentucky, 2008.
Cowan, Ruth Schwartz. *More Work for Mother: The Ironies of Household Technology from the Open Hearth to the Microwave*. New York: Basic Books, 1983.
Cowan, Ruth Schwartz and Francesca Bray, Organizers. "Presidential Roundtable: Diversity as Method in the History of Technology." Presentation, Society for the History of Tech-nology Annual Meeting, Albuquerque, NM, October 9, 2015.

Cramer, Katherine J. *The Politics of Resentment: Rural Consciousness in Wisconsin and the Rise of Scott Walker*. Chicago, IL: University of Chicago Press, 2016.

Cross, Al. "Stop Overlooking Us!:' Missed Intersections of Trump, Media, and Rural America." In *The Trump Presidency, Journalism, and Democracy*, 231–256. New York: Routledge, 2018.

Danbom, David B. "Romantic Agrarianism in Twentieth-Century America." *Agricultural History* 65, no. 4 (1991): 1–12.

———. *Born in the Country: A History of Rural America*. Baltimore, MD: The Johns Hopkins University Press, 1995.

Daston, Lorraine and Peter Galison. *Objectivity*. New York: Zone Books, 2010.

"Dennis Brinkman Posing with Case International Combine." Author's personal records. Greene, IA, 2000.

De Zoysa, Richard. "America's Foreign Policy: Manifest Destiny or Great Satan?" *Contemporary Politics* 11, no. 2–3 (2005): 133–156.

Dimitri, Carolyn, Anne Effland and Neilson Conklin. "Economic Research Services/USDA." In *The 20th Century Transformation of U.S. Agriculture/EIB3*, 1–12. Washington, DC: US Government Publishing Office, 2005.

Downey, Gary. "What Is Engineering Studies For? Dominant Practices and Scalable Scholarship." *Engineering Studies* 1, no. 1 (2009): 55–76.

Edgerton, David. *Shock of the Old: Technology and Global History Since 1900*. Oxford: Oxford University Press, 2007.

Elliott, Carl. *Better than Well: American Medicine Meets the American Dream*. New York: W.W. Norton & Co., 2003.

Essig, Mark. *Lesser Beasts: The Snout to Tail History of the Humble Pig*. New York: Basic Books, 2015.

"Farmer and Family Posing with Threshing Machine." Author's personal records. Greene, IA, 1940.

Federico, Giovanni. "Not Guilty? Agriculture in the 1920s and the Great Depression." *The Journal of Economic History* 65, no. 4 (December 2005): 949–976.

Fitzgerald, Deborah. "Blinded by Technology: American Agriculture in the Soviet Union, 1928–1932." *Agricultural History* 70, no. 3 (1996): 459–486.

———. *Every Farm a Factory: The Industrial Era in American Agriculture*. New Haven, CT: Yale University Press, 2003.

Foucault, Michel. *Discipline and Punish: The Birth of the Prison*. New York: Vintage Books, 1979.

———. *Security, Territory, and Population: Lectures at the Collége de France 1977–1978*. New York: Palgrave Macmillan, 2007.

Fouché, Rayvon. "Say it Loud, I'm Black and I'm Proud: African Americans, American Artifactual Culture, and Black Vernacular Technological Creativity." *American Quarterly* 58 (2006): 639–661.

Fridlund, Mats. "Buckets, Bollards and Bombs: Towards Subjective Histories of Technologies and Terrors." *History and Technology* 27, no. 4 (2011): 391–416.

Fry, John J. "'Good Farming-Clear Thinking-Right Living:' Midwest Farm Newspapers, Social Reform, and Rural Readers in the Early Twentieth Century." *Agricultural History* 7, no. 1 (2004): 34–49.

Giddens, Anthony. *The Consequences of Modernity*. Stanford, CA: Stanford University Press, 1990.

Gjerde, Jon. *The Minds of the West: Ethnocultural Evolution in the Rural Midwest, 1830–1917*. Chapel Hill: University of North Carolina Press, 1997.

Goffman, Erving. *The Presentation of Self in Everyday Life*. New York: Anchor, 1959.

Gutman, Amy, ed. *Multiculturalism: Examining the Politics of Identity*. Princeton, NJ: Princeton University Press, 1994.

Haecker, A.L. "'Do I Need a Silo?' Half a Million Farmers in the U.S. Have Answered 'Yes.'" *Better Farming* 43, no. 6 (June 1920): 6.

Hall, Stuart. "Ethnicity: Identity and Difference." *Radical America* 23, no. 4 (1987): 9–20.

Hamilton, David. *Deep River: A Memoir of a Missouri Farm*. Columbia: University of Missouri Press, 2001.

Hamilton, Shane. "Agribusiness, the Family Farm, and Politics of Technological Determinism in the Post-World War II United States." *Technology and Culture* 55, no. 3 (2014): 560–590.

Haraway, Donna. "A Cyborg Manifesto: Science, Technology, and Social-Feminism in the Late Twentieth Century." In *Simians, Cyborgs and Women: The Reinvention of Nature*, edited by Donna Haraway, 149–181. New York: Routledge, 1991.

Harding, Sandra. *Whose Science? Whose Knowledge?: Thinking from Women's Lives*. Ithaca, NY: Cornell University Press, 1991.

Hughes, Thomas P. "Technological Momentum." In *Does Technology Drive History? The Dilemma of Technological Determinism*, edited by Merritt Roe Smith and Leo Marx, 101–113. Cambridge, MA: MIT Press, 1994.

Johnson, Hank, Enrique Larana and Joseph R. Gusfield. "Identities, Grievances and New Social Movements." In *New Social Movements: From Ideology to Identity*, edited by Larana, Johnston and Joseph R. Gusfield, 3–35. Philadelphia, PA: Temple University Press, 1994.

Jones, Rodney H. "Creativity and Discourse." *World Englishes* 29, no. 4 (2010): 467–480.

Jones-Imhotep, Edward. "Malleability and Machines: Glenn Gould and the Technological Self." *Technology and Culture* 57, no. 2 (2016): 287–321.

Jørgensen, Marianne and Louis Phillips. *Discourse Analysis as Theory and Method*. London: Sage, 2002.

Kline, Ronald R. *Consumers in the Country: Technology and Social Change in Rural America*. Baltimore, MD: Johns Hopkins University Press, 2000.

———. "Resisting Development, Reinventing Modernity: Rural Electrification in the United States before World War II." *Environmental Values* 11, no. 3 (August 2002): 327–344.

Larsen, Jonas. "Families Seen Sightseeing: Performativity of Tourist Photography." *Space and Culture* 8 (2005): 416–434.

Lash, Scott, Ulrich Beck and Anthony Giddens. *Reflexive Modernization. Politics, Tradition and Aesthetics in Modern Social Order*. Cambridge: Polity Press, 1994.

Latour, Bruno. *Science in Action*. Cambridge, MA: Harvard University Press, 1987.

Lerman, Nina E. "Categories of Difference, Categories of Power Bringing Gender and Race to the History of Technology." *Technology and Culture* 51, no. 4 (October 2010): 893–918.

Lipsitz, George. *Time Passages: Collective Memory and American Popular Culture*. Minneapolis: University of Minnesota Press, 1990.

Livingston, John A. "Other Selves." In *Rooted in the Land: Essays on Community and Place*, edited by William Vitek and Wes Jackson, 134–139. New Haven, CT: Yale University Press, 1996.

Ludwig, Arnold. *How Do We Know Who We Are? A Biography of the Self*. Oxford: Oxford University Press, 1997.

MacLeod, Douglas. "The Corn Belt: An Exercise to Define the Limits of a Region." *Journal of Geography* 110, no. 1 (2011): 32–46.

Malcolm, Andrew H. "Problems on Farms Take Toll on Family Life." *New York Times*, November 20, 1984, A1, A17.

MNB Farms, Ltd. Webpage, http://bormannag.com/history (accessed 12/1/14).

"Mr. Germeroth, Farm Security Administration Client Sitting on Tractor Which Was Bought by Means of FSA Loan, Sheridan County, Kansas," *Library of Congress*. https://www.loc.gov/pictures/item/2017740694/ (accessed 8/9/23).

Noble, David F. "Social Choice in Machine Design: The Case of Automatically Controlled Machine Tools." In *The Social Shaping of Technology*, edited by Donald MacKenzie and Judy Wajcman, 161–176. Buckingham: Open University Press, 1999.

Nordin, Dennis S. and Roy V. Scott. *From Prairie Farmer to Entrepreneur: The Transformation of Midwestern Agriculture*. Bloomington: Indiana University Press, 2005.

Nye David E. *Electrifying America: Social Meaning of a New Technology*. Cambridge, MA: MIT Press, 1992.

Olmstead, Alan L. "The Mechanization of Reaping and Mowing in American Agriculture, 1833–1870." *The Journal of Economic History* 35, no. 2 (June 1975): 327–352.

Oudshoorn, Nelly and Trevor Pinch, eds. *How Users Matter: The Co-Construction of Users and Technology*. Cambridge, MA: MIT Press, 2003.

Pacey, Arnold. *Meaning in Technology*. Cambridge, MA: MIT Press, 1999.

Peterson, Tarla Rai. "Jefferson's Yeoman Farmer as Frontier Hero: A Self Defeating Mythic Structure." *Agriculture and Human Values* 7, no. 1 (1999): 9–19.

Ramsey, Jr., Guthrie. *Race Music: Black Cultures from Bebop to Hip Hop*. Berkeley: University of California Press, 2003.

Rasmussen, Wayne D. "The Impact of Technological Change on American Agriculture." *Journal of Economic History* 22 (1992): 578–599.

Rhode, Paul W. *Arresting Contagion: Science, Policy, and Conflicts over Animal Disease Control*. Cambridge, MA: Harvard University Press, 2015.

Ringer, Benjamin B. and Elinor R. Lawless. *Race-Ethnicity and Society*. New York: Routledge, 1989.

Rubin, Avi. *Ottoman Nizamiye Courts: Law and Modernity*. New York: Palgrave Macmillan, 2011.

Schatzberg, Eric. "Ideology and Technical Choice: The Decline of the Wooden Airplane in the United States, 1920–1945." *Technology and Culture* 35, no. 1 (1994): 34–69.

Schneider, Keith. "New Product on Farms in Midwest: Hunger." *New York Times*, September 29, 1987, A1, B24.

Shchafft, Kai A. "Rurality and Crises of Democracy: What Can Rural Sociology Offer the Present Moment?" *Rural Sociology* 86, no. 3 (2021): 393–418.

Shortridge, James R. "The Emergence of the 'Middle West' as an American Regional Label." *Annals of the Association of American Geographers* 74, no. 2 (1984): 209–220.

Spencer, J.E. and Ronald J. Horvath. "How Does an Agricultural Region Originate?" *Annals of the Association of American Geographers* 53, no. 1 (1963): 74–92.

Swidler, Ann. "Culture in Action: Symbols and Strategies." *American Sociological Review* 51 (1986): 273–286.

Taylor, Charles. *The Ethics of Authenticity*. Cambridge, MA: Harvard University Press, 1991.

Veblen, Thorstein. *The Theory of the Leisure Class*. New York: Mentor Books, 1953, originally published in 1899.

Wall, Joseph Frazier. *Iowa: A Bicentennial History*. New York: W.W. Norton & Company, 1978.

Warntz, William. "An Historical Consideration of the Terms 'Corn' and 'Corn Belt' in the United States." *Agricultural History* 31, no. 1 (1957): 40–45.

Weinberg, Albert K. *Manifest Destiny: A Study of Nationalist Expansionism in American History.* Gloucester, MA: Peter Smith, 1958.

Weiss, Gilbert and Ruth Wodak. "Introduction: Theory, Interdisciplinarity and Critical Discourse Analysis." In *Critical Discourse Analysis*, edited by Gilbert Weiss and Ruth Wodak, 1–32. New York: Palgrave Macmillan, 2003.

Whittlesey, Derwent. "Major Agricultural Regions of the Earth." *Annals of the Association of American Geographers* 26, no. 4 (1936): 211–212.

Wiebe, Robert H. *The Search for Order, 1877–1920*. New York: Hill and Wang, 1967.

Williams, Rosalind. *The Triumph of Human Empire*. Chicago, IL: University of Chicago Press, 2013.

Winner, Langdon. "Do Artifacts have Politics?" In *The Whale and the Reactor: A Search for Limits in an Age of High Technology*, edited by Langdon Winner, 19–39. Chicago, IL: University of Chicago Press, 1986.

1

SETTING THE STAGE

The Genealogy of Contemporary Rural Identity in the Midwest

> Agriculture is, perhaps, of all the useful arts that which improves most slowly amongst democratic nations. Frequently, indeed, it would seem to be stationary, because other arts are making rapid strides towards perfection. On the other hand, almost all the tastes and habits which the equality of condition engenders naturally lead men to commercial and industrial occupations.
>
> *Alexis de Tocqueville*, Democracy in America, *1840*[1]

Growing up in the suburbs of Washington, D.C., few of my friends had even seen a "real" farm. These eastern urbanites vocally looked down on rural Americans as backward. In the summers, many of my classmates would travel to Europe, the Caribbean, and other exotic places. I went to Iowa or Missouri. In a family with a father who became the first son to ever leave the farm and a mother with trepidations about air travel, we actually *drove* to Iowa or Missouri. The family packed into a car and for hours and passed field after flat field of corn and soybeans from eastern Ohio to the middle of America. When we finally turned off the main highway and onto the road to my grandparents' and uncle's Iowa farms, the landscape did not change although my boredom had started to evolve into restrained anticipation as I looked out over the horizon. Finally, I received the annual announcement I had waited for over two long days of monotony when my father said in a clear voice with pride, "Son, see that field? That's Brinkman land!" While my father meant the phrase to carry a hint of sarcasm, I must admit that the landscape literally changed for me. Rows and rows of drab green corn on a flat landscape became bucolic plants majestically waving in the wind to announce a clear blue sky above. I no longer saw corn, I imagined my grandfather starting out in his early twenties with oxen chasing the American dream, my father driving the old Farmall M as

DOI: 10.4324/9781032637952-2

a teenager with the whole world in front of him. I saw everything I was taught I should be proud to be.

Of course, the field was, in reality, no different than the rows of corn in eastern Ohio (hence my dad's joke), but it looked different to me. My perceptions were confirmed over the next few weeks of my "vacation" when I worked on the farm. My grandfather had moved on from the oxen and the Farmall M to huge combines and tractors guided by global positioning systems, but the ox harness was still in the barn and the Farmall still parked in the machine shed in pristine condition so visitors could drive it, imagine what it was like in the "old days," and marvel at the fact that it still ran. Upon arrival, the family had to thoroughly inspect the car that we drove to Iowa. Years later when I could drive my own car to the farm, I felt ashamed to arrive in a Honda thinking that my conservative grandfather would look down on an imported car. To my surprise he met me with "Honda, those are great cars. I have one of those, too." When I pressed further and asked him why he drove a Japanese car he responded proudly, "They buy my corn."

Arriving at the farm where my dad grew up, I delighted in walking through the huge dairy barn, no longer in use, with my father who explained how he worked alongside his father and brother as a child.

The cows were here in front of this trough and we would pull this trip rope and bales of hay would fall into the hay mow. Here is the water tank where we kept the milk cool. The cows always kept it warm in here even in the winter.

Following the barn tour, I would go to the huge machine shed to sit on the most recent Case International tractors and combines. The door would open to reveal these huge beautiful red machines that seemed poised as if ready to leap into action. An assuring aroma of old corn and hay filled the air. I would climb up the side of the largest combine I could find and open the door. To a 12-year-old, the inside of the cab looked like something out of Star Wars. An impressive array of buttons, lights, levers, screens, and gadgets surrounded a comfortable leather chair with arms. The whole cabin leaned forward, looking out through a huge slanted window in a way that made you feel like a god perched on top of the world. The same experience came when climbing into the enormous tractors.

I would ask my dad "How many acres do they farm, dad?" and he would boast "about 1,500." My uncle would invariably arrive in a large semi-truck with "Brinkman Farms, not for hire" painted on the side and brimming to the top with corn. Sometimes, he arrived with an empty truck and would pull up to the large grain bins to the right of the barn and pull a corn auger into position. In this case, we climbed the ladder on the side of the semi-truck and watched a steady yellow stream of corn shoot into the enormous truck bed and think, "this corn is going all over the world!" We knew the racing bushels above our heads were destined for Asia or Europe because they had already been sold as futures before they were even planted. While I knew the corn kernels had grown as a plant, I saw them more as dollar bills than natural objects and the hum of the auger seemed to groan "money, money, money...." Indeed, dipping my hand into the warm pile of grain and letting it flow through my fingers, it amazed me how uniform and identical each kernel was to the point that they seemed more like mass-produced manufactured artifacts than anything produced by nature. Again, I felt proud that a few of us Brinkmans could produce so much so "perfectly" and feed people all over the globe. Yet, when I returned home to suburban Washington, D.C. and grocery store clerks found out we moved there from Kansas City we would get questions like, "did you have electricity?" and "I love the Wizard of Oz, what is it like to be in a tornado?"

Compared to my suburban surroundings back east, everything was big, loud, manly, and, somehow more "real." I had thoughts of my family emigrating from Germany with nothing but the clothes on their backs (at least that's what the narrative said) and my grandfather starting out with only a few acres and an ox. My imagination turned to a heroic story about my great, great grandfather who faced persecution as a "Volga German" forced to live in Russia and escaped on a bicycle to seek freedom in America just before a Bolshevik take-over. "Look how far we've come," I thought. While I did not own a single piece of farm equipment, these monuments of steel somehow made me feel good about myself. I felt as if I had come from a long line of tenacious scrappers who made it, a self-image I confirmed with several days of what many would consider unglamorous and back-breaking labor. I picked up large rocks in a field behind one of these dust-producing machines in the middle of the summer, shoveled corn into a dangerous auger in a hazy corn bin, and cleaned out a pigsty in the frigid Iowa December. While my classmates played video games back home, I shoveled corn next to an auger that would literally tear my leg off had I made a wrong step. Yet, these dirty and dangerous jobs would always leave me in a good mood. I got the sense that I had found something inherent to being a Brinkman although I could not really articulate what that "something" was. Next, the family would eat lunch in the house, a hearty German lunch of cheese and meats, with the radio blaring crop reports. The conversation revolved around politics, crop prices, hauling grain, and hedging. My uncle would sometimes mention the names of men I had never met, but who formed a large team of advisors about taxes, chemical applications, and commodity markets. After two days in the car from Washington, D.C., I was on the other end of the world.

German Origins of Midwest Farmers' Relationship with Technology

The rural modern identity farmers developed in the Corn Belt in the early part of the 20th century originated in German and Jeffersonian agrarianism. This chapter describes these two origin identities as they existed in 18th- and 19th-century rural America. The agrarian ethos that German immigrants brought to the Midwest in the 19th century became particularly important for how 20th-century farmers would use technology. As one historian described the typical immigrant to frontier

Iowa in the 1850s, "he would more likely have come from one of the west German states than from any other part of Europe" and many of the American-born arrivals immigrated from Pennsylvania, Ohio, and Indiana where German-Americans comprised the largest ethnic group.[2] As late as the 2021 census, 31.3% of Iowans reported German heritage, more than double the number of Iowans reporting heritage from any other single European nationality, including Irish-Americans at only 12.7%.[3] Other Midwest states reported similarly high rates of German ancestry, with Wisconsin reporting between 37.05 and 40.5%, North Dakota at 35.77%, South Dakota at 33.25%, and Minnesota at 30.41%.[4] In the 2010–2014 census, virtually every county in each Corn Belt state reported German American as the largest ethnic group.[5]

American observers of German farmers in the 18th and early 19th centuries noted several unique aspects of their brand of agrarianism. These characteristics include the view of manual labor as a religious act, the eschewing of hired labor or slaves in favor of family-based production, the obsession with obtaining personal property and control over work processes, the inclusion of women in production processes and land inheritance, and the distrust of outsiders as a threatening "other."[6] Many migrants to the Midwest in the mid-19th century exhibited the German distrust of slavery as an "abhorrent" institution of the upper classes similar to peasantry under the old European monarchy. Thus, many Corn Belt settlers, "would be highly suspicious of bankers, lawyers, and most politicians," in other words, distrustful of elites.[7] The German farmer even in the 18th century thought himself more moral than members of the upper classes holding government positions. Indeed as early as 1794, one observer noted that most German farmers distrusted government officers as "too many and overpaid," an attitude held by Midwest farmers toward government bureaucrats in the 20th century.[8]

Most importantly, German rural identity relied on the display of success through material objects of production. Conversely, German agrarianism viewed demonstrations of wealth through non-productive artifacts as frivolous and immoral. Indeed, by the time Germans began moving into the Midwest in the 19th century, they had participated in *"a pattern of audience."*[9] In this repeated cultural practice, a "modern" social elite would observe and evaluate a rural person. The farmer uses material objects to perform an agrarian identity. This pattern of observation and performance continues an unarticulated debate over the modernity and morality of the rural dweller.[10]

Benjamin Rush's 1789 *Account of the Manners of the German Inhabitants* opens a window into the culture of German agrarianism and serves as an early example of the pattern of audience. Often, the account highlights differences between traditional peasant German and American/English aristocratic agrarian traditions.[11] Rush hailed from a group of American elites dedicated to the broader Enlightenment project of perfecting man and promoting progress through greater rationality, quantification, and scientific knowledge.[12] For gentlemen such as Rush in the 18th century, this "modern" worldview extended beyond natural philosophy to a wide variety of domains, including agriculture. Rush and his elite colleagues formed the geographic outer edge of a broader informal network of elites centered in Europe, the "Republic of Letters," who "submitted themselves to a moral economy of obligations and responsibilities that transcended the boundaries of nation."[13] Thomas Jefferson, for instance, personified these gentlemanly farmers.[14]

On February 11, 1785, Rush and other men of letters in Pennsylvania and surrounding states formed the Philadelphia Society for the Promotion of Agriculture (the Society). The twenty-three founding members included four signers of the Declaration of Independence (including Rush) and two future members on the Convention that drafted the United States Constitution. Rush has served as Surgeon General of the Continental army beginning in 1777. When he embarked on his study of "German inhabitants" in 1789, he served as the Treasurer of the U.S. Mint, a position he held since 1779. Benjamin Franklin, who had founded the American Philosophical Society in 1744, later became a member of the Society as did the author of *Common Sense*, Thomas Paine. Samuel Powel, the mayor of Philadelphia, served as the Society's first president. While the Society failed to organize enough financial resources to publish its own journal, unlike some of the European agricultural societies, its Committee of Correspondence published articles and letters in local periodicals and newspapers including the *Pennsylvania Gazette*, the *Pennsylvania Mercury*, the *Independent Gazetteer*, the *Pennsylvania Packet* and *Daily Advertizer* [*sic*], and the *Delaware Gazette*.[15]

The Society conceptualized "agriculture" in narrow terms to mean anything pertaining to practical improvements in growing crops, raising animals, producing products, or farm infrastructure such as farm design, fencing, or barn construction. The Society did not concern itself with improvements in business practices, although it did consider mill construction within the purview of agriculture.[16] Rush and the elites of the Society attempted to direct agricultural improvement and experimentation by announcing prizes for very specific activities. One announcement for prizes on April 5, 1785 read,

> For the greatest quantity of ground, well fenced, in locus trees or poles, of the sort used for posts or trunnels, grown in 1789, from seed sown after Feb. 8, 1788, not less than one acre, nor fewer than 1500 per acre – a gold medal.[17]

Several prize announcements called for agricultural experiments according to detailed specifications such as, "For an account of a vegetable food that may be easily procured and preserved, and that best increases milk in cows and ewes, in March and April, founded on experiment – a gold medal."[18] In experimenting on using oxen, the Society typically sought to incorporate the Enlightenment ideal that *everything* should be noted and measured.[19]

In awarding prizes, the Society held a very high standard for originality that almost amounted to arrogance and condescension toward contributors. In the January 12, 1790 minutes, the Society notes a rejection of an application by "a Southern

Farmer" for methods preventing damage of crops by insects because in his notes he "does not distinguish between the Hessian fly…and the common wheat fly" and because his technique of destroying the eggs of the insects "is not of his invention."[20] Overall, the Society sought to use Enlightenment experimentation to make American agriculture and the American countryside "look" English, from making cheeses in the "Cheshire" style to experiments erecting hedgerows across Pennsylvania.[21] For the enlightened mind, the model for modernity was the reason and rationalism emanating from gentlemanly Europe whether in natural philosophy or in agriculture.[22]

While the members of the Society lamented the anti-modern American dirt farmer who chose subsistence farming over enlightened agriculture, they encountered another "other" whose agrarian tradition achieved a high level of efficiency while similarly rejecting the Enlightenment ideals of Rush and his fellow savants.[23] As early as 1743, Governor George Thomas of Pennsylvania estimated that German immigrants comprised 60% of the European population, and Rush noted that, "they have, by their industry, been the principle instruments of raising the state to its present flourishing condition."[24] The success and "industry" of the German farmer in Pennsylvania, which by 1751 had led all colonies in agricultural exports, posed a challenge to Rush and the elites of the Society. If the basis of the new American agricultural identity was to be the enlightened modernity of gentlemen, how did the Germans, who formed an efficient agrarianism through following tradition rather than enlightened experimentation, fit into this "modern" American identity?[25] Rush responded to this quandary the way any enlightened gentlemen of the 18th century would. He set out in the 1780s to conduct a natural history of these strange, but productive people and record everything in great detail from their farming methods to their views on marriage and their construction of houses.[26]

Rush gives a positive account of the Germans from the beginning to end of his account, partially because other gentlemen of his rank, such as the French aristocrat Théophile Cazenove, often mocked the German-American farmer for being ignorant, cheap, and tied to Old World superstitions, while maintaining an insularity that rejected a more civic-minded republicanism.[27] One gets a sense that Rush aimed to rehabilitate a group that he viewed as making his commonwealth so productive. Perhaps more importantly, Rush saw the German immigrant narrative as contributing to the formation of a broader identity regarding America as a land of opportunity. While one may expect this "underdog" immigrant identity more in a later 20th-century mythos, it surprisingly appears in the early pages of Rush's account where he notes that most Germans came to the colonies as indentured servants, even having to sell their children. Rush further observes, "Many, who at home, had owned property, and converted it into money, were robbed in tran-situ, [sic] by ship owners, importers, and sea captains." Despite these hardships, Rush concludes, "'yet…from this class have sprung some of the most reputable and wealthy inhabitants of this province.'"[28]

Many other Germans had literally been sold by German princes to Hessian and other mercenary units in the Revolutionary War.[29] This immigrant narrative reinforced later associations between machinery, independence, and a family legacy of "innovative" technology use. Such stories about immigration would contribute to farmer's ideas of themselves as "inborn innovators" that created successful farms through hard work and technological acumen not inheritance or hereditary title. This type of "bootstrap" narrative, as journalist Alissa Quart calls it, became an important aspect of 20th-century rural globalized ultramodernity in the Midwest.[30]

Rush's positive impression, however, primarily from the fact that the Germans had rendered the Pennsylvania countryside into the kind of ordered, rational, and efficiently productive place that Rush and other gentlemen strived for, but without the natural philosophy and experimentation of the Society. As Rush states

> The Germans taken, as a body, especially as farmers, are not only industrious and frugal, but skillful cultivators of the earth…The German farm was easily distinguishable from those of others, by good fences, the extent of orchard, the fertility of the soil, productiveness.

Rush specifically noted, "The Germans seem more adapted to agriculture and improvement of the wilderness; and the Irish, for trade."[31] Rush commented that the Germans took over failed farms from Scotch-Irish settlers, "and often double[ed] the value of an old farm in a few years."[32]

Rush recorded specific farming and cultural practices which, he believed, made traditional German agriculture particularly productive. First, the Germans' culture of utility led to investment in productive infrastructure rather than in the home. The German farmer showed status through the size of his barn, the condition of his fences, and the size of his animals, not through a "frivolous" display of wealth through his house or his clothing. For the German-American farmer, technologies of utility and productivity were moral ways of displaying wealth and achieving what historian Harold Cook calls "an objectification of self."[33] Rush noted that while many American farmers invested in their homes and allowed animals to range freely, the Germans first invested in large impressive barns, which protected animals from the weather, conserved their energy for harvest, facilitated collection of manure for fertilizer, and allowed for greater control over feeding. This use of manure combined with plaster of Paris or lime, made with kilns on the farm itself, rendered the German farmer's land much more productive.[34] In contrast, Cazenove, a French elite, did not conceptualize this German cultural practice of displaying wealth through utilitarian objects of production as a virtuous or moral display of prosperity. In describing the small and cramped German home, Cazenove stated, "Probably one of the causes of this slovenliness and lack of comfort is that they do not know any better."[35]

Second, the Germans modified the American "girdle and belt" method for clearing new land. Instead of removing a ring of the outer bark layer, resulting in tree death within a year, the Germans physically pulled the tree out by the roots, which allowed earlier cultivation, and then burned them, leading to longer fertility and softer soils. These more pliable fields also avoided expensive plow repair.[36] In choosing new land, Rush indicates the Germans had some kind of folk or tacit knowledge that allowed them to accurately evaluate the future productivity of soils. The German method of predicting soil fertility, therefore, remained a mystery to Rush. Third, the Germans saved their best grain and distilled beverages for sale rather than personal consumption, preferring instead what Rush referred to as "Sourcrout" and other vegetables derived from kitchen gardens.[37] Fourth, as Cazenove pejoratively described, the Germans "were thrifty to the point of avarice" and according to Rush, "are afraid of debt."[38] For Rush, the German's thrifty nature led to increased productivity in a number of ways. For example, the Germans acted as "great economists of their wood," a practice made easier using porcelain stoves and the design of homes with the fireplace in the middle of the structure.[39] This thrifty use of firewood conserved the energy of draft horses used to haul and cut the wood. The German draft horse, therefore, was better bred, fed, and rested than the Anglo-American counterpart.[40]

Fifth, Rush noticed that "unlike their English and Irish neighbors, they [the German farmers] never, as a general thing, had colored servants, or slaves."[41] In addition, the "Germans seldom hire men to work on their farms" in spite of their wealth.[42] The German farmers satisfied labor needs instead by suspending gender roles with women working alongside men doing heavy labor in the fields, by having large families, and by maximizing the production value of draft animals. Sixth, the Germans designed the Conestoga wagon, made famous by Westward expansion in the 19th century, pulled by a renowned breed of draft horse (the Conestoga horse). This wagon and horse combination allowed the German farmer to carry up to 3,000 pounds of product to markets in urban areas over rough roads. As Rush described "in the months of September and October, it is no uncommon thing… to meet in one day fifty or one hundred of these wagons, on their way to Philadelphia."[43] Seventh, Rush noted a work ethic so strong and so engrained that while English and American men inquired as to the dowry of future brides or the wealth of their families before the marriage, the German suitor concerned himself only with the young woman's work habits. Rush described several maxims governing German behavior including "Eine fleissige Hausfrau ist die beste Sparbuechse" (an industrious housewife is the best money safe) or "To fear God, and to love work."

Finally, the Germans placed great value on ownership of personal property and the ability to give children land at marriage whereas the English-American population at the time featured a large number of landless young English-American men.[44] Rush does not give a detailed explanation of why many English-American families did not pass land onto all of their sons. While Pennsylvania and most northern states had multigeniture intestate inheritance laws in the late 18th century, most southern states retained the English common law practice of primogeniture until the late 1780s or early 1790s.[45] Therefore, Rush may have been comparing German inheritance practices to American-English families in the South in this portion of his account. Alternatively, he may intend to indicate that many Anglo-Pennsylvanians retained primogeniture in their wills notwithstanding state law allowing for inheritance by multiple children in the absence of a valid will.[46] Most likely, Rush means to indicate that German farmers gifted land to all of their children, both sons and daughters, well before the father's death allowing for more children to farm their own land sooner than their English neighbors who had to wait for testate succession.[47] In any event, this German cultural practice of property gifting to all of their heirs, sons and daughters, rather than just a few male heirs allowed the accumulation of traditional tacit knowledge over several generations. German farmers preserved agricultural knowledge through land gifts that would motivate their children to continue farming. In addition, Rush's account relates how German agrarianism views women as part of important production processes outside the home as well as family land accumulation.

While Rush found the productivity and efficiency of German farming impressive and hailed the German-American countryside as exemplifying the ordered, rational, aesthetic that Rush and his fellow elites found appealing, he struggled with determining whether this agrarian culture was truly modern or enlightened. Rush noted the complete rejection among the Germans of any attempts at formal secular education, and acknowledged a reputation that "the Germans are deficient in learning." Cazenove more bluntly described "the total lack of education of the farmers."[48] Even more surprising to Rush was the German farmers' reliance on what to an enlightened mind could only be described as ancient superstition rather than reasoned rationalism. Rush wrote, "The German farmers are very much influenced in planting and pruning trees, also in sowing and reaping, by the age and appearance of the moon" as well as a theory "that wood not felled at a full moon, is very soon attacked by worms and soon rots."[49]

Rush therefore describes two distinct agrarian cultures that would form the genealogical basis for the later rural-urban conflict. Early in the 20th century, modernity would become a prominent ingredient in a broader American identity and discourse, although taking different forms than Rush's aristocratic 18th-century conception. Rush and his Society proffered an agrarianism marked by an elite Enlightenment culture which embraced the experimental method and progress but with a clear aristocratic paternalism and view that the reasonable rich could best show everyone else how to conduct agriculture in a properly modern way. In contrast, German agrarianism features a traditional peasant culture with an association between work and independence reinforced by what we would call today a "rags to riches" immigration narrative and agricultural practices that would be seen as efficient and productive by their aristocratic contemporaries, but perhaps still not truly "modern."

Natural histories of the German farmer conducted by 18th century elites and savants such as Rush and Cazenove established a *pattern of audience* that would repeat itself through the rest of American history whereby the elite and "modern" observer would judge the prosperous peasant farmer as either "modern" or backwards. The farmer would, in turn, use methods and technologies *performatively* to establish his own identity. In the 20th century, farmers would reassert their own notions of modernity more overtly (Figure 1.1). Such a pattern of observation and performance would remain an important unspoken mechanism by which identities could be formed and conveyed across the rural-urban divide. In fact, Rush frequently noted negative views of German-American farmers by others of his social rank as anti-modern, foreshadowing later rural yokel stereotypes employed by urban reformers, academics, bureaucrats, and writers. While Rush seeks to counter these negative views, the fact that he cited them and feels the need to argue so forcefully against them, speaks to their prevalence and strength among 18th-century Anglo-Americans. For example, Rush declares, "our Germans are stigmatized as dolts" and at one point refers to the German farmers as "our much-abused 'Pennsylvania Dutchmen.'"[50] He cites a Buffalo newspaper article describing the "German Farmers of Pennsylvania" as a "fragment of the middle ages, uneducated, and uncultivated."[51] Rush also cites a letter from Benjamin Franklin to Peter Collinson on May 9, 1753, in which Franklin characterizes the German farmers in Pennsylvania as "the most stupid of their own nation."[52] This idea that those who

FIGURE 1.1   German-born American artist Lewis Krimmel seeks to paint a positive image of the 19th-century German-American farmer as a prosperous but humble patriarch. By associating the farmer with financial success, modesty, and family he depicts the German agrarian as moral. Lewis Krimmel, *Country Wedding*, 1820. Oil, 16 3/16 in. × 22 1/8 in. Pennsylvania Academy of Fine Arts, Philadelphia, PA. https://www.pafa.org/museum/collection/item/country-wedding-bishop-white-officiating (accessed 10/27/23).

toil on farms are the "stupid ones" foreshadows later 20th-century discourse viewing Midwest farmers who chose to *stay* on farms rather than migrate to urban centers as "the stupid ones who were left behind."[53]

Even when attempting to rehabilitate the German-American farmer, Rush often
does so with a backhanded compliment such as "there are other things, besides
political soundness, valuable in a citizen" and "there is no false mental glitter about
them: in a word, they are rather men of sound judgment, than brilliant rhetori-
cians or one sided ideologists." Rush encourages his enlightened reader to "learn
to respect the excellent sense they display in the ordinary concerns of life."[54] Yet,
the German-American farmers of Pennsylvania were the first Americans to produce and read what amounted to a 20th-century-style agricultural journal. While
Rush and the members of the Society were infrequently publishing advances in
agricultural techniques in local newspapers, the Pennsylvania German paper *Die
Germantauner Zeitung* published a whole series of articles in 1787 and 1791, with
detailed descriptions of how to prepare land with artificial fertilizers during plowing, care for orchards, employ manure more effectively, improve the quality of
potato crops, and make meadows more productive. The journal also proposed new
and improved methods on cultivating fodder and feeding stalled cattle. Other articles detailed the entire process of fruit farming or manufacturing artificial fertilizer. Another German newspaper in Pennsylvania, *Neu Unpartheyische Lancäster
Zeitung* contained similar articles on agricultural methods while yet another paper,
Straatsbote, featured numerous advertisements for chaff separators, harvest cra-
dles, scythes, and sickles designed by German inventors similar to ads in later
20th-century American farm journals for agricultural technologies.[55] In 1790, the
paper *Neu Unpartheyische Readinger Zeitung* even published a paper by George
Morgan of the Society who received a prize for model barn and barnyard design.[56]
In spite of Rush's account in 1789, and notwithstanding the impressive productive capacity of German farming methods, few Society members showed a similar
interest in gaining knowledge from the backward, anti-modern, German-American
farmer. A thorough search through the minutes and writings of the Society shows
little mention of this substantial body of published and experienced German agricultural knowledge existing right in Philadelphia's backyard. One notice produced
by the Society to instruct young farmers on how to best start a farm in 1819 cited
several methods "to be found in European books" but nothing about their productive German neighbors.[57]

To the Germans, the "other" of the 18th century had been the British and
the German princes who interfered with their property, family, and work. The
German farmers were "among the first to shoulder the gun" in the Continental
Army not because of the kind of patriotism that inspired Rush and his colonial
colleagues, but because the British represented a threat to German private property.
The 18th-century German agrarian viewed work as much more than a prideful
occupation, but as a sacred act that reinforced bonds between God, the family, and
the land itself. Rush indicates that the Germans avoided the use of slaves because

slavery was irreconcilable with religion.[58] In addition to violating scripture, having another person that did not belong to your family work on your land, according to the German agrarian ideal, was an immoral way to make a living. Hence, the Germans also avoided using hired labor. Perhaps slavery also reminded Germans of indentured servitude and military impressment, the opposite of the free independent German morally cultivating his own land with his own family. Rush states that for the German farmer "His first object, is to become a freeholder. The highest compliment that can be paid to them, on entering their house, is to ask them: Is this house *yours*? [Emphasis added]."[59] In this sense, Germans used barns, fences, and well-managed fields to perform the farmers' work ethic, independence, and morality.

After interviewing several German farmers, Rush concluded, "I am persuaded, that no chains would be able to detain them from sharing in the freedom of their Pennsylvania friends and former fellow subjects 'We will assert our dignity' (would be their language)."[60] Rush's quote indicates a strong cultural practice of performing a "dignified" self-image through work and improvements in personal property. The prosperous look of their property confirmed the moral righteousness of the German-American farming way of life.[61]

Other observers of German agrarians in early America noted the contrast between the German agrarian's view of work and property and that of lower-class English immigrants. The trustees of the Georgia colony, led by James Edward Oglethorpe, for example, actively sought to draw German immigrants in the early 18th century in an effort to establish a large class of working-class whites as a way of resisting the importation of a slave-based plantation economy from South Carolina. As Rush also noted in his treatise on Germans in Pennsylvania, many poor Englishmen would simply squat on inferior land and barely break the soil "preferring to steal and starve rather than work in the fields," lest they might "look like slaves." If Englishmen in early Georgia had means, they would often flee to South Carolina where they could purchase slaves to perform farm labor. The German and Scandinavian immigrants, on the other hand, would gladly "dirty their hands" by engaging in farm labor and make poor land productive. As accounts of elites such as Rush and Oglethorpe suggest, these two peasant cultures not only had different methods of production, but also vastly divergent views of the relationship between work, property, status, and morality.[62]

Letters from Germans immigrating to Iowa in the 19th century also display a German rural identity. For example, the German teacher Johannes Gillhoff compiled stories from letters by his former students who had immigrated to rural Iowa in the late 19th century into one fictitious character, Jürnjakob Swehn, in *Letters of a German American Farmer*.[63] While some regard Swehn as a composite of several different farmer immigrants from Gillhoff's Mecklenburg, Germany to rural Iowa, other scholars have argued that Swehn is actually Carl Wiedow from Glaisin, Germany, who settled in Clayton County, Iowa in the 1860s.

Written in Low German, students in northern Germany read Gillhoff's stories widely in schools through the early 20th century. The book reveals that several aspects of German agrarianism appeared in frontier Iowa in the 19th century. First, Swehn still sees the world largely through ancient folk and religious lenses originating in his small German village of peasants and shows little modern sensibility even when discussing the acquisition of new farm machinery. When mentioning new equipment, Swehn tends to simply list items rather than discuss greater "efficiency" or present artifacts as a means of achieving "progress." In recalling his attendance at the 1893 Chicago World Fair, Swehn notices the personalities of certain people at the exhibits rather than new artifacts or methods presented.

Swehn may represent a transitional figure between traditional German agrarianism and rural capitalistic modernization in that he tends to self-reflect on his farming methods, a feature of modernity. Swehn changes when he finds better work processes or farming techniques that differ from those used in his native Mecklenburg, but he does not appear to define himself in terms of innovativeness, mechanical savvy, or business acumen. "I said to myself," Swehn wrote "you have been a real dummkopf... You have to get a different schooling to work in America. Jürnjakob, you have to put your brain to work, otherwise nothing is ever going to come of you." However, Swehn did not seek to adopt the modern techniques or scientific methods embraced by Iowans in the 20th century such as experimenting with seed hybridization or finding novel uses for the latest machinery.

As with the German farmers observed by Rush, Swehn sees his farm and his productivity as the only moral displays of success. When, for example, he discovered that his son in college at the University of Iowa had started wearing "a gold watch, a gold ring with a big stone, a gold tiepin, and all such things as that," Swehn scolded him with a long immigrant narrative about how the family originated as impoverished peasants and that the jewelry "doesn't fit into our family." As with his rationale for many of his production methods, Swehn also saw the jewelry through a religious lens as violating the "ninth and tenth commandments."[64] Instead of these immodest ways of dressing, he boasted of success in his letters by recalling the number of cattle he owned, the amount of his acreage, and the machinery he had accumulated, always reminding Gillhoff, "And it all costs plenty of sweat."[65]

Still, Swehn does not view machinery in the same way modern capitalistic farmers would in the early 20th century. While serving as productive symbols of success, Swehn still sees the machines as tools just like other object of production on the farm. He does not regard more advanced objects as special or attach to them a privileged status among his collection of artifacts. Swehn never uses the term "technology" or "innovative" or "progressive" or even "newest" or "up-to-date" to describe his farm machinery. Swehn does not define himself in terms of using the most current technology. He displays no tendency to fetishize mechanical farm equipment by discussing it more than other tools, land, or buildings. Nor does Swehn show any desire to learn how his machinery works and he does not redesign or repurpose hardware to gain a competitive advantage over his neighbors. Thus, Swehn adheres to the traditional German agrarian practice of displaying wealth through productive tools, but he still has not adopted a modern capitalistic identity.

Third, Swehn exhibits a fierce independence and distrust of authority resulting from his experience as, first, a serf in Germany and then as a hired man in a noble system of land ownership that prevented him from acquiring his own property. Although serfdom ended in Mecklenburg in 1820, local dukes still controlled almost every aspect of Swehn's life, including what tools he could use and what church he had to attend.[66] Swehn continuously reminds Gillhoff that he owns his own land, farm buildings, animals, and equipment.[67]

Fourth, Swehn and his wife seem to farm together and share decisions about the production processes on the farm. The marital relationship strikes the 21st-century reader as surprisingly equal. Swehn ultimately decided to not feed pigs to chickens after consulting with his wife who scolds him as an equal "Jürnjakob, you are one dummkopf of a Deutschman. And if you get another idea like that, make sure you don't tell me."[68] Fifth, Swehn shows distrust of people from outside the farm and experiences awkward discomfort in downtown Chicago, though he had not yet acquired the same sense of alienation that Corn Belt residents would express toward their urban cousins in the 20th century as the rural urban conflict started to emerge. Therefore, Swehn offers a picture of some features of agrarianism that would later play important roles in the ways rural Americans interacted with technology as the 20th century progressed. Contextual factors such as modernity and the rural-urban conflict would later reach Iowa and other parts of the Corn Belt, reshaping German agrarianism.

Swehn's experience as an immigrant also shows views of German farmers as ignorant and anti-modern when observed by Americans occupying higher social ranks. Reminiscing about his arrival in the U.S., Swehn portrayed how most Americans regarded Germans in the late 19th century as, "Here comes another one of those dumb Deutschmans, not a red cent to his name."[69] Swehn often sees success in agriculture as the most effective way to overcome these stereotypes.[70]

By the early 20th century, when money and power had concentrated in cities and within central government, the decedents of German immigrants had settled in the Midwest Corn Belt. Drawing on their own German-American agrarian tradition as well as Jeffersonian agrarianism emanating from the American Southeast, farmers reinforced the cultural values of personal property, family, and control over work processes. In addition, these Midwesterners resisted the urban version of modernity, urban industrialism, with their own discourse of rural capitalistic modernity. Added to barns and fences as representing their work ethic and rural independence were tractors, cream separators, and combines. The German agrarians in the Corn Belt came to define themselves in terms of these objects in a new way as reinforcing not only their capacity to succeed but to *be modern*.

The events that most re-shaped the identities of German Midwesterners into a modern capitalist sense of self were the rise of the progressive reform movement, and the rural-urban conflict beginning in the early 20th century and intensifying in the 1920s. The pattern of audience in the 1920s featured the advent of the "rube" stereotype as a significant motivation for farmers to develop their own rural modern identity.

Jeffersonian Agrarianism

In addition to German agrarianism, the Anglo-American discourse-identity bundle of Jeffersonian agrarianism migrated into the rural Midwest during the 19th century. Scholars have debated the proper definition of traditional Jeffersonian agrarianism, but the one I use in this book is the view of farmers in virtuous and heroic terms as people chosen by God to civilize and "save" society by planting the idealized "pastoral garden."[71] Jefferson viewed his farming activities through the lens of both the European Enlightenment and the strict English class system. As historian Mark Essig explains, this Jeffersonian view arises from "the Englishman's idea of himself" described by "Neat fields, tidy hedgerows, healthy animals-all spoke of his virtue and godliness." Essig points out, the Anglo-American Protestant practice of using agricultural technologies to reinforce the moral self has a long history pre-dating Thomas Jefferson. Essig argues that English settlers in the colonial world reacted to the lack of gold in the English colonies by seeing colonization through farming as more virtuous and more Christian than the Spanish use of soldiers and mines. "Instead of killing Indians," Essig writes "they would convert them to Christianity and train them as farmers." English settlers went as far as seeing themselves as fulfilling God's decree in Genesis: "Let them have dominion over the fish in the sea, and over the fowl in the air, and over the cattle, and over all the earth."[72]

Thomas Jefferson adopted these English views of order to uphold small yeoman farmers as chosen by God to civilize the American continent and preserve democratic ideals. The prosperity of industry and survival of the cities, according to Jefferson and subsequent 20th-century populists like William Jennings Bryan, rested on this group of independent small yeomen braving the frontier. Further, Jeffersonian agrarianism ascribes to a pastoral ideal that romanticizes the farmer's work with nature.[73] Barbara J. Scot, in her recollection of visiting her Iowa farm family, described the spiritual elements of Jeffersonian agrarian myth eloquently as,

> The most dignified and worthy of all callings was the land – a land rightfully assumed because of a promise to people far distant in time and place. Once on the land, you and your progeny were part of a succession that was not only grounded in the Bible, it was meant to endure until Judgment Day. On the family farm.[74]

William Jennings Bryan secured the Democratic nomination for the presidency in 1896 by appealing to Jeffersonian agrarianism with his "A Cross of Gold" speech,

> The hardy pioneers who have braved all the dangers of the wilderness, who have made the desert to blossom as the rose – the pioneers away out there [pointing to the West], who rear their children near to Nature's heart… It is for these that we speak.[75]

Unlike German agrarianism, Jeffersonian ideals include clearer gender roles in the production process and embrace having laborers outside the family, such as slaves or hired hands, work the land.[76] Jefferson hired working-class whites or used slaves to work his land while he experimented with new techniques taken from readings on agronomy and husbandry published in Europe and kept meticulous records. He even designed a new plough in 1794 featuring a less resistant moldboard, but he never actually used it as he considered such manual labor as inappropriate for a gentleman of his high social rank.[77]

Some Midwesterners dispute the idea that pure traditional Jeffersonian agrarianism still drives farmers in forming discourses or identities. For example, in a personal essay written in 1985 about growing up on a family farm in Kansas, historian Thomas D. Isern noted the lack of a pastoral on modern farms by asking

Where is the yeoman farmer of Jeffersonian myth today? I know that last winter Bill Short on Diamond Creek in Chase County fed quail in his barn and allowed wild turkeys on his silage, but otherwise I see little resemblance between the mythical Jeffersonian yeoman and the contemporary farmer of my acquaintance.[78]

Sociologists and communications scholar Tarla Rai Peterson argues that Jeffersonian agrarianism creates a persistent and dysfunctional frontier myth in which the farmer holds onto a "self-image of a hero victimized by circumstances," but who overcomes these obstacles through independence and grit and the inventive use of resources.[79] Peterson, argues that pure and unaltered Jeffersonian yeoman frontier myth dominates rural discourse so completely, that it creates a monolithic rural identity in which farmers engage in a constant battle with technology as a threat to the pastoral ideal. As such, according to Peterson, farmers view themselves as victims of technology growing beyond their control preventing "positive adaptation" to social and economic change.[80] I dispute that farmers in the Corn Belt still retain anti-technological aspects of Jeffersonian agrarianism as well as the pastoral while retaining the heroic yeoman myth. Jeffersonian agrarianism, I argue, combined with German agrarianism creating a hybrid rural identity in the rural Midwest on the eve of the 20th century and rise of rural modernity.

Notes

1 Alexis de Tocqueville, "What Causes Almost All Americans to Follow Industrial Callings," in *Democracy in America*, ed. Richard D. Heffner (New York: Penguin Books, 1984), 213–216.
2 Wall, *Iowa*, 54–55.
3 "Iowa Quick Facts: Social Characteristics," *Iowa.gov State Data Center*, https://www.iowadatacenter.org/index.php/quick-facts/iowa-quick-facts. (accessed 10/09/23).
4 Tom Kertschner, "Home away from the Homeland: Why so Many German Immigrants Chose Wisconsin," *Wisconsin Public Radio* (May 18, 2023); "German Population by State [Updated January 2023]," *World Population Review*, https://worldpopulation review.com/state-rankings/german-population-by-state (accessed 10/9/23).
5 U.S. Census Bureau, "German Roots," https://www.census.gov/content/dam/Census/library/visualizations/2016/comm/german_roots.jpg (accessed 1/21/17); "German Americans-The Silent Minority." *The Economist*, February 7, 2015, http://www.

economist.com/news/united-states/21642222-americas-largest-ethnic-group-has-assimilated-so-well-people-barely-notice-it (accessed 3/31/16).

6 For an observation of these characteristic of German agrarianism, see Benjamin Rush, *An Account of the Manners of the German Inhabitants of Pennsylvania, Written 1789*, ed. Israel Daniel Rupp (Philadelphia, PA: Samuel P. Town, 1875).

7 Wall, *Iowa*, 54–55.

8 Rush, *An Account of the Manners of the German Inhabitants of Pennsylvania, Written 1789*, 34–35.

9 My notion of "pattern of audience" adds to recent works by scholars on class and rural resentment in America by introducing technology as a crucial tool of rural resistance to a perceived urban threat. I also suggest that using technology to perform for urban observers is a repeated cultural practice in rural America with historical continuity. See Nancy Isenberg, *White Trash: The 400-Year Untold History of Class in America* (Viking: New York, 2016), xvi–xvii, 1–42; Cramer, *The Politics of Resentment*.

10 Again, see Rush.

11 Rush.

12 Alyn Brodsky, *Benjamin Rush: Patriot and Physician* (New York: St. Martin's Press, 2004).

13 Martin Rudwick, *Bursting the Limits of Time: The Reconstruction of Geohistory in the Age of Revolution* (Chicago, IL: University of Chicago Press, 2005), 31–32.

14 Isenberg, *White Trash*, 85–88.

15 Manuela Albertone, "The American Agricultural Societies and the Making of the New Republic, 1785–1830," in *The Rise of Economic Societies in the Eighteenth Century*, ed. Koen Stapelbroek and Jani Marjanen (New York: Palgrave Macmillan, 2012), 339–369; Lucius F. Ellsworth, "The Philadelphia Society for the Promotion of Agriculture and Agricultural Reform, 1785–1793," *Agricultural History* 42, no. 3 (1968): 189–200.

16 *Minutes of the Philadelphia Society for the Promotion of Agriculture*, from its Institution in February, 1785, to March, 1810 (Philadelphia, PA: John C. Clark & Son Printers, 1854), 71.

17 Ibid., 46; a "trunnel," according to the *Collins English Dictionary*, is a dowel used for pinning planks of timbers together. *Collins English Dictionary*, s.v. "trunnel." http://www.collinsdictionary.com/dictionary/english/trunnel (accessed 5/3/16).

18 *Minutes of the Philadelphia Society for the Promotion of Agriculture*, 71.

19 Ibid., 47.

20 Ibid., 57.

21 Ibid., 47.

22 For a detailed and well-written work on the importance of 18th century France to 18th-century enlightenment culture including natural philosophy, reason, and rationalism see Jessica Riskin, *Science in the Age of Sensibility: The Sentimental Empiricists of the French Enlightenment* (Chicago, IL: University of Chicago Press, 2002), Chapter 1, Chapter 3.

23 For a description of how members of the Society looked down on American "dirt farmers" as stalling the social and economic development of the new nation, see Ellsworth, "The Philadelphia Society for the Promotion of Agriculture," 189–200.

24 Rush, 5.

25 Sociologist Harry M. Collins describes Michael Polyani's conception of tacit knowledge as an art that cannot be transmitted through transcription but only passed on by example or by doing from master to apprentice. The person holding tacit knowledge may not even consciously know he or she possesses it. See H.M. Collins, *Changing Order: Replication and Induction in Scientific Practice* (Chicago, IL: Chicago University Press, 1992), 77, n.5.

26 While Germans at the time occupied many occupations, an 1838 survey found that the number of farmers doubled all other occupations combined. Rush, 11.

27 Although Cazenove's account was written in 1794 after Rush's document, his negative impression of German farmers as backwards exemplified some gentlemanly accounts of the period. Théophile Cazenove, *Cazenove Journal, 1794*, ed. Rayner Wickersham Kelsey (Haverford: The Pennsylvania History Press, 1922).

28 Rush, 6–7.

29 Ibid., 56–57.

30 Alissa Quart, *Bootstrapped: Liberating Ourselves from the American Dream* (New York: Ecco Press, 2023).

31 Rush, 10–11.

32 Ibid., 13.

33 Cook explores social and cultural conceptions of moral displays of wealth in Dutch consumer and merchant culture in the 16th and 17th centuries. Harold J. Cook, *Matters of Exchange: Commerce, Medicine, and Science in the Dutch Golden Age* (New Haven, CT: Yale University Press, 2007), 14–15, 43, 68–69.

34 Rush, 12, 14, 16.

35 Cazenove, *Cazenove Journal*, 84.

36 Rush, 14–15.

37 Ibid., 20–24.

38 Ibid., 23; Cazenove, 34.

39 For a more thorough discussion on the centrality of stoves in the household economy of pre-20th-century America as well as design differences between German/Scandinavian and English/American stoves see Cowan, *More Work for Mother*, 20–26, 54–62.

40 Rush, 17–18.

41 Ibid., 23.

42 Ibid., 24.

43 Ibid., 26–27.

44 Ibid., 30; Cazenove, 44. For a discussion of how the English and American custom of holding onto land until death resulted in large numbers of landless young men, see James Henretta, "Families and Farms: Mentalité in Pre-Industrial America," *William and Mary Quarterly* 3rd ser., 35, no. 1 (1978): 6–7.

45 Lee J. Alston and Morton Own Schapiro, "Inheritance Laws Across Colonies: Causes and Consequences," *The Journal of Economic History* 44, no. 2 (1984): 277–287.

46 "Primogeniture" is the common-law right of the firstborn son to inherit his ancestor's estate, usually at the expense of his younger siblings. "Multigeniture" is the division of wealth among all sons, or perhaps all children. "Intestate succession" is the method used to distribute property owned by a person who has died without a valid will. Bryan A. Garner, *Black's Law Dictionary* (St. Paul, MN: West Group, 1999), s.v. "intestate succession," "primogeniture;" Alston and Schapiro, 277.

47 See for example, Henretta, "Families and Farms."

48 Rush, 40; Cazenove, 34.

49 Rush, 31.

50 Referring to a German as a Dutchman may have been a racist term in the 18th century.

51 Rush, 65.

52 Ibid., 63.

53 For a discussion of early 20th century rural stereotypes see James H Shideler, "'Flappers and Philosophers,' and Farmers: Rural-Urban Tensions in the Twenties," *Agricultural History* 47, no. 4 (1973): 289 and Randall Patnode, "'What these People Need Is Radio,' New Technology, the Press, and Otherness in 1920s America," *Technology and Culture* 44, no. 2 (2003): 286.

54 Rush, 67–69.

55 Leo Bressler, "Agriculture Among the Germans in Pennsylvania During the Eighteenth Century," *Pennsylvania History* 22, no. 2 (1955): 103–133; *The Pennsylvania-German Society*, Vol. 29 (Philadelphia: Pennsylvania-German Society, 1922), 131.

56 *The Pennsylvania-German Society*, 129–130.
57 "Agriculture: Notices for a Young Farmer, From the Memoirs of the Philadelphia Agricultural Society," *The American Farmer, Containing Original Essays and Selections on Rural Economy and Internal Improvement* 7 (June 25, 1819): 1, 13.
58 Rush, 63.
59 Ibid., 32.
60 Ibid., 56.
61 While this association between farming and morality mirrors the view of the Jeffersonian Democrats of the era, it had not yet gained a Jeffersonian association with patriotism or ideas about democratic citizenship. This agrarian morality grew out of a completely independent German agrarian tradition.
62 German agrarianism is the author's term and concept. Isenberg only notes that these German, Swiss, French Huguenot, and Scottish Highlander immigrants in Georgia "seemed prepared for lives of hardship, arriving as whole communities of farming families." Isenberg, *White Trash*, 58.
63 Johannes Gillhoff, *Letters of a German American Farmer*, trans. Richard Lorenz August Trost (Iowa City: University of Iowa Press, 2000).
64 Ibid., 47.
65 Ibid., 40.
66 Ibid., vii–viii.
67 Ibid., 5, 25, 40.
68 Ibid., 29.
69 Ibid., 6.
70 Ibid., 23.
71 Thomas Jefferson, *Notes on the State of Virginia* 1787 (Chapel Hill, NC: University of North Carolina Press, 1995).
72 Essig, *Lesser Beasts*, 131–132.
73 For example, see Wall, *Iowa*, 136.
74 Barbara J. Scot, *Prairie Reunion* (New York: Farrar, Straus and Giroux, 1995), 54. Certainly, this version of the pure Jeffersonian yeoman frontier hero still appears in popular treatments of American farming, such as author Richard Rhodes' acclaimed 1989 book *Farm: A Year in the Life of an American Farmer*. Richard Rhodes, *Farm: A Year in the Life of an American Farmer* (New York: Simon & Schuster, 1989).
75 William Jennings Bryan, "Democratic National Convention Address-A Cross of Gold," *American Rhetoric*, 1896, https://www.americanrhetoric.com/speeches/williamjenningsbryan1896dnc.htm (accessed 7/8/23).
76 For example, see Benjamin Rush's comparison of German and English farmers in the 18th century with respect to gender and inheritance as well as slavery in Rush, 15–63.
77 Isenberg, 85–88.
78 Thomas D. Isern, "The American Dream: the Family Farm in Kansas," *Midwest Quarterly* 26 (1985): 359–361.
79 Tarla Rai Peterson, "Jefferson's Yeoman Farmer as Frontier Hero," 9–10.
80 Scot, *Prairie Reunion*, 9, 15–17.

Bibliography

"Agriculture: Notices for a Young Farmer, From the Memoirs of the Philadelphia Agricultural Society." *The American Farmer, Containing Original Essays and Selections on Rural Economy and Internal Improvement*, June 25, 1819.
Albertone, Manuela. "The American Agricultural Societies and the Making of the New Republic, 1785–1830." In *The Rise of Economic Societies in the Eighteenth Century*, edited by Koen Stapelbroek and Jani Marjanen, 339–369. New York: Palgrave Macmillan, 2012.

Alston Lee J. and Morton Own Schapiro. "Inheritance Laws Across Colonies: Causes and Consequences." *The Journal of Economic History* 44, no. 2 (1984): 277–287.

Bressler, Leo. "Agriculture Among the Germans in Pennsylvania During the Eighteenth Century." *Pennsylvania History* 22, no. 2 (1955): 103–133.

Brodsky, Alyn. *Benjamin Rush: Patriot and Physician.* New York: St. Martin's Press, 2004.

Bryan, William Jennings. "Democratic National Convention Address-A Cross of Gold." *American Rhetoric,* 1896. https://www.americanrhetoric.com/speeches/williamjennings-bryan1896dnc.htm (accessed 7/8/23).

Cazenove, Théophile. *Cazenove Journal, 1794,* edited by Rayner Wickersham Kelsey. Haverford, PA: The Pennsylvania History Press, 1922.

Collins English Dictionary, s.v. "trunnel." http://www.collinsdictionary.com/dictionary/english/trunnel (accessed 5/3/16).

Collins, H.M. *Changing Order: Replication and Induction in Scientific Practice.* Chicago, IL: Chicago University Press, 1992.

Cook, Harold J. *Matters of Exchange: Commerce, Medicine, and Science in the Dutch Golden Age.* New Haven, CT: Yale University Press, 2007.

Cramer, Katherine J. *The Politics of Resentment: Rural Consciousness in Wisconsin and the Rise of Scott Walker.* Chicago, IL: University of Chicago Press, 2016.

Ellsworth, Lucius F. "The Philadelphia Society for the Promotion of Agriculture and Agricultural Reform, 1785–1793." *Agricultural History* 42, no. 3 (1968): 189–200.

Essig, Mark. *Lesser Beasts: The Snout to Tail History of the Humble Pig.* New York: Basic Books, 2015.

Garner, Bryan A. *Black's Law Dictionary.* St. Paul, MN: West Group, 1999.

"German Americans-The Silent Minority." *The Economist,* February 7, 2015. http://www.economist.com/news/united-states/21642222-americas-largest-ethnic-group-has-assimilated-so-well-people-barely-notice-it (accessed 3/31/16).

"German Population by State [Updated January 2023]." *World Population Review.* https://worldpopulationreview.com/state-rankings/german-population-by-state (accessed 10/9/23).

Gillhoff, Johannes. *Letters of a German American Farmer.* Translated by Richard Lorenz August Trost. Iowa City: University of Iowa Press, 2000.

Henretta, James A. "Families and Farms: Mentalité in Pre-Industrial America." *William and Mary Quarterly* 3rd ser., 35, no. 1 (1978): 6–7.

"Iowa Quick Facts: Social Characteristics." *Iowa.gov State Data Center.* https://www.iowadatacenter.org/index.php/quick-facts/iowa-quick-facts (accessed 10/9/23).

Isenberg, Nancy. *White Trash: The 400-Year Untold History of Class in America.* Viking: New York, 2016.

Isern, Thomas D. "The American Dream: the Family Farm in Kansas." *Midwest Quarterly* 26 (1985): 357–367.

Jefferson, Thomas. *Notes on the State of Virginia* 1787. Chapel Hill, NC: University of North Carolina Press, 1995.

Kertschner, Tom. "Home away from the Homeland: Why so Many German Immigrants Chose Wisconsin." *Wisconsin Public Radio,* May 18, 2023.

Krimmel, Lewis. *Country Wedding.* 1820. Oil. 16 3/16 in. × 22 1/8 in. Pennsylvania Academy of Fine Arts, Philadelphia, PA. https://www.pafa.org/museum/collection/item/country-wedding-bishop-white-officiating. (accessed 10/27/23).

Minutes of the Philadelphia Society for the Promotion of Agriculture, from its Institution in February, 1785, to March, 1810. Philadelphia, PA: John C. Clark & Son Printers, 1854.

Patnode, Randall. "'What these People Need Is Radio:' New Technology, the Press, and Otherness in 1920s America." *Technology and Culture* 44, no. 2 (2003): 285–305.

Peterson, Tarla Rai. "Jefferson's Yeoman Farmer as Frontier Hero: A Self Defeating Mythic Structure." *Agriculture and Human Values* 7, no. 1 (1999): 9–19.

Quart, Alissa. *Bootstrapped: Liberating Ourselves from the American Dream*. New York: Ecco Press, 2023.

Rhodes, Richard. *Farm: A Year in the Life of an American Farmer*. New York: Simon & Schuster, 1989.

Riskin, Jessica. *Science in the Age of Sensibility: The Sentimental Empiricists of the French Enlightenment*. Chicago, IL: University of Chicago Press, 2002.

Rudwick, Martin. *Bursting the Limits of Time: The Reconstruction of Geohistory in the Age of Revolution*. Chicago, IL: University of Chicago Press, 2005.

Rush, Benjamin. *An Account of the Manners of the German Inhabitants of Pennsylvania, Written 1789*, edited by Israel Daniel Rupp. Philadelphia, PA: Samuel P. Town, 1875.

Scot, Barbara J. *Prairie Reunion*. New York: Farrar, Straus and Giroux, 1995.

Shideler, James H. "'Flappers and Philosophers,' and Farmers: Rural-Urban Tensions in the Twenties." *Agricultural History* 47, no. 4 (1973): 283–299.

The Pennsylvania-German Society, Vol. 29. Philadelphia: Pennsylvania-German Society, 1922.

Tocqueville, Alexis de. "What Causes Almost All Americans to Follow Industrial Callings." In *Democracy in America*, edited by Richard D. Heffner, 213–216. New York: Penguin Books, 1984.

U.S. Census Bureau. "German Roots." https://www.census.gov/content/dam/Census/library/visualizations/2016/comm/german_roots.jpg (accessed 1/21/17).

Wall, Joseph Frazier. *Iowa: A Bicentennial History*. New York: W.W. Norton & Company, 1978.

2

"ARE WE READY FOR THIS?"

Urban Industrialism, Rural Resistance, and Rural-Urban Conflict

> Owning two thirds of the personal property of the country, evading payment of taxes wherever possible, the corporations throw almost the whole burden up on the land, upon the little homes, and the personal property of the farms.
>
> Robert M. La Follette, "The Danger Threatening
> Representative Government," 1897[1]

In the fall of 1935, college professors at Harvard University commenced a study of American accents. Harvard Professor Harry H. Hall had invented a device in which a person would talk into a microphone connected to a sound filter and a cathode ray tube, which produced a visual representation of the person's voice on a piece of sensitized paper. With the contraption, Professor Hall intended to take a "photograph" of different accents to study their characteristics. The study went smoothly until December of 1935 when Professor Hall used his machine to photograph the voice of Professor Miles L. Hanley from the University of Wisconsin. Hall classified Hanley's upper Midwest accent, particularly the way he said the letter "s," as "random noise" and likened it to the rating he got when recording a vacuum cleaner. Soon, Midwest professors "hustled … to defend the Midwest accent" at a meeting in Kansas City, Missouri where they "scoffed at the 'nasal tricks' of New Englanders." Professor H. Miles Heberer from Kansas State railed, "the speech of the down east New Englanders is much easier to satirize. The old-time hick comedies were based on the language of New England. No Kansas farmer talks like that…" The Kansan went on, "If ours [accent] sounds like a vacuum cleaner, some of the nasal tricks they do with an 'r' liken it to a buzz saw." Professor J.S. Smith from Sioux Falls, SD declared, "A Harvard accent? Why, that sounds much like the purring of a bobcat."[2] Marquette Physics Professor A.E. Elo, based in Milwaukee,

DOI: 10.4324/9781032637952-3

Wisconsin, agreed. Hall's insult of the Wisconsin accent doomed the acceptance of his invention among his Midwest colleagues. Clearly, rural self-consciousness about rube stereotypes could even derail some of the nation's intellectuals.

The Rise of "Rube" Stereotypes in the Midwest and the Rural-Urban Conflict

Negative urban views of rural America, along with German and Jeffersonian agrarianism, strengthened due to the rural-urban conflict between Midwest farmers and "city folk" intensifying in the 1920s and continuing into the 1930s. These negative urban discourses increased, in part, to alleviate ambivalence many urbanites felt over their decision to leave rural life. Modernity became strongly associated with urban living when new mass communications technologies combined with the accelerated pace of migration of rural denizens to cities in the early 20th century. Closer historical analysis also reflects contesting and shifting agrarian discourses and identities in which many farmers embraced modern sensibilities while giving them a distinctly rural flavor as a way of responding to negative urban stereotypes.

If one uses the push-pull metaphor commonly deployed to conceptualize immigration and migration, a discourse of urban superiority strengthened the pull of the city beginning in the early 20th century. Negative views of farmers by urban Americans had begun to intensify in the early 20th century when high farm commodities prices caused manufacturers to accuse farmers of keeping food supplies low through laziness and inefficiency.[3] When commodities prices fell in the 1920s and rural migration to cities increased, these negative attitudes accelerated to assuage regrets many new urbanites had over abandoning rural life. A new vocabulary of condescension, in which farmers were known as bumpkins, hicks, yokels, and rubes reflected this notion of urban superiority as did "rube songs" popular in the 1920s mocking farmers as stupid, closed-minded, and unsophisticated.[4]

Although encountered less in contemporary rhetoric, the "rube" served as a particularly important icon of early 20th-century America. The word originated in England or America as a shorthand version of the surname "Reuben" to denote someone unsophisticated and countrified. In Genesis 29:32, Reuben, or the Hebrew version Rĕ'ūḇēn, is the name of the firstborn son of the patriarch Jacob and his wife Leah and the English began using it as a surname during the Protestant Reformation. It is unclear why the Anglo-Americans associated the surname Reuben with a rustic lifestyle. One theory is that Reuben denoted a countrified person due to the supposed widespread occurrence of biblical forenames in rural communities in the late 19th century.[5] The rural rube icon may go even further back to the earliest years of colonial New England as a stock character of the British theater named "Hodge" described by one historian as "an archetypal Yorkshire countryman characterized by a mix of provincialism, cunning calculation, and quick but rude rejoinders." Hodge could have morphed into several "Yankee" stereotypes in rural

New England depicted in plays, painting, almanacs, cartoons, newspapers, and journals from 1820 to 1850. This Yankee took a variety of names from Yankee Doodle, Brother Jonathan, Jack Downing, Deuteronomy Dutiful, Salon Shingle, and Rip Van Winkle, but they all featured a rural person with a gangly body, ill-fitting and old-fashioned cloths, a simple or foolish grin, a particular dialect, and a hidden shrewdness and tenacity that could not be trusted. This northern precursor of the rural working-class "rube" split off from the "hillbilly" stereotype as the 19th century progressed with the later representing poor whites in the South or in mountain regions.[6]

In any case, both the rube and the hillbilly functioned as ways for urban elites to justify the changing social order that accompanied rapid industrialization by denigrating the rural way of life. Indeed the 1920 census was the first in American history in which urban dwellers outnumbered their rural cousins.[7] At the same time, the cunning and "get ahead at any price" imagery of the rube allowed pernicious negative images of rural people to persist without alienating the common man's American identity associated with a rugged individualism, strength, and courage.[8]

The rural rube also has links to 19th-century racial stereotypes about rural African Americans. Namely, the rube stereotype as it appeared in the 1920s emerged when urbanites adapted European folk stock characters for vaudeville, burlesque, and minstrel shows in cities beginning in 1870. In this racial context, the rube again served an important role in working out "what it means to be urban" in the U.S. by contrasting city living with rural whites and African Americans. After the Civil War, a new "bias of the modern and urban, drew lines between the progress of the city and what was often depicted as its opposite, beyond the city limits" that had not existed in prior European or Yankee depictions of the rural.[9]

Writers employed the rube stereotype about Corn Belt farmers from the later nineteenth into the early 20th century. Although many authors lionized farmers in 18th-century America in heroic terms, they wrote favorably about mostly the Anglo-American farmer in the North who comprised the population's majority.[10] Once these Anglo-American farmers joined German and other immigrant farming groups to become a minority "other," observers assigned them pejorative characteristics in the late 18th century. For example "Wal, I Swan (Git Up, Napoleon)" written in 1907 by Benjamin Hapgood Burt, depicts an uneducated farmer who becomes drunk at a state fair and gives away his bull while expressing ignorance of his son's more glamorous life in Philadelphia.[11] This view of farmers as unsophisticated and anti-modern also pervaded the work of urban writers and scholars in the 1920s. In "The Husbandman," first published in the *American Mercury* in 1924, Henry L. Mencken described the farmer as "a serious fraud and ignoramus, a cheap rogue and hypocrite, the eternal Jack of the human pack," unable to understand any political and economic issue beyond his own self-interests.[12] In the *Wireless Age* magazine, one urban actress described rural America with "its warped outlook on life, its ignorance of current events, its mean and petty superstitions."[13]

Midwest farmers reacted to these rube stereotypes with attempts to create a discourse of condescension about urban Americans. These attempts to parry insults against rural life reveal that the rural-urban conflict existed not only in the halls of Congress and in debates over tariff policies, but as a lived experience of farmers. Rural denizens carried self-consciousness about the rube as a dominant image on almost a daily basis and it had seeped into even rural folk culture by the mid-1920s. During a speech to a group of Corn Belt teachers in St. Paul, Minnesota in 1921 by the managing editor of the *Farm Journal, Farmer's Wife*, Ada Melville Shaw recounted how mainstream non-farm-journal media

> Took account of farms and farmers chiefly from the standpoint of condescension toward men and women who made daily close contact with the dirt of the field, the barn, the chicken yard, the hog pen; or, from the standpoint of the farmer as a comic or picturesque contribution to a certain type of fiction.[14]

Rural Americans in the Midwest, therefore, knew they needed to perform in ways that combated these negative images of the farm coming out of urban America. One Corn Belt farmer and journal editor for *Better Farming*, Fred L. Chapman, for example, wrote a column in 1923 entitled "The Real Rube" in which he recalls,

> I can remember the time when I was a little boy, I almost dreaded a trip to the city because I might appear green to the more sophisticated lads who were better acquainted with street cars and tall buildings. I did not know then that there is a big world of things which the country boy and girl knows but which are utterly foreign to his city cousin. We have all heard of the town girl visiting the country who confessed she did not know the difference between a wild tree and an apple tree.

Chapman concludes, "I'll say that the laugh is about the city kid. He is the real 'rube.'"[15]

The clash over the urban yokel stereotype of rural Americans stood as the cultural manifestation of a broader rural-urban conflict in the 1920s that also presented political and economic disputes. The falling prices for farm products alongside the rising prices of manufactured goods farmers had to buy from urban producers created a tangible struggle between urban and rural interests. This conflict manifested itself politically in the debate over the McNary-Haugen bills and other measures to raise prices of commodities.[16] In 1919, farmers formed the American Farm Bureau Federation (AFBF), a political advocacy organization, and congressmen joined in the "Farm Bloc" to reverse this price disparity. Farm journals and farmers in the 1920s exhibited an obsession with the McNary-Haugen Bill and other legislation intended to raise prices for farm commodities, but which never passed into law. President Calvin Coolidge twice vetoed the McNary-Haugen Bill on February 25, 1927 and May 23, 1928. Editorialists in farm journals vehemently attacked

Coolidge for killing the bills while leaving similar tariff protections for urban big business in place.[17] For example, one writer mocked the President writing, sarcastically "Ah, the courage of Coolidge is vastly underestimated. He bares his breast to the arrows of outrageous fortune in defense of the steel trust and the banker with an unparalleled calmness and remains Cool!"[18]

The farmer needed a means other than the ineffective Farm Bloc and the federal government to compete with the "outrageous fortune" of these urban interests. Many farmers expressed the idea that technological use represented a more moral means of solving the farmer's financial problems in the 1920s than legislation like the McNary-Haugen Bill because the former promised to enhance the farmers' independence. For many of these farmers, technology promised a means of modernizing and competing with industrialism without depending on politicians whose farm blocs seemed ineffective to combat urban interests in the pro-business political environment of the 1920s. Harry Rehm of North Dakota, for example, wrote into *Wallaces' Farmer* stating that the only way tariff legislation could help hog farmers "would be to appoint guardians for those of us who can not think for ourselves. Personally, I like to handle my own affairs as far as possible, even if I go bankrupt doing it."[19]

Harold Rohwer of O'Brien County, Iowa opposed the McNary-Haugen Bill because

> If the new Ford is not a success will Henry shout 'rotten legislation?' Indeed not. He will say, 'You have made a mistake somewhere, Henry.' And furthermore he won't ask congress to find it and remedy it; he will do it himself; and if the farmers can't pull themselves out of the rut of their own making they are a sorry lot indeed.

Rohwer then listed several items that he believed would help the farmer more than legislation from Washington including, "More following of daily markets and less following of murder trials; More tractors and less automobiles; More early morning rises, less midnight frolicking; More and longer working days in a year." The farmer, according to Rohwer, must abandon the image of "the old one-crop farmer" and embrace productive technology, hard work, and education to combat urban dominance.[20] In short, the farmer must take matters into his own hands by modernizing. As *Better Farming* proudly exclaimed as early as 1925,

> Farm work with modern machinery has come to be more and more of an engineering job. The farmer who handles the binder, a tractor, or a truck, who installs a drainage system on his farm, or who builds his small farm building in his spare time, must be somewhat of an engineer as well as a farmer.[21]

Midwesterners not only perceived of their money flowing into cities but their labor force too, as young men and women abandoned the farm for urban jobs. The farmers of the Corn Belt found themselves in an awkward position of working as both capitalist owners and laborers without identifying with either urban capital *or* labor. The urban industrialists feared that farmers in the early 20th century would join radical labor movements (absent farm bureau organizations) a concern reflecting yet another urban misunderstanding of rural America. The farmer not only had substantial amounts of capital invested in property, he or she felt superior to labor.[22] At the same time, farmers saw urban business as a threat both economically and culturally. Organizations such as the Farmer's Union consciously regarded any farm organization perceived to align with the Chamber of Commerce, big business, or agricultural colleges with suspicion.[23] While not all farmers viewed urban interests in such radical terms, they still retained the German agrarian view of urban actors outside the farm as a dangerous "other."

People throughout the rural Corn Belt realized that "the roaring 20s" had passed them by, and they blamed urban interests, both labor and capital, for their unequal standing.[24] Farmer J.F. Murphy of Clark County, Illinois wrote into *Wallaces' Farmer* in 1924 the tongue-and-cheek suggestion of fixing the farmer's price problems by abandoning the McNary-Haugen bills altogether and take a "short cut – one that congress cannot vote on. Organize the farmer into a federation and cut the working day to an eight-hour day; the week into five and one-half days." After all, Murphy asked "Why should the farmer work more hours per day than any other people?"[25] Murphy's sarcasm reveals an underlying serious truth that most farmers felt slighted by both urban businesses and labor living the modern "good life" on the shoulders of low food prices made possible by the overworked farmer.

Farmer Blanche Stein wrote into *Better Farming* angrily criticizing the U.S. government for preaching equality and bringing peace to Europe in World War I while allowing farmers to be "victims of distressing turmoil and unrest within our own borders." After lamenting rural-to-urban migration, Stein argued, "Our country will never be a democracy until we extend equality from the city to rural areas – until we give to the farmer equal rights socially, recreationally, educationally, religiously, and economically." This rural-urban inequality, according to Stein arose from a mix of high wages in the city and the drudgery of farm work.[26]

As Stein and others in the Corn Belt saw the Farm Bloc in Congress as ineffective and failing to push through agricultural legislation, technology became a way farmers could take it upon themselves to level the economic and cultural playing field. Never mind that farmers would later gain substantial support from the federal government through Franklin Delano Roosevelt's New Deal programs such as the Agricultural Adjustment Act and the Rural Electrification Administration (REA). Even prior to the New Deal, rural Midwesterners often benefitted from working with progressive reformers and extension programs.[27] While farmers gained from such government and university efforts and supported Roosevelt, the identities of Midwesterners as independent agrarians standing up to city dwellers through technological use had also been established by the 1930s. The view of urbanites and a distant government as a threat fit the Jeffersonian notion of farmers heroically overcoming obstacles and the German distrust of an outside other, regardless of how many rural homes were lit by the REA. At the same time, the active presence of progressive reformers and extension services in the rural Midwest in the early 1900s created a social landscape ripe for the rise of a uniquely rural version of modernity that incorporated both Jeffersonian and German elements.

Rural denizens in the 1920s knew they had to modernize or risk losing the contest with urban America or, even worse, submitting to urban reformer's view of what rural America should look like. Some farmers prior to the New Deal such as E.S. Murphy from Fremont County, Iowa even suspected a conspiracy of urban interests in which the federal government's farm investigators colluded with speculators to construct crop reports that drove down commodity prices. As Murphy wrote into *Wallaces' Farmer*, "Can you explain why the speculators' reports coincide with the government reports, no matter whether the latter is accurate or inaccurate? Do you not think it is a fact that they get their information from the same source?"[28]

Farmers wrote into farm journals suggesting all kinds of measures for solving the price problem, such as Midwest states forming marketing associations to fix prices or farmer-owned banks.[29] Others formed cooperatives to set higher prices in an attempt to combat what rural Americans saw as large urban corporations colluding to drive down the value of farm commodities.[30] But a discourse and identity of Jeffersonian and German agrarianism mandated that farmers act independently to uplift their own production processes and create their own modernization rather than rely on outside reformers. As Joseph Frazier Wall stated in writing the bicentennial history of Iowa in the late 1970s, Iowans saw their state's history as one in which farmers tried "to live and let live" unbothered by "eccentric purveyors of panaceas" conspiring to "disrupt his life, and drag him, willy-nilly, into the whirlwind of ideological dispute."[31] With this cultural conception of fierce independence threatened from outside forces, the farmer needed to conceive of himself as developing his own modern rural identity through technological use.

To combat the myriad of threats farmers perceived as swirling around them in the 1920s from urban interests and confronted by stereotypes that affronted their moral sense of self, Midwest agrarians needed to develop and perform a new rural identity. Through technological use, male and female farmers would help to devise a distinctly rural modernity beginning in the early 20th century that preserved aspects of German and Jeffersonian agrarianism. More than any economic or functional motivation, this desire to form and perform identity drove technological change in the Corn Belt.

Urban Industrialism and Rural Resistance

The rube stereotype that viewed farmers as a social problem formed just one feature of a more comprehensive bundle of discourses and identities, I have named *urban industrialism*. Farmers in the Corn Belt did not oppose modernity per se, only urban industrialism proffering a specific urban version of modernity. Additionally, this urban industrialism, as it existed within the social milieu of 1920s America, motivated the formation of a competing rural discourse identity bundle of *rural capitalistic modernity*. I urge the reader to view the history of rural resentment of urban America, including the rural-urban conflict of the 1920s, as not just a struggle over competing tangible interests such as economic power. Instead, the contest between rural and urban America results from a clash of two opposing visions of what it means to be "modern." Technology plays a central role in this conflict because users reinforce their competing identities and send signals to the other side through performative use of material objects. People use technology both to realize the type of person they think they should be and to perform this identity for an outside "other" regarded as less moral. Since the 1920s, the pattern of audience has taken a similar shape in the form of farmers using technology to combat rube stereotypes and urban versions of modernity.

I seek to see the meaning of technology through the eyes of the farmers using the artifacts, rather than from the perspective of the outside observer. A careful reading of the discourse of farmers in the Corn Belt early in the 20th century taking their points of view seriously reveals that they sought to modernize, but they rejected visions from "others" in academia, business, or government to literally "make every farm a factory."

To many Midwesterners, a diverse group of cultural, business, and commercial actors in early 20th-century America echoed the myth that American farmers saw themselves as old fashioned and opposed to progress and modernism. Many supporters of this notion of a persisting Jeffersonian yeoman ideal hailed from industrialized urban areas or from the federal government. Rural groups or institutions also critiqued urban life by juxtaposing it with an outdated and idealized agrarian imagery. Farmers saw two groups of progressive reformers perpetuating this anti-modern view of farmers prior to World War I. The first, the Country Life Movement, consisted entirely of urban-based educators, social scientists, religious leaders, and philanthropists concerned that increased rural-to-urban migration had caused a drain of intelligent and ambitious people in rural communities, potentially leading to a fall in U.S. agricultural productivity. In 1907, President Theodore Roosevelt lent legitimacy to the movement by forming the Country Life Commission to study and write reports on rural problems. This commission of experts, however, did not contain a single farmer. Its chair, Liberty Hyde Bailey, worked as a horticulturalist at Cornell University.

Following the issuing of questions, receiving letters from farmers, and holding hearings and conducting interviews with farmers in forty states, the Commission's 1909 report concluded that farmers had failed to keep pace with urban Americans in terms of standard of living, education, and methods of productivity. What rural America needed, among other reforms according to the report, were urban education models to teach children agricultural skills that many rural denizens thought they had already learned on the farm.[32] "Life on the farm," the report stated, "must be made permanently satisfying to intelligent, progressive people." Since the

Commission determined that rural America lacked these types of "progressive" people, they framed their task as, "The work before us, therefore, is nothing more or less than the gradual rebuilding of a new agriculture and new rural life."[33] Many on the Commission even viewed farmers as less intelligent and enterprising than urban dwellers because urban migration resulted in a thinning out of the rural gene pool, suggesting that the popularity of eugenics among some progressive reformers informed their view of rural Midwesterners.[34]

The Country Life Movement exemplified the traditional notion (which my theory of performative use disputes) that farmers only have agency in determining their lifestyles, work processes, and uses of technology to the extent that urban experts and activists have shown them the efficient or best way of doing things. Indeed, the Commission's report elicited defensive attitudes from many farmers in rural America who saw it as evidence of an attack on the rural way of life by urban voices (Figure 2.1). In Figure 2.1, a cartoon published for a rural audience in a February 1928 issue of the *Farm Journal*, three intellectuals, a historian, a psychologist, and an economist dressed in suits worn by the urban upper classes sling handfuls of mud at a woman labeled "agriculture." The woman dressed in a robe and wearing a laurel wreath with a cornucopia overflowing with food next to her, represents the moral purity of agriculture depicted for years under Jeffersonian agrarianism.

FIGURE 2.1   A *Farm Journal* editorial reflects the rural-urban conflict of the 1920s as well as the alienation felt by some in rural America toward academics and reformers. Walter Burr, "Rural Bunk: Favorite Indoor Sport is Slandering Agriculture: Why do American Farmers Continue to Stand for It?," *Farm Journal* (February 1928): 9–10.

She covers her face in a defensive position with her arm to shield herself from the filth thrown by the scowling and aggressive intellectuals. The bountiful cornucopia highlights the injustice, from a rural point of view, of an urban-dominated economic and social system that flings mud at the very people who feed and support it, another Jeffersonian idea. The artist of the cartoon, entitled "Rural Bunk," clearly intends to reflect rural resentment and show support for resistance that already existed, evidenced by the drawing's subtitle which asks "Favorite indoor sport is slandering agriculture: Why do American farmers continue to stand for it?"

Farmers both worked with progressive reformers during this period while also resenting them and challenging their claims, hence my argument that the rural-urban conflict involved contesting notions of modernity. As David Danbom states in *Born in the Country: A History of Rural America*, "rural people did not agree that there was a crisis in their schools or even that there was anything very wrong with them." The backlash by rural America only confirmed the urban misconception that farmers were backward and that they opposed modernity. Stories in the urban-based popular press and reports by progressive reformers following the Country Life Commission report fanned rural resentment. For example, the *Ladies Home Journal* in 1909 criticized farm husbands for purchasing labor saving equipment in the fields before the home, where women supposedly work. This report alienated many farm women who viewed themselves as partners in the farming operation rather than as homemakers (see Chapter 4). The founding editor of Wallaces' Farmer, Henry Wallace, also characterized the *Ladies Home Journal* article as slanderous to rural men.

The second group of urban progressive reformers resented by some farmers were economists and scientists led by experimental agriculturalist such as Seaman Knapp. Beginning in 1902, Knapp started promoting agricultural extension services, whereby county agents would demonstrate innovative farming methods. The extension system gained federal support with the passage of the Smith-Lever Act in 1914, which created bureaus organized by state land-grant agricultural and mechanical colleges. While the extension system became popular in the rural South, where farmers saw it as a means to end their dependence on cotton, it often met with resistance in the Midwest because many farmers saw the extension agents as further evidence of unnecessary meddling by urban reformers. Agents also alienated some women by focusing on domestic skills such as cooking even though for many Midwest farm families, women participated in farm production.[35] As with negative reactions toward the Country Life Commission Report, this inability of some Midwest farmers to embrace the extension program served as further evidence of farmers' anti-modern Jeffersonian yeomanism from the point of view of urban dwellers.

Rural objections to schemes implementing urban industrialism had little to do with economics. Many American agricultural economists, led by M.L. Wilson, for example, attempted to implement an urbanized model of industrialized wheat production modeled on his work with collectivized Soviet peasants in the 1920s and 1930s.[36] This industrialization never actually occurred in American agriculture, even though Wilson and others attempted to implement it in rural Montana, because it clashed with other aspects of rural modernization such as individual ownership of property, competition, individualized incentive, and the value of the nuclear family. The farmer does not see himself as relinquishing control and ownership to the same extent as urban factory laborers even in contemporary agriculture.

Another model of urban industrialism in the 1920s appeared in the form of Hawthorn Farm, a 4,200-acre estate owned by the electricity mogul Samuel Insull, on which 35 men operated 17 tractors and 24 horses. Insull bought 24 separate family farms over a 17-year period and combined them into one huge operation. The agricultural engineer E.R. Wiggins celebrated Hawthorn Farm at the annual meeting of the Society for Agricultural Engineers in Lincoln, Nebraska on June 18, 1924. In an article published in *Better Farming,* Wiggins described Insull's creation as the wave of the future:

[T]here is no farm that could be selected to show better the saving in operating expenses brought about by mechanical power and machinery. It is so organized as to bring man labor to a minimum, and have the work done in its proper season. That the plan is right is shown by the fact that since 1917, the year the largest number of men were employed, the farm has been greatly increased in size, yet the number of men has not increased. The efficient application of labor-saving machinery is the answer.

Further, Hawthorn Farm "solved many agricultural engineering problems" such as arranging all fields at more than 100 acres with as few fences as possible "to make straight lines of field sides."

So given the efficiency and innovations from Wiggins' perspective as an agricultural engineer, why did Hawthorn Farm *not* become the model of agriculture in the Corn Belt? A traditional view that technology and methods appear in farming only for functional or economic reasons, as Wiggins presumes, would suggest that the entire Corn Belt should have become like Hawthorn Farm in the coming years. After all, modern farming, according to Wiggins and the traditional view of agriculture, is simply the rational application of labor and capital to increase efficiency and profit and Hawthorn Farm carried this task out with greater effect than any other model at the time.

Wiggins' description of the relationship between labor and technology on Hawthorn Farm provides the answer of why its blueprint failed to spread throughout the Corn Belt. Wiggins writes:

In the general labor and horses are centralized in one place, and the mechanical power in another place. There are two main boarding houses, and it has been found that it pays to keep the men concentrated. "It's best" said Mr. J. C. Reuse, Manager, "to keep the gangs of workers together, as it saves time. It is cheaper to haul the men around in a bus than to have them scattered all over the place."

None of the 35 men owned any of the land or equipment or farmed with the family at the center. They worked under the direction of a manager and lived in communal boarding houses rather than the family farm home. The manager conceived of the workers as "gangs" rather than individuals because "the idea of concentration of effort prevails. As many workers as possible work on one job at the same time." The workers labored under a strict hierarchy of managers. A head serviceman and his assistant under the manager had "entire charge of all tractors and equipment" and a field boss directed all use of these technologies. In fact, this efficient centralized management, according to manager Reus, allowed for *less* use of modern equipment. Wiggins noted that the workers received high wages to ensure the most efficient operation of machinery.[37]

Given the complaints of farmers in farm journals of a lack of a regular paycheck and work that often led to no profit, one would expect that any rational economic actor would choose the high wages on farms like the Hawthorn Farm over the unstable earnings on a family farm. Further, one would expect farmers to enthusiastically sign up for such work as laborers who would be paid regardless of the well-known risks inherent to agriculture such as poor weather, price fluctuations, and crop diseases, particularly in an era before crop insurance. Unlike many farmers in the Corn Belt in the 1920s for whom electricity represented an expense, the workers on Hawthorn Farm even enjoyed electric lighting as part of their monthly salaries.[38]

Similarly, Scott Rowley, a corporate law professor, made the argument in favor of reimaging farms as carbon copies of urban factories in a 1928 *Wallaces' Farmer* article entitled "Do We Need Corporation Farming? An Argument for Putting Corporate Methods into Agricultural Production." Rowley begins his essay with language subtly suggesting that he does not identify with the previous dominant discourse of Jeffersonian agrarianism that moralized the yeoman farmer by addressing only "those engaged in farm management, and also of landlords who have been unable to secure satisfactory tenants." Rowley then suggests a plan to implement "modern business practices" by organizing farmers in the same geographical vicinity to form a large corporation by which:

Each owner and stockholder might turn over personal property on his farm-horses, stock, equipment, crops – to the corporation, in payment for his shares of stock. A competent manager, preferably a graduate of an agricultural college, could be hired at a fair salary, and a certain percentage of the corporation profits, perhaps with an opportunity to purchase a certain amount of stock in the corporation. Each of the landowners could rent his farm to the operating corporation, for a cash rental and could also, should he so desire, hire his services to the corporation, at the prevailing scale of wages for farm-hands. Each of such owners could retain his farm residence for his own use, and his services could, primarily, be utilized on his own farm. In case he might not live on the farm, the farm residence could be occupied by one of the employes [*sic*]. The idea would be that at least one man should, primarily, live on and work one farm, but that he should be subject to the direction of the manager, who might send him, temporarily, to work elsewhere.

Given the economic advantages that this arrangement would provide in the form of economies of scale, greater specialization of labor, and regular wages, Rowley could not understand why any farmer would resist his plan. After all, manufacturers often sold their plants or combined with parent companies in the 1920s to gain economic advantages, so why would owners of capital and property in the country not do the same?

As with Wiggins' evaluation of the benefits of Hawthorn Farm, Rowley reasoned that farmers would surely welcome his more rational, efficient, and modern corporate reorganization. A response essay by an unnamed farmer and journal editor entitled "Are We Ready for This?," however, reflects a long and widespread resistance to plans like Rowley's scheme and Hawthorn Farm. The rebuttal cynically begins, "Corporation farming has had a lot of advertising lately. Here is some more," as if to say "here we go again!" The farmer, in claiming to speak for the majority of farmers then stated, "We doubt if corporation farming can or should succeed. It has yet to prove its efficiency. With few exceptions, corporation farms have lost money." More importantly, the farmer argued,

> Even if it were efficient, corporation farming would still be socially undesirable. We need the family farm; we need the social qualities it generates. Love of the land is something too precious in the life of a farmer, and in the long life of a nation, to be traded for stock in a corporation-farm.[39]

The reason models like Hawthorn Farm or Rowley's corporation farming failed to become the model of agriculture in the coming years lies in the fact that despite the potential economic advantages of reorganizing every farm as a literal factory, such urban-derived creations threatened the farmer's identity as an independent, family-oriented, producer in control of work and technology. Wiggins' training as an agricultural engineer and Rowley's education as a lawyer, which taught them to view the world in terms of rational economic processes, blinded them to the fact that farmers imbed their identities and ideas about morality within the equipment they use and the fields they plowed. While promising to make the farmer financially richer, schemes of urban industrialism also sought to take away from the farmer and the farm wife the unspoken things that were ultimately more important to them: the farm home, the family, independence, and the Jeffersonian and German view of themselves as moral agrarians. Throughout this book, the reader will discover that rural Americans in the Corn Belt would see any agricultural arrangement that took authority over production processes away from the farmer and removed the farm family from the center of life as a threat to rural identities from an outside "other."

Thus, from the 1920s onward, farmers did not object to modernity per se but to conceptions or forms of modernity that threatened agrarian identities developed by Jeffersonian and German agrarianism with deeply ingrained rural notions of morality. Farmers in the 1920s looked at their inferior position within the rural-urban conflict and realized, even without articulating it, that they needed to form their own version of modernity and use the latest technologies, or men like Insull and Rowley would end up doing it for them. From the view of people in the Corn Belt in the 1920s, the Jeffersonian yeoman must not be reduced to the status of the "mindless" urban factory worker. Nor could the old Jeffersonian yeoman, with his small dirt farm, his horse-drawn plow, his one-room schoolhouse, and his wife raising a few chickens win this conflict between urban and rural America. The yeoman farm family needed something else very quickly while not losing their moral selves. They needed an updated yeomanry. They needed technology, and they needed it on their own terms. Farmers would use technology to combat elements of rural industrialism that threatened their control of work while adopting other elements such as faith in progress and scientific rationalism. Rural Midwesterners blended these progressive features with Jeffersonian ideas like the farmers as a hero facing obstacles and the superior morality of rural life with German elements such as using objects to perform identity and family-based production processes. The resulting rural identity was rural capitalistic modernity and the following chapter details how farmers used technology to create and perform this new rural identity.

Notes

1 Robert M. LaFollette, "The Danger Threatening Representative Government," Mineral Point, 1897, *Robert M. LaFollette Papers* (Wisconsin Historical Society).
2 "Hear Bobcat's Purr in Broad 'A' of Harvard," *Chicago Daily Tribune* (December 2, 1935), 1.
3 Theodore Saloutos and John D. Hicks, *Agricultural Discontent in the Middle West: 1900–1939* (Madison: University of Wisconsin Press, 1951), 21–22.
4 Shideler, "'Flappers and Philosophers,' and Farmers: Rural-Urban Tensions in the Twenties," 289.
5 *Oxford English Dictionary* (Oxford: Oxford University Press, 2017), s.v. "rube;" *Oxford English Dictionary*, s.v. "Reuben."
6 Anthony Harkins, *Hillbilly: A Cultural History of an American Icon* (Oxford: Oxford University Press, 2004).
7 Hal S. Barron, "Rural America on the Silent Screen," *Agricultural History* 80, no. 4 (2006): 384.
8 Harkins, *Hillbilly*, 5–15.

9 Mark Evans Bryan, "Yeoman and Barbarians: Popular Outland Caricature and American Identity," *The Journal of Popular Culture* 46, no. 3 (2013): 463–480.

10 Patnode, 286–287.

11 *The Folklorist*, http://www.folklorist.org/song/Wal_I_Swan_(Giddyap_Napoleon,_ Ebenezer_Frye). (accessed 3/17/16).

12 Henry L. Mencken, "The Husbandman," *A Mencken Chrestomathy* (New York: Knopf, 1924), 360–364.

13 Mona Morgan, "Small Town's Radio Receivers Will Emancipate It, Actress Predicts," *Louisville Courier Journal* (March 1923), Section 2, quoted in Patnode, 286.

14 Ada Melville Shaw, "The Country Child's Schooling: An Answer to the Question, 'How Much Education Does the Rural Child Need?'" *Farmer's Wife* 23, no. 8 (January 1921): 288, 308.

15 Fred L. Chapman, "The Real Rube," *Better Farming* 46, no. 12 (December 1923): 2.

16 George N. Peek, "The McNary-Haugen Plan for Relief," *Current History* 29, no. 2 (November 1, 1928): 273; Darwin N. Kelley, "The McNary-Haugen Bills, 1924–1928: An Attempt to Make the Tariff Effective for Farm Products," *Agricultural History* 14, no. 4 (October 1940): 170–180.

17 "Farm Bureau Holds Convention: Iowa Farmers Renew Demands for McNary-Haugen Bill-Hearst Re-Elected." Wallaces' Farmer 53, no. 3 (January 20, 1928): 90 (7, 14).

18 Saloutos and Hicks, *Agricultural Discontent in the Middle West*, 399–400, quoting *Farmer's Union Herald*, March, 1927.

19 Harry Rehm, "'What Farmer Can Kick Now,'" *Wallaces' Farmer* 53, no. 3 (January 20, 1928): The Voice of the Farm, 94 (10).

20 Harold Rohwer, "Helping the Farmer," *Wallaces' Farmer* 53, no. 3 (January 20, 1928): The Voice of the Farm, 94 (10).

21 M.M. Jones, "Farm Shop will Save Money," *Better Farming* 48, no. 1 (January 1925): 14.

22 Saloutos and Hicks, 259–260.

23 Ibid., 283–284.

24 See for example, Thomas E. Wilson, "Farmers Are in a Fix," *Better Farming* 44, no. 1 (January 1921): 4.

25 J.F. Murphy, "Eight Hour Day for the Farmer," *Wallaces' Farmer* 49, no. 46 (November 14, 1924): Voice of the Farm, 1479 (11).

26 Blanche Stein, "Building a Rural Civilization: It Must be Based Upon the Highest American Ideals, Expressed in Terms of Christianity, Recognizing the Farmer's Right to Economic Welfare," *Better Farming* 45 no. 1 (January 1922): 4, 16, 19.

27 For historians discussing the importance of the Progressive Reform movement and extension service in the modernization of Midwest agriculture see Dorothy Schwieder, *75 Years of Service: Cooperative Extension in Iowa* (Ames: Iowa State University Press, 1993); W.D. Rasmussen, *Taking the University to the People. Seventy-Five Years of Cooperative Extension* (Ames: Iowa State University Press, 1989); Laurie M. Carlson, *William J. Spillman and the Birth of Agricultural Economics: Missouri Biography Series* (Columbia: University of Missouri Press, 2005); Nancy K. Berlage, *Farmers Helping Farmers: The Rise of the Farm and Home Bureaus, 1914–1935* (Baton Rouge: Louisiana State University Press, 2016); Elizabeth Sanders, *Roots of Reform: Farmers, Workers, and the American State, 1877–1917* (Chicago, IL: University of Chicago Press, 1999); Catherine M. Stock, *Rural Radical: Righteous Rage in the American Grain* (Ithaca, NY: Cornell University Press, 1996); Gabriel N. Rosenberg, *The 4-H Harvest: Sexuality and the State in Rural America* (Philadelphia: University of Pennsylvania Press, 2015); Victoria Saker Woeste, *The Farmer's Benevolent Trust: Law and Agricultural Cooperation in Industrial America, 1865–1945* (Chapel Hill: The University of North Carolina Press, 1998); Charles Postel, *The Populist Vision* (Oxford: Oxford University Press, 2007).

28 E.S. Murphy, "Government Crop Reports," *Wallaces' Farmer* 53, no. 16 (April 20, 1928): Voice of the Farm 638 (6).

29 John H. Bundy, "Grain Marketing," *Wallaces' Farmer* 46, no. 2 (January 14, 1921): 54 (14); W. S. Short, "The Banker and the Farmer," *Wallaces' Farmer* 46, no. 2 (January 14, 1921): 54 (14).
30 "Individualism and Collectivism," *Wallaces' Farmer* 46, no. 1 (January 7, 1921): 5; "Co-Operative Legislation for Iowa," *Wallaces' Farmer* 46, no. 1 (January 7, 1921): 8.
31 Wall, *Iowa*, 66.
32 *Report of the Commission on Country Life* (Washington, DC: GPO, 1909), 1–9, 25.
33 Ibid., 17.
34 Kline, *Consumers in the Country*, 91.
35 Danbom, *Born in the Country*, 174–175.
36 Deborah Fitzgerald, "Blinded by Technology," 459–486; Fitzgerald, *Every Farm a Factory*.
37 E. R. Wiggins, "Growing 1,000 Acres of Corn on Hawthorn Farm," *Better Farming* 47, no. 7 (July 1924): 3–4, 7.
38 Ibid., 3.
39 Scott Rowley, "Do We Need Corporation Farming? An Argument for Putting Corporate Methods into Agricultural Production," *Wallaces' Farmer* 53, no. 10 (March 9, 1928): 376 (6); "Are We Ready for This?" *Wallaces' Farmer* 53, no. 10 (March 9, 1928): 376 (6).

Bibliography

"Are We Ready for This?" *Wallaces' Farmer* 53, no. 10 (March 9, 1928): 376 (6).

Barron, Hal S. "Rural America on the Silent Screen." *Agricultural History* 80, no. 4 (2006): 383–410.

Berlage, Nancy K. *Farmers Helping Farmers: The Rise of the Farm and Home Bureaus, 1914–1935*. Baton Rouge: Louisiana State University Press, 2016.

Bryan, Mark Evans. "Yeoman and Barbarians: Popular Outland Caricature and American Identity." *The Journal of Popular Culture* 46, no. 3 (2013): 463–480.

Bundy, John H. "Grain Marketing." *Wallaces' Farmer* 46, no. 2 (January 14, 1921): 54 (14).

Burr, Walter. "Rural Bunk: Favorite Indoor Sport Is Slandering Agriculture: Why do American Farmers Continue to Stand for It?" *Farm Journal*, 52, no. 2 (February 1928): 9–10.

Carlson, Laurie M. *William J. Spillman and the Birth of Agricultural Economics: Missouri Biography Series*. Columbia: University of Missouri Press, 2005.

Chapman, Fred L. "The Real Rube." *Better Farming* 46, no. 12 (December 1923): 2.

"Co-Operative Legislation for Iowa." *Wallaces' Farmer* 46, no. 1 (January 7, 1921): 8.

Danbom, David B. *Born in the Country: A History of Rural America*. Baltimore, MD: The Johns Hopkins University Press, 1995.

"Farm Bureau Holds Convention: Iowa Farmers Renew Demands for McNary-Haugen Bill-Hearst Re-Elected." *Wallaces' Farmer* 53, no. 3 (January 20, 1928): 90 (7, 14).

Fitzgerald, Deborah. "Blinded by Technology: American Agriculture in the Soviet Union, 1928–1932." *Agricultural History* 70, no. 3 (1996): 459–86.

———. *Every Farm a Factory: The Industrial Era in American Agriculture*. New Haven, CT: Yale University Press, 2003.

Harkins, Anthony. *Hillbilly: A Cultural History of an American Icon*. Oxford: Oxford University Press, 2004.

"Hear Bobcat's Purr in Broad 'A' of Harvard." *Chicago Daily Tribune*, December 2, 1935, 1.

"Individualism and Collectivism." *Wallaces' Farmer* 46, no. 1 (January 7, 1921): 5.

Jones, M.M. "Farm Shop Will Save Money." *Better Farming* 48, no. 1 (January 1925): 14.

Kelley, Darwin N. "The McNary-Haugen Bills, 1924–1928: An Attempt to Make the Tariff Effective for Farm Products." *Agricultural History* 14, no. 4 (October 1940): 170–180.

Kline, Ronald R. *Consumers in the Country: Technology and Social Change in Rural America*. Baltimore: Johns Hopkins University Press, 2000.

LaFollette, Robert M. "The Danger Threatening Representative Government." Mineral Point, 1897. *Robert M. LaFollette Papers* (Wisconsin Historical Society).

Mencken, Henry L. "The Husbandman." In *A Mencken Chrestomathy*, edited by H.L. Mencken, 360–364. New York: Knopf, 1924.

Morgan, Mona. "Small Town's Radio Receivers Will Emancipate It, Actress Predicts." *Louisville Courier Journal* (March 11, 1923): sec. 2.

Murphy, E.S. "Government Crop Reports." *Wallaces' Farmer* 53, no. 16 (April 20, 1928): Voice of the Farm 638 (6).

Murphy, J.F. "Eight Hour Day for the Farmer." *Wallaces' Farmer* 49, no. 46 (November 14, 1924): Voice of the Farm, 1479 (11).

Oxford English Dictionary. Oxford: Oxford University Press, 2017.

Patnode, Randall. "'What These People Need Is Radio:' New Technology, the Press, and Otherness in 1920s America." *Technology and Culture* 44, no. 2 (2003): 285–305.

Peek, George N. "The McNary-Haugen Plan for Relief." *Current History* 29, no. 2 (November 1, 1928): 273.

Postel, Charles. *The Populist Vision*. Oxford: Oxford University Press, 2007.

Rasmussen, Wayne D. *Taking the University to the People. Seventy-Five Years of Cooperative Extension*. Ames: Iowa State University Press, 1989.

Rehm, Harry. "What Farmer Can Kick Now." *Wallaces' Farmer* 53, no. 3 (January 20, 1928): The Voice of the Farm, 94 (10).

Report of the Commission on Country Life. Washington, DC: GPO, 1909.

Rohwer, Harold. "Helping the Farmer." *Wallaces' Farmer* 53, no. 3 (January 20, 1928): The Voice of the Farm, 94 (10).

Rosenberg, Gabriel N. *The 4-H Harvest: Sexuality and the State in Rural America*. Philadelphia: University of Pennsylvania Press, 2015.

Rowley, Scott. "Do We Need Corporation Farming? An Argument for Putting Corporate Methods into Agricultural Production." *Wallaces' Farmer* 53, no. 10 (March 9, 1928): 376 (6).

Saloutos, Theodore and John D. Hicks. *Agricultural Discontent in the Middle West: 1900–1939*. Madison: University of Wisconsin Press, 1951.

Sanders, Elizabeth. *Roots of Reform: Farmers, Workers, and the American State, 1877–1917*. Chicago, IL: University of Chicago Press, 1999

Schwieder, Dorothy. *75 Years of Service: Cooperative Extension in Iowa*. Ames: Iowa State University Press, 1993.

Shaw, Ada Melville. "The Country Child's Schooling: An Answer to the Question, 'How Much Education Does the Rural Child Need?'" *Farmer's Wife* 23, no. 8 (January 1921): 288, 308.

Shideler, James H. "'Flappers and Philosophers,' and Farmers: Rural-Urban Tensions in the Twenties." *Agricultural History* 47, no. 4 (1973): 283–299.

Short, W.S. "The Banker and the Farmer." *Wallaces' Farmer* 46, no. 2 (January 14, 1921): 54 (14).

Stein, Blanche. "Building a Rural Civilization: It Must be Based Upon the Highest American Ideals, Expressed in Terms of Christianity, Recognizing the Farmer's Right to Economic Welfare." *Better Farming* 45, no. 1 (January 1922): 4, 16, 19.

Stock, Catherine M. *Rural Radical: Righteous Rage in the American Grain*. Ithaca, NY: Cornell University Press, 1996.

The Folklorist. http://www.folklorist.org/song/Wal_I_Swan_(Giddyap_Napoleon,_Ebenezer_Frye). (accessed 3/17/16).

Wall, Joseph Frazier. *Iowa: A Bicentennial History*. New York: W.W. Norton & Company, 1978.

Wiggins, E.R. "Growing 1000 Acres of Corn on Hawthorn Farm." *Better Farming* 47, no. 7 (July 1924): 3–4, 7.

Wilson, Thomas E. "Farmers are in a Fix." *Better Farming* 44, no. 1 (January 1921): 4.

Woeste, Victoria Saker. *The Farmer's Benevolent Trust: Law and Agricultural Cooperation in Industrial America, 1865–1945*. Chapel Hill: The University of North Carolina Press, 1998.

3

"THE FUTURE OF AN IDEA"

Farmer's Use of Technology to Perform Rural Capitalistic Modernity

> People were crazy to buy a new car and B.D.'s rash of car salesmen were just as crazy to sell. Oscar Price sold four in one day by walking along with the farmers talking as they cultivated.
>
> Ruth Wilson, Buffalo Chips: The History of a Town, year unknown[1]

In the late summer of 1923, hundreds of farm families loaded into their new Ford and Packard automobiles and headed for DeKalb, Illinois. Other farm families piled into trains and some even walked, increasing the small town's population of 5,000 to over 30,000 people. The hundreds of families came not for a sporting event or to see a famous orator speak, but to help celebrate the tenth anniversary of the DeKalb County Farm Bureau. The organization boasted of being the first county farm bureau in the U.S. "marking a new scientific and business era in farming" and leading to the founding of similar "movement[s]" in "1,600 other countries and to 46 other states." The celebration included "bands, and parades, and exhibits of modern farm implements and methods," but the main attraction featured a pageant involving over four thousand people representing every county in Illinois. Almost everyone in DeKalb County helped with the play directed by Nina Lamkin, a dramatic organizer of Community Service, a national organization helping communities organize large-scale theatrics.

The pageant entitled "Forward! Farm Bureau" contained four episodes – "The Birth of an Idea," "The Growth of an Idea," "The Development of an Idea," and "The Future of an Idea" in which the stars of the show, a farmer, his wife, and two sons experienced the modernization of agriculture firsthand. Action took place around the farm home and went through "a series of changes and improvement until finally it evolved into the most up-to-date of 1923 farm homes." The real stars in the play were evolving technologies and production methods.

DOI: 10.4324/9781032637952-4

The show concluded with over 60 floats demonstrating the DeKalb Farm Bureau's vision of an even more modern future. As one observer, Mabel Travis Wood, exclaimed, "A pageant has always been the best way to illustrate progress." Nor was the DeKalb production unusual, as many smaller towns had organized similar demonstrations throughout the Midwest.[2] But Mabel Wood was wrong in her assessment that pageants of the kind in DeKalb, Illinois would prove the best method of demonstrating "progress." Farm families in DeKalb, Illinois and thousands of other places throughout the Corn Belt would cast themselves in the role of the performers in "Forward! Farm Bureau." Their farms, homes, barnyards, and machine sheds would serve as their stage with tractors, cream separators, radios, hybrid corn, and electric appliances as their props. These farm families, through the mundane everyday use of material objects rather than grand theatrical productions, would stage a "pageant" even more impressive and more significant to the future of the way people in the developed world would eat than Nina Lamkin could possibly have envisioned. These "actors" are, in fact, not finished with the performance.

A competing ideology influencing the farmer's relationship with agricultural technology, which I call *rural capitalistic modernity*, came to prominence in the 1920s and gained strength in the 1950s and the 1960s as a result of Cold War ideologies. This sense of modernity viewed mechanization as a symbol of American cultural values and discounted the notion of farms as industrialized and centrally managed businesses. It encompassed a culture in which farmers' identities as modern, progressive, independent capitalists (as opposed to urbanized industrialists) became embedded in the technologies they employed. The rural discourse that developed in the Midwest Corn Belt embraced progressive modern ideas about the faith in technology, rationalism, modern business practices, and capitalist competition in bringing about progress while injecting traditional notions about morality, prosperity, and ethical ways of living with origins in Jeffersonian and German agrarianism.

The mistake many farm memoirists and historians make is taking a technological determinist view of technological change. For example, commentators on agriculture often assume that the shift from a communal to an individual experience occurred because of new farming technologies such as tractors that did not require collective production processes.[3] However, nothing mandated that the Corn Belt farmer would use the new tractors in a non-communal way. Farmers could have very easily decided that only one farmer would purchase a tractor and contract with adjacent farms as they had done with threshing machines. Similarly, one could imagine a production process whereby farmers would communally cultivate one another's land with multiple tractors to complete the work even faster with lower capital investment. The tractor's design itself did not dictate its individualistic use. An unarticulated identity of rural capitalistic modernity, which introduced capitalist ideas about competition into Jeffersonian and German agrarianism, in combination with the design of the latest tractors explains this abandonment of communalism.[4] Identities and technologies co-constructed one another.

Similarly, one cannot understand this rural need for modernization wholly through rational economic explanations. In fact, one could argue that new farm machinery and methods, which promised to increase production, would hurt farmers economically by leading to even lower commodity prices. Thus, only within the cultural context of urban-rural contestations over the meaning of modernity and rural identities can one fully understand why farmers found increased supply as a solution to low prices. An economist may point out that a new tractor promises to increase an individual farmer's profits through taking greater advantage of economies of scale. Farmers may not internalize the effect that the increase in his individual production may have on prices more broadly. This economic "moral hazard" argument, however, fails to realize that this individualist capitalist way of thinking about production and technology, the very mindset that causes the farmer to ignore externalities, is itself a feature of the new rural capitalistic identity that formed in response to the rural-urban conflict. Farmers prior to the 1920s, while they valued independent property ownership, also viewed themselves as part of communities rather than as fierce individual capitalist competitors against their neighbors. The rural landscape of the Corn Belt in the late 19th century featured roads full of people walking between fields to work communally. The communal nature of threshing in the 19th century and the accompanying "threshing dinners" have become part of the mythology of the Midwest. Thus, the change from a communal to an individual identity marked a significant shift in the way farmers thought of themselves. In fact, many farmers after mechanization would cite not drudgery or poverty as the most negative aspects of a rural lifestyle but, rather, social isolation.[5]

Further, farmers in the Midwest Corn Belt helped to develop a discourse based on rural identity containing seemingly opposing views and positions that differed from urban industrialism. This rural modern discourse became remolded from 1920 to the present in a complex negotiation between rural identity and modernity that influences the ways farmers use technologies and think about themselves (Figure 3.1). In this way, farmers use both technologies and discourses to work

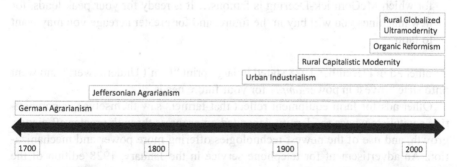

FIGURE 3.1 Timeline of rural identities in the Corn Belt. The discourse-identity bundles overlap and often inform one another. The start dates of each identity are approximate.

on the rural moral self. When Iowa or Ohio farmers adopt the scientific method to breed a new seed variety, or exhibit sophisticated abilities to understand and alter technology, or welcome a new road for reasons such as gaining a competitive advantage in a national market, they express a modern sensibility. For example, in 1920, *Better Farming* described Walter Tomlinson, a farmer in Defiance County, Ohio as "one of Defiance county's progressive, scientific farmers and stock raisers." Tomlinson had used "up-to-date dairy equipment" and the latest stock breeding methods to report an annual profit of $8,500, a large sum at the time.[6]

Quotes from farmers, farm journal advertisements, photos of farmers, and even the actions taken by farmers in the 1920s reveal that rural capitalistic modernity already existed by the end of the 1920s.[7] Advertisements in the *Farm Journal* in the 1920s for items used in work processes also show that farmers had already adopted their own version of modernity based on faith in capitalism and progress through use of technology, yet retaining a rural flavor that preserved the nuclear family and kept the farmer in control of work processes. Ads for tractors especially drew on modern sensibilities and reflected the latest capitalistic ideologies. These advertisements suggest that farmers had already embraced growth, the association of time with money, and capitalistic competition with their neighbors through the adoption of the latest technologies.[8] For example, an ad for McCormick-Deering Farmall tractors in the February, 1928 issue of the *Farm Journal* read

> Prosperity is rare whenever time is being wasted on a large scale. Time is the most valuable thing we have. It is the very essence of life itself. That is where the great value of *machines* and *power* and *planning* comes in. These factors help a man to *multiply his work, his production, and his profit*." The ad concludes "Time is always money on a well-managed farm.[9]

Another McCormick-Deering ad in September of 1928 read,

> Remember that in this tractor you will own the *modern-4-cylinder power plant* for which McCormick-Deering is famous... It is ready for your peak loads, for new machines you will buy in the future, and for greater acreage you may want to farm.[10]

Another ad in December 1928 stated in large print "Don't Underpower!" and went onto state "Invest in power *ample* for your future needs."[11]

Other ads for farm equipment reflect that farmers saw themselves as competitive capitalists whose modernity depended on progress through greater efficiency, growth, and use of the newest technologies offering more power and mechanization. An advertisement for telephone service in the January, 1928 edition of the *Farm Journal* contained discourse of this rural capitalistic modernity. The ad tells the story of farmers who used phones to sell to far-off agricultural markets early in the morning and deliver the product "at night when my neighbors are sleeping."

The ad goes on to read, "The work of a whole year may hinge on the result of a few days. It's easy to lose $10 a steer by selling at the wrong time… The modern farm home has a telephone."[12]

As a McCormick-Deering advertisement from 1928 shows, these marketing messages rejected the kind of industrialized modernity advocated by Rowley and Hawthorn Farm (see Chapter 2) by retaining the nuclear family with the farmer himself in control of work processes even while constructing a new rural identity around modernity and mechanization. The ad shows an idealized winter scene in which a father and a son haul a Christmas tree. In the meantime, other workers efficiently use the tractor in the background to do winter work with a prosperous family farm in the background. (Figure 3.2). The ad reads

The American farmer is always improving things he has done. He goes on compounding farm science and knowledge, motive power and mechanical equipment, managing his farm-factory with greater ease and efficiency. The more he works with his brains, intelligently, the less he works with his hands…. He is abreast with the best of men and competes with the world…. More than ever it is a certainty that "Good Equipment Makes a Good Farmer Better.[13]

FIGURE 3.2 "McCormick-Deering advertisement," *Farm Journal* 52, no. 1 (January 1928): 25.

Ads for electricity and consumer products in the 1920s also indicate that farmers had adopted their own rural modern identity incorporating capitalism, faith in progress and science, and mechanization while rejecting the type of urbanized industrialization that saw farmers simply as factory workers. In one ad published in the October 1928 issue of *Farm Journal*, a farmer stands confidently talking to a team of electricians running power lines to his large and prosperous barn and farmhouse as if giving them directions on how to wire his property (Figure 3.3). The drawing associated masculinity with strength, control over work, progress, technology, and the moral raising of male children. The farmer's haymow on the barn overflows with hay and he has the latest grain silo between the barn and the farmhouse. The farmer's son of about six years old stands next to him listening to his father directing the electricians. The artist placed the child prominently above the graph showing the growth of electricity in rural districts as if to associate the introduction of new technology with, literally, the family's future. Everything in the advertisement draws the eyes upward as if to subtly convey that a family "on the rise" has electricity. One of the electricians climbs an electric line and the wires seem to shoot up into the sky toward the peak of the barn. Birds and tall trees in the sky further focus the observer's gaze upward into the future.

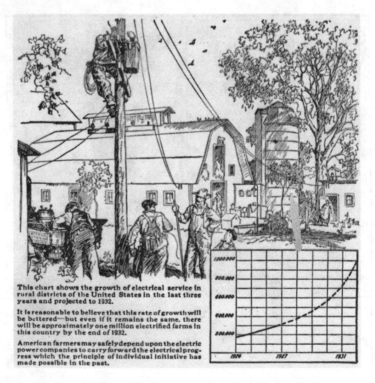

This chart shows the growth of electrical service in rural districts of the United States in the last three years and projected to 1932.

It is reasonable to believe that this rate of growth will be bettered—but even if it remains the same, there will be approximately one million electrified farms in this country by the end of 1932.

American farmers may safely depend upon the electric power companies to carry forward the electrical progress which the principle of individual initiative has made possible in the past.

FIGURE 3.3 "National Electric Light Association advertisement," *Farm Journal* 52, no. 10 (October 1928): 45.

The tendency of scholars to attribute increased mechanization or electrification of the farm to purely economic motives ignores the fact that early in the 20th century, it was not so clear to many farmers that new technologies actually increased profit. Because of the complexity of calculating the costs and benefits of many agricultural inputs and outputs, one could show, for example, that horses outperformed tractors in the 1920s or vice versa. Farm journals contained debates between agrarians that reveal this uncertainty. While one farmer may note the tendency of horses to break down in the heat, another may note the penchant for early tractors to malfunction frequently. One farmer may cite the savings allowed by the tractor in the form of not paying a "hired man" to drive horses while another farmer may point out the gasoline savings allowed by animal power.[14] In addition, farming with motors required an investment in time and effort to learn new mechanical skills needed to maintain them.[15]

Other mechanized farm equipment in the 1920s and 1930s brought similar technical disincentives to their adoption. For example, using the new mechanical chemical sprayers required changing the oil every 40 to 60 days and the farmer had to flush out the tank, pump, nozzles and all connecting hoses each day during spraying. In addition, the user had to keep all fittings and screw joints tight throughout the engine and pump, as well as check all the "plungers and packing to see that they are not leaking." A farmer neglecting any items in this long list of arduous maintenance tasks would result in a dysfunctional sprayer and a net loss to the already cash-strapped agrarian. Yet, farmers adopted mechanical sprayers in such large numbers that by 1930, the National Association of Farm Equipment Manufacturers received enough concerned letters from users of the fickle devices that they ran a whole series of farm journal articles about how to care better for the new technology. Farmers continued to adopt mechanical sprayers because doing so made them modern, even though the new machines often performed unreliably in the field and many agrarians had little idea of how to properly maintain them.[16]

Similarly, when farmers praised the advantages of installing their own electric power plants on their farms, they often justified the expense for reasons that had little to do with adding articulable production value. Rural denizens of the Corn Belt, in the 1920s for example, rarely spoke of electrification in the kind of rational economic discourse proffered by University of Wisconsin agricultural engineer Floyd Waldo Duffee. In 1924, Duffee wrote into *Better Farming* urging farmers to electrify their farms because

> Experiments show that in the process of stabling cows, cleaning mangers, weighing, and feeding grain, hay and silage, that a half hour or 34.9 percent of the time was saved by the use of electric lights when compared with coal oil lanterns.[17]

In contrast, farmers themselves rarely employed Duffee's rational economic quantifications of work hours to justify their decision to electrify. Instead, they used discourse reflecting a desire to achieve a more modern image and lifestyle. One

1926 advertisement for the Colt Light Plant in the journal *Farmer's Wife* declared "Colt Light has brought modern convenience to hundreds of thousands of farm women," followed by a letter written by the farming couple "Mr. and Mrs. S.B. Rudicil" of New Trenton, Indiana. In the March 15, 1926 letter, the Rudicils describe the electric light as,

> It is always clear and bright, and the fixtures so beautiful that they never fail to attract attention. In our living room we have a handsome central fixture and a floor lamp with a silk shade-in our parlor also, a nice central fixture.

Mrs. Rudicil also noted her use of the Colt iron and hot plate not because they brought economic benefits to the farm but because "They have lighted my work so much that I would be willing to pay the price of the whole plant for them if it were necessary." The husband and wife concluded that, "We have never made any investment that has been of more benefit to the entire family. Every one of us is proud of our Colt Light Plant."[18] Hence, the Rudicils felt that the Colt Light Plant made them modern and believed the light signaled their identities to others.

When farmers did cite economic reasons for electrification, they rarely articulated the same advantages noted by progressive intellectuals. In a January 1922 issue of *Farmer's Wife*, Minnesota farmer Harper Christensen, for example, cited as the primary economic benefits of electricity the time she saved milking cows and the fact that light in her hog house prevented predators from killing sows.[19] Just two years later, Duffee would note neither of these economic benefits.[20] In fact, Duffee's argument in favor of rural electrification focused on avoiding the risk of fire posed by oil lanterns.[21]

Similarly, ads for isolated generation facilities by General Motor's Delco-Light subsidiary used modern discourses praising open-mindedness while emphasizing farmer's independence from the city and the benefits to the farm family. Many other advertisements associated light with prosperity, a happy home, and other symbols of modernity such as automobiles. In one advertisement, entitled "Delco Light *and* the Open Mind," the farmer stands next to his garage door with a modern car in front of it (Figure 3.4). He turns toward a large and prosperous farmhouse as if talking to his wife who stands in the open door of the house. The entire farm is dark except for the farmer and his wife, both of whom are bathed in light. The farmer stands in a swath of light emitted by a flood light at the front of the garage. The wife stands in a house in which light pours out of every window and door. The ad, in using the term "open mind," implies that light does not bring simply economic benefits but also symbolizes that the mind of its user is also bright and forward-looking.[22]

The Delco ads sometimes told the origin story of the generator-battery technology, in which a farmer wrote to General Motors to say that he had creatively powered his home using his Cadillac's built-in equipment to light his house. Such stories emphasized the notion that innovation began on the farm, not in urban areas.[23]

DELCO-LIGHT
and The Open Mind

FIGURE 3.4 "General Motors Delco Light advertisement," *Farm Journal* 52, no. 10 (October 1928): 39.

Another Delco ad in March of 1928 promised shorter hours for labor as well as larger profits, indicating that farmers, unlike factory workers, saw technology as giving them greater control over time as well as work processes.[24]

Farmers themselves desired to electrify their homes and barns and many took independent action to install small electric power plants on their properties. Victor Brown, for example, a Somers, Wisconsin farmer installed a plant to light his house and all his work buildings as early as 1924. The article reflected that Brown, "typical of many farmers," saw electrification as one technology in a group of artifacts making his farm modern including a "two-plow tractor, self-lift plow, double disk harrow, stationary gas engine," and many other artifacts. The author notes, "he [Brown] says he will not be completely satisfied until he had every possible necessity and convenience."[25]

Therefore, as with tractors, farmers did not ultimately install costly power plants on their farms because of an agreed-upon set of rational economic advantages brought by electricity. People on farms who had come to think of themselves as modern became more likely to emphasize the advantages of these technologies and underplay their costs or inconveniences. The Rudicils, for instance, characterized the expense of ten cents per day to operate their Colt Power Plant as low "since we used to pay nearly as much for oil lamplight," but they overlooked the cost of installation and maintenance.[26]

Further, farmers encountered several new technologies early in the 20th century and continuously had to make decisions on which ones to adopt.[27] Of course, a rational choice would have been to reject all of them, particularly when faced with uncertainty about the economic viability of many new artifacts. One farmer writing into *Wallaces' Farmer* in 1925 lamented that ideals like "progress" and "efficiency" had become so widely adopted among farmers in the Midwest that farming had become unrecognizable from the "peace and quiet" of his "childhood days" and lamented, "How much progress can we stand?" The farmer noted that the newest technologies did not lead to economic advantages but, rather, financial strain declaring "How long can we pursue progress with our limited means?"[28]

Farmers, therefore, needed an identity such as rural capitalistic modernity to see themselves in a way that favored the adoption of the tractor, the mechanical sprayer, and the power plant. The promise of uncertain economic advantages was not enough to nudge them toward adopting and using new technologies. When one reads accounts of rural residents first encountering automobiles, for example, they cite many disadvantages of the new technology such as the danger that turning a crankshaft would recoil and break the user's arm or the high volume of dust that became sucked inside the cab on unpaved roads. Others recall batteries constantly dying and the lack of gas stations in the country, requiring the driver to often borrow the nearest farmer's kerosene. The user could only add this borrowed kerosene after draining the remaining gas left in the gas tank. Other users recall the inability of early cars to scale even the small hills of the Corn Belt. Rural denizens early in the 20th century nevertheless embraced automobiles, as Matt DeVries of Buffalo Center, Iowa recalls because "Cars were an exciting novelty." DeVries continues,

> Dad came in and yelled 'Come on out.' There's a car coming down the road.' It was D.B. Sterling's brother driving a car with a chain drive. We all ran out and stood on the hill to watch it go by.

Many people in the Corn Belt even invested in an additional expense of a car garage because nitrogen emitted by horse manure would damage the finish on cars kept in barns. Few of these accounts justify this increased cost in economic terms or noted that automobiles, in fact, increased the speed of travel.[29] Another Buffalo Center resident, John Howe, recalling his father bringing home a phonograph on their farm expressed similar sentiments "We were so tickled, we had a phonograph! We really appreciated something like that and thought it was mysterious as heck!"[30]

Hence, a purely "rational" analysis of accounting cannot explain early 20th-century modernization. In addition, the function of the device cannot fully account for its use. Rather, rural capitalistic modernity gave farm families the push to adopting newer artifacts. While some farmers resisted modernization, many others adopted new technologies wholesale and with enthusiasm notwithstanding uncertainty about their economic viabilities.

An analysis of the *Farm Journal* in the 1920s also shows that rather than seeing themselves as backward agrarians, farmers in the Midwest already embraced modern business practices and the need for higher education to learn the newest business and farming techniques. While these farmers may have articulated a desire to make more "rational" decisions as the reason for adopting these business practices, an equally significant unstated motive was to *be* modern. Articles reflect a balance between urban academics and traditional rural knowledge that formed part of the uniquely rural discourse of modernity. One article entitled "Orderly Farm Business" highlighted farmers' efforts to learn new accounting techniques developed in agricultural schools while another article entitled "College on the Farm" discussed the value of both going to college and "learning at home" through practical experience and 4H clubs.[31]

Further, by 1940, farmers had already begun to take photos of themselves and their families proudly posing with machines, the symbols of this independent rural modernity. As the 1920s progressed, both men and women on Midwest farms would construct and perform distinctly rural modern self not by discussing it directly, but by simply working and using technology.

Evidence from farm journals indicates that many urban dwellers in the early 1920s, including advertisers and manufacturers, greatly underestimated this modern sensibility as already existing in rural America. *Better Farming*, published in Chicago, Illinois, for example, declared "It has become almost proverbial to say 'The farmer ought to be a businessman.' This is a phrase that is loosely bandied about by those who do not realize that the farmer is a businessman." M.V. Casey noted the inability of urban automobile advertisers or dealers in 1920 to recognize the farmer as a modern capitalist and address how farmers would use motorized trucks. The advertisers appealed only to urban users even though many farmers had already gladly adopted motorized vehicles over horses. Casey also noted that dealers did not offer instructions on how farmers could repair or modify trucks themselves.[32] In other words, the truck industry failed to appeal to rural modernity. In fact, the same issue of *Better Farming* containing Casey's article reported that farmers had already widely adopted truck use despite the inability of advertisers to appeal to them. According to 1919 statistics from B.F. Goodrich, farmers used 78,789 trucks in hauling grain and livestock compared with 65,928 used by manufacturers and 64,486 used by retailers.[33]

Several articles in *Wallaces' Farmer* in the 1920s confirm how farmers themselves had also already adopted "scientific" methods to develop new corn strains and seed inoculation techniques.[34] One 1925 piece recounts how farmers in Downy, Iowa organized a competition comparing crossbred corn strains that they developed as a way of improving and standardizing production.[35] Similarly, when hybrid corn first appeared on the market in 1932, farmers enthusiastically adopted it even though they had to purchase it from seed companies. By 1939, for example, three-fourths of all Iowa corn acreage was hybrid corn.[36]

Other articles and letters exhibited farmers' enthusiastic adoption of manufactured animal feed, new uses for mechanized equipment, and novel crop fertilization methods.[37] Thus, farmers performed their modern identities through technology use. Coen Belt agrarians increasingly thought of themselves as entrepreneurs acting on their own and adopting modern concepts like efficiency and cost accounting. One Corn Belt farmer, R.B. Rushing, for example, wrote to *Better Farming* in 1925 explaining

My experience is that the farmer should never be afraid to buy and invest his money in new and improved labor saving and money making tools. Just now, as the snows and blusters of winter have passed away, and the spring and working season is over, it is a good plan for the farmer to carefully consider the proposition of buying any Implements that will increase the farm's efficiency.[38]

Another 1921 article in *Wallaces' Farmer* entitled "The Best Corn in Iowa" discussed new corn breeds not originating from universities or extension stations, but developed by farmers growing corn, comparing strains, and keeping records using the experimental method.[39] One Corn Belt farmer, Henry Lunz, wrote a letter to *Better Farming* in 1920 to share new seed-selection methods he developed with the newest fanning mills writing "The modern fanning mill is almost a wizard when it comes to cleaning and separating seeds. The right mill in skilled hands will do seemingly impossible things."[40] Similarly, in a 1928 article, Illinois farmer Charles D. Kirkpatrick described a sophisticated use of phosphate to fertilize his 500-acre farm that included research of experimental results and the use of a technoscientific network of experts. Kirkpatrick writes,

> Because several British experiment farms have proved that fineness of grinding [of phosphate] is important, we were content to take the manufacturer's guarantee. We employed a firm of commercial engineers to make private tests for us. At our request several comparative tests were made for neighboring state experiment stations. By their tests, our purchases have ranged from 94.8 to 87.5 per cent [*sic*] thru a standard dry screen. This same firm made screen tests on materials which another mill had furnished the Iowa and Missouri experiment stations. The rock phosphate used at Missouri tested. 78 per cent and at Iowa 83.7 per cent. In another test, the rock phosphate used at Ohio tested 67 per cent.[41]

Other Corn Belt farmers in 1924 formed "cow testing associations" to incorporate the scientific method and modern business accounting more thoroughly into milk production by hiring testers to work with each farmer. The tester

> Spends one day every month on the farm of each member where he weighs and tests the milk from each cow, computes its value and estimates the total cost of feed for the month. When this amount has been charged against the value of the milk and fat produced for that month, the owner is able to determine the profit or loss on each cow in his herd.

The tester served as a sort of expert employee of the farmer's as he also "assists in selecting feeds and figuring the most profitable rations. He also offers suggestions on breeding, management, and other phases of the business as deemed advisable by the attitude of the member."[42]

By the 1920s, at least some farmers found original uses for new technologies on their own and understood the modern business concept of creating more output from invested capital. For these rural denizens, rural capitalistic modernity had clearly taken hold. For example, Donald H. Wells, a corn farmer in Croton, Ohio gave similarly detailed accounts of his use of a tractor to double his land holding in 1924 with numerical accounts of tractor hours, fuel costs, and yields to calculate efficiency and profit.[43] Concern with efficiency and personal ownership of technology motivated Cyrus H. Lancaster of Indiana to abandon communal threshing for a small grain thresher.[44] Another 1930 piece reports that

> [W]ork for the tractor in winter months is no problem for J.J. Zeman, Winneshiek County, Iowa. He has bought a stone-crusher which he is using to turn a rocky corner of this farm into limestone that is sold to other farmers for sweetening sour soil. Some stone is sold for road building. There's an example of real farm management.

The article further explains that, "balling hay is the winter job that keeps N. Schaub's tractor busy."[45]

One proposed article in a press release by Robert A. Jones of the National Association of Farm Equipment Manufacturers also suggested that farmers had associated increased use of tractors and cars to economies of scale as early as 1930. In the article, Jones described one farmer's use of a tractor in Wisconsin who used a tractor

> To take on an additional 40 acres, half of which was four miles from the house and the other 20 about two miles away. The car was used to drive back and forth to work which would have been impossible with horses.

Jones went onto describe how the Wisconsin farmer "utilized his tractor for custom work [work under contract on another farmer's farm], taking in $200 this last fall in about two weeks by filling silos. By filling his own silo he figured that he saved a thirty-dollar expense." Jones describes how another Wisconsin farmer found increased uses of the tractor to increase his land size,

> Operating pea viners in three years brought in $1,370 to one man who reports that gas and oil for this work cost him $195. On the average, this tractor grinds 900 bushels of grain each year, a saving of $45 in cash outlay not to mention the saving in the long haul to town and waiting at the mill. Because of time saved over former methods in putting in his crops, this farmer had the opportunity to tile drain most of his farm over a six year period, with a saving of $380 which he would have otherwise expended in outside labor. This tractor is also used to run the threshing machine and kill weeds in the late summer with a harrow.[46]

Farmers Zeman and Schaub also saw the new roads spreading into their communities, not as a threat to the farm by urban encroachment, as some rural people feared, but as a profit opportunity.[47] Other farmers such as Daire Cobbler of Hedrick, Iowa, even used his tractor in off-peak seasons for roadwork.[48] In another article by extension agent F.S. Wilkins recalls how he and an investigator with the U.S. Department of Agriculture (USDA) surprisingly found a town in Washington, Iowa

> Where soybeans are not a novelty, where they are a common and extensively grown crop on practically every farm, where the farmers know how to grow them with a minimum of expense, and where the yields have been going over twenty bushels per acre.

Thus, farmers, not the USDA "soybean investigator," initiated the growing of new crops and invented their own techniques and uses for technology.[49]

Letters to magazine editors from Corn Belt farmers also demonstrate their embrace of mechanized machinery and the newest accounting methods. For example, farmer F.W. Hawthorn's 1930 letter to the editor states,

> My seven-year cost-account record tells me the average fuel cost on a three-plow tractor is 38 cents per hour of operation… The depreciation charge was $1,020 for 2,688 hours of work, or exactly 38 cents per hour, the same as the fuel costs.

In another article, W.J. Breakenridge, a farmer in Palo Alto County, Iowa, described his management practices in similarly modern terms: "[w]e departmentalize the farm… Accounts are kept separately for the beef cattle, the cows, hogs, horses, chickens, and for various crops." A "close student of markets," the author noted that he "tries to buy on the breaks and sells on bulges."[50]

Similarly, farmer A.L. Haecker wrote into *Better Farming* in 1920 analyzing evidence from experimental stations all over the Corn Belt on the productivity and profitability of the new silos spreading through the Midwest. Haecker stated that over a half million farmers had already done such a detailed analysis of data in ways resembling a scientific method and had chosen to install the new silos on their farms.[51] These voices hardly portray backward farmers resisting mechanization and change until ushered into the modern age by progressive reformers. Nor do they reflect the pre-modern sensibility of Swehn and other German immigrants in which religion and folk knowledge drive production decisions. Rather, these farmers took it upon themselves to use the latest technology and business practices to construct their own rural modern identities and discourses.

Other letters to farm journals by Corn Belt farmers reflect the enthusiastic adoption and modification of the newest technologies with an underlying motive to equal or exceed the modernity of urban residents. Farmer Rich Lucas wrote into *Better Farming* in 1925 urging other farmers to enhance the lives of their families by buying a radio. Lucas begins by stating

> We farmers situated miles from our home town or many miles from large cities are not able to secure much of the pleasures our city cousins do as to operas,

band concerts, lectures, etc., by attending them personally, but this pleasure can be secured by all farm folks by the use of radio.

Lucas not only bought a radio to equal the modern lifestyle of his "urban cousins," he modified it to make it even better by repurposing parts from a phonograph. Lucas writes,

> The horn or tone carrier on a good phonograph is an excellent loud speaker so I bought a phonograph attainment and attached our radio onto our Victor and it makes us a perfect loud speaker. All I have to do is slip off the microphone and slip on my radio attachment so we use only one set of head phones to tune in by.[52]

Thus, Lucas demonstrates the enthusiastic adoption and modification of the latest technology.

Evidence also confirms that farmers viewed financial or business decisions through a lens of shifting identities toward a self-image of modern technological users and innovators. Illinois farmer W.F. Nagel, for example, declared in 1923 that "[i]f power farming means the most economical application of mechanical power to the operation of an individual farm, then I am a power farmer." As a result of his newfound realization about himself, Nagel was ready to completely abandon horse farming as old-fashioned. Nagel concluded "I do not believe it is a question of whether power farming will pay or not. It is more a question of whether it will pay to keep as many horses as formerly were thought necessary."[53] Nagel's letter appears alongside an article about how an unprecedented number of farmers in 1923 had sought courses in gas engine management and basic mechanical engineering.[54]

Farmers not only expressed rhetoric, but also pursued activities consistent with rural-capitalistic modernity starting after World War I. In the fall of 1919, farmers helped establish their own national advocacy organization, the American Farm Bureau Federation (AFBF), for the purpose of giving farmers more control in modernizing agricultural life. By 1920, the AFBF claimed to represent more than 100,000 members. Farmers flocked to the AFBF because, as one newspaper noted, "it starts with the farmer, not from any social or political standpoint, but as a producer and distributer, and the idea of production, and of increasing it by scientific methods, is carried along from the bottom to the top."[55] Reflecting the perceived importance of the AFBF, one farmer stated, "[w]hile farmers might hold off from joining any of the other organizations, no up-to-date farmer could afford to keep out of this one."[56] Many urban financial and commercial interests supported the AFBF because it offered a more conservative organization to counter more radical rural movements such as The National Farmers Union. Farmers joined for purposes of increasing the power of rural residents through modernization.[57] In fact, the AFBF and many state farm bureaus tended to frame farmers as rural capitalists to satisfy urban industrialist's desire to prevent a more radicalized farmer-labor alignment. Thus, farm bureaus and a discourse of rural capitalism satisfied two opposing interests at the same time. *From the farmer's point of view*, farm bureaus gave them a way of competing with strengthening urban interests; for urban industrialists, it offered a means of avoiding, in the words of Illinois Agricultural Association president Henry J. Scone, "any policy that will align organized farmers with the radicals of other organizations."[58] In this way, modernity offered social consensus while still satisfying the need of opposing parties to engage in conflict.[59]

Quotes from farmers in magazines in the 1920s reflect this same modern sensibility when discussing specific methods of production. Farmer G.O. Merryman of Grundy County, Iowa observed in a 1925 editorial in *Wallaces' Farmer*,

> [T]o those that are still skeptical [of using the newest machinery], let us say you will come to use the two-row cultivator just as surely as you have the self-binder, the automobile, hay loader, tractor and other modern tools and conveniences that reduce time and labor and production costs.[60]

Another farmer in O'Brien County, Iowa wrote into *Wallaces' Farmer* stating that by 1928, he already owned a truck and two tractors, which he used for "planting, disking, harvesting and the fifty other jobs it can be used for." The farmer discounted the use of horses because they "break down from the heat," while overlooking mechanical problems with tractors. The farmer also bought his own truck even though trucks for hire in many Iowa towns "haul as cheap or cheaper than one can afford to own a truck."[61] Many other farmers discussed tractor use in a similar way that only recognized the benefits of new machinery compared to horse-powered equipment and reflected a willingness to make the new equipment profitable by finding new uses. Earl Hill of Low Moore, Iowa, for example, stated in 1924 "We have done light belt work with our motor cultivator, pulled an 8-foot disk, 6-foot pending on the size... It does a good job of cultivating [corn] as any horse-drawn machine I ever used." According to the article, many farmers in the Corn Belt were enthusiastically adopting tractors for cultivation on their own without the influence from advertisers, university experts, or government assistance.[62] *Better Farming* reported that these novel uses for tractor belts had become widespread among Midwest farmers as early as 1923.[63] Further, by 1940, farmers had already begun to take photos of themselves and their families proudly posing with machines, the symbols of this independent rural modernity (Figure 3.5).

Nor did farmers view modernity in terms of urban industrialism. In a study of rural Illinois families, Jane Adams found that none of the farm families she interviewed or lived with thought of their farms as factories either in the 1920s or the 1990s.[64] Similarly, the January 20, 1928 issue of *Wallaces' Farmer* reported how eleven "Master Farmers" from Iowa addressed a rural audience over the WHO radio station out of Des Moines, Iowa. These Master Farmers all shared common practices consistent with rural capitalistic modernity. As the article explains,

> These men have, since 1921, invested freely in limestone, improved machinery and in better farm buildings. Three have built new homes since then, and nearly

FIGURE 3.5 "Farmer Posing with Farmall F-20 Tractor with a Wind-Powered Water Pump in Background." Author's personal records. Powersville, IA, 1940.

all the others have spent a considerable amount in making the homes better adapted to the needs of the family. These improvements include remodeling, additions of porches, heating systems, electric lights, and bathrooms.[65]

Farmers, therefore, saw themselves as modern businessmen, but not as equivalent to urban industrialists running a factory because the family and the farmer as controlling productivity remained central moral components of their sense of modernity. One farmer, George Steen, speaking at the radio address linked efficiency and profit with moral ways of living in his speech:

> We may apply limestone and phosphate to all our land, and rotate crops to the best advantage; we may convert bumper crops into livestock with the greatest efficiency and market them to the best advantage; we may accumulate a good property; yet we can still be failures as farmers. Unless we can transform the profits of good farming into homes and neighborhoods where people get more of the worth while things that come with good schools and good churches and a healthy social life, we are poor farmers.[66]

These "Master Farmers," such as 1926 winner Lewis Morris of Grimes, Iowa, served as experts and teachers at Iowa State University in short courses for farmers to learn better "money making methods." The article proudly declares "Farmers who are anxious to learn the results of new experiments in an effort to make as much money as possible on their own farms are business men."[67] The farmers, therefore, saw themselves as modern businessmen, but not as equivalent to urban industrialists running a factory. The family and the farmer still controlled productivity and remained central moral components of rural modernity. It was not only men who have resisted urban industrialism through technology use. Women on Midwest farms have also used technology to contest urban ideas about what counts as a "modern wife" beginning in the early 20th century.

Notes

1 Ruth Wilson, *Buffalo Chips: The History of a Town* (unpublished book, Buffalo Center, IA, year unknown), 31.
2 Mabel Travis Wood, "A Pioneer Farm Bureau Celebrates," *Better Farming* 46, no. 4 (August 1923): 3.
3 See for example Hamilton, *Deep River*, 157.
4 Many historians have noted the transition of American farmers to capitalism in the early 20th century. See for example Christopher Clark, et al., "The Transition to Capitalism in America: A Panel Discussion," *The History Teacher* 27, no. 3 (May 1, 1994): 263–288; Steven Hahn and Jonathan Prude, eds., *The Countryside in the Age of Capitalist Transformation: Essays in the Social History of Rural America* (Chapel Hill: University of North Carolina Press, 1985); James A. Henretta, *The Origins of American Capitalism: Collected Essays* (Boston, MA: Northeastern University Press, 1991); Allan Kulikoff, *The Agrarian Origins of American Capitalism* (Charlottesville: University Press of Virginia, 1992); Charles Sellers, *The Market Revolution: Jacksonian America, 1815–1846* (New York: Oxford University Press, 1991); Michael Merrill, "Cash Is Good to Eat: Self-Sufficiency and Exchange in the Rural Economy of the United States," *Radical History Review* 1977, no. 13 (January 1977): 42–71; Henretta, "Families and Farms;" Clark, *The Roots of Rural Capitalism: Western Massachusetts, 1780–1860* (Ithaca, NY: Cornell University Press, 1990); Clark, "Household Economy, Market Exchange and the Rise of Capitalism in the Connecticut Valley, 1800–1860," *Journal of Social History* 13, no. 2 (December 1, 1979): 169–189; Nancy Grey Osterud, "Gender and the Transition to Capitalism in Rural America," *Agricultural History* 67, no. 2 (April 1, 1993): 14–29.
5 Harold F. Breimyer, *Over-Fulfilled Expectations: A Life and an Era in Rural America* (Ames: Iowa State University Press, 1991), 49.
6 "Aged Farmer Makes Over Eight Thousand," *Better Farming* 43, no. 8 (March 1920): 10.
7 See, for example, "Why We Made Our Home Modern," *Farm Journal* 54, no. 2 (February 1930): 32, 89.
8 The prevailing marketing theory is that advertising reflects cultural values (albeit in a distorted way) rather than completely creating such views. As a result, advertising may serve as evidence of ideas already existing within its target audience, particularly when it conforms with other contextual evidence. See Roland Marchand, *Advertising the American Dream: Making Way for Modernity, 1920–1940* (Berkeley: University of California Press, 1985); Richard Pollay and Katherine Gallagher, "Advertising and Cultural Values: Reflections in the Distorted Mirror," *International Journal of Advertising* 9 (1990): 359–372.
9 "McCormick-Deering advertisement," *Farm Journal* 52, no. 2 (February 1928): 21.
10 "McCormick-Deering advertisement," *Farm Journal* 52, no. 9 (September 1928): 35.
11 "McCormick-Deering advertisement," *Farm Journal* 52, no. 12 (December 1928): page unknown.
12 "American Telephone and Telegraph Co. advertisement," *Farm Journal*, 52, no. 1 (January 1928): 53.
13 "McCormick-Deering advertisement," *Farm Journal* 52, no. 1 (January 1928): 25.
14 For an example of two Iowa farmers debating the merits of adopting tractor farming over horses in the 1920s, see Reader, O'Brien County, Iowa, "Truck and Tractors," *Wallaces' Farmer* 53, no. 12 (March 23, 1928): 480 (10), The Voice of the Farm.
15 "Motor Skill in Demand: Many Farmers Seek Instruction in Gas Engine Management," *Better Farming* 46, no. 11 (November 1923): All Around the Farm, 9.
16 Robert A. Jones to "The Editor," National Association of Farm Manufacturers, Farm Press Release #76, "Care Lengthens Sprayer's Life," January 2, 1930, Box 24, Record Group Number 503, Accession Number 79–001, Ag Engineering Records, Auburn University Archives, 5.

17 F.W. Duffee, "Modern Lights Save Farmers Work Hours," *Better Farming* 47, no. 6 (June 1924): 8.

18 Mr. and Mrs. Rudicil, "Colt Light advertisement," *Farmers' Wife* 26, no. 12 (December 1926): 611.

19 Harper Christensen, "Electricity on Our Farm," *Farmer's Wife* 33, no. 8 (January 1922): 677.

20 Duffee, "Modern Lights Save Farmers Work Hours," 8.

21 Ibid.

22 "General Motors Delco Light Advertisement," *Farm Journal* 52, no. 10 (October 1928): 39.

23 Ibid.

24 "General Motors Delco Light Advertisement," *Farm Journal* 52, no. 3 (March 1928): 67.

25 E.R. Wiggins, "Comfort, Convenience, and Economy with Electric Power and Light Plants: Thousands of Famers and Farmers' Wives Point Out Many Advantages of Electricity Both in the Home and the General Farm Activities," *Better Farming* 47, no. 5 (May 1924): 4.

26 Mr. and Mrs. Rudicil, "Colt Light Advertisement," 611.

27 "Motor Skill in Demand," 9.

28 "The Voice of the Farm: How Much Progress Can We Stand?" *Wallaces' Farmer* 10 (March 27, 1925): 461.

29 Wilson, *Buffalo Chips*, 31–34.

30 Ibid., 37; it should be noted that John Howe is the author's great grandfather.

31 George Price, "College on the Farm: 4-H Clubs Provide Opportunity to 'Learn by Doing' at Home,'" *Farm Journal* 52, no. 4 (April 1928): 24; Walter Burr, "Orderly Farm Business," *Farm Journal* 52, no. 1 (January 1928): 38–39.

32 M.V. Casey, "What Kind and Size Truck Shall I Buy?: Some Practical Ideas for the Prospective Truck Buyer," *Better Farming* 43, no. 2 (February 1920): 5.

33 "Farmers Used 78,789 Trucks During 1918," *Better Farming* 43, no. 2, (February 1920): Farm Mechanics, 16.

34 Henry A. Wallace, "Fifth Iowa Corn Yield Contest," *Wallaces' Farmer* 50, no. 7 (February 13, 1925): 1, 18; Russell H. Beck, "The Voice of the Farm: Inoculating Soybeans," *Wallaces' Farmer* 50, no. 9 (February 27, 1925): 298 (10).

35 B.N. Stephenson, "A Community Checks up its Corn," *Wallaces' Farmer* 50, no. 1 (2 January 1925): 8 (8).

36 Wall, *Iowa*, 131.

37 "Cultivation that Kills Weeds," *Wallaces' Farmer* 50, no. 19 (8 May 1925): 676 (6); "Too Busy to Milk," *Farm Journal* 54, no. 1 (January 1930): 40; Warner, H.W. "Getting More Mileage on the Manure-Spreader," *Farm Journal* 54, no. 2 (February 1930): 38; "Topics in Season," *Farm Journal* 54, no. 2 (February 1930): 44; Grif McKay, "Good Cows, Well Fed"; "Topics in Season," (March 1930): 28; "Why Some Farms Pay," *Wallaces' Farmer* 50, no. 12 (March 20, 1925): 440 (28); Miller Purvis, "New Ideas in Feeding Poultry," *Wallaces' Farmer* 53, no. 11 (March 16, 1928): 431 (7); J.G. Haney, "Making Alfalfa and Clover Hay," *Better Farming* 46, no. 7 (July 1923): 3.

38 R.B. Rushing, "Buying Farm Implements: Good Tools and Implements Pay for Themselves," *Better Farming* 48, no. 1 (January 1925): 13.

39 Henry A. Wallace, "The Best Corn in Iowa," *Wallaces' Farmer* 46, no. 2 (January 7, 1921), 43 (1).

40 Henry Lunz, "Better Farmers See Need of Fanning Mills," *Better Farming* 43, no. 8 (March 1920): 28.

41 Charles K. Kirkpatrick, "Using Rock Phosphate How the Walden Farm Has Profited by it," *Wallaces' Farmer* 53, no. 14 (April 6, 1928): 551 (9).

42 E.M. Harmon, "The Cow Testing Association: How it Works for the Herd Owner and the Community," *Better Farming* 47, no. 10 (October 1924): 7.

43 E.R. Wiggins, "How Tractors Make Farms Pay Profits: Many of the Best Farmers Have Found that Ample Power Provides the Means of Doing Their Work at the Right Time," *Better Farming* 47, no. 3 (March 1924): 6–7, 11.

44 E.R. Wiggins, "The Small Grain Thresher-a Profitable Farm Machine: Farmers Save Money Threshing Their Own Grain and Earn Extra Money Threshing," *Better Farming* 47, no. 5 (May 1924), 3.

45 See also Fred W. Hawthorn, "More Efficient Tractor Farming," *Farm Journal* 54, no. 3, (March 1930): 41–42.

46 Robert A. Jones to "The Editor," National Association of Farm Manufacturers, Farm Press Release #76, "Increasing the Farm Business," January 2, 1930, Box 24, Record Group Number 503, Accession Number 79–001, Ag Engineering Records, Auburn University Archives, 2; for another farmer in the Corn Belt associating increased tractor use with economies of scale in the 1920s, see L.T. Woods, "I Use Big Machinery and a Little Pencil," *Tractor Farming* (May-June 1926): page unknown.

47 "Topics in Season," (February 1930): 36.

48 S.E. Gamble, "Reaping Large Rewards From Good Roads," *Better Farming* 48, no. 1 (January 1925): 12.

49 F.S. Wilkins, "Where Soybeans Replace Oats: Wapello County, Iowa, Community Finds Soy Yields More and Pays Better," *Wallaces' Farmer* 53, no. 12 (March 23, 1928): 477(7).

50 D.F. Marlin, "Pulling Back to Prosperity," *Wallaces'Farmer* 50, no. 1 (January 2, 1925): 1 (1), 12 (12).

51 Haecker, "'Do I Need a Silo?" 6, 14.

52 Rich Lucas, "Radioize the Farm Home," *Better Farming* 48, no. 3 (February 1925): All Around the Farm, 10.

53 "255 Acres: 4 Horses; A Tractor; A Truck," *Better Farming* 46, no. 11 (November 1923): All Around the Farm, 9.

54 "Motor Skill in Demand," 9.

55 "Farm Federation's Power: Life Topsy, Bureau 'Jes' Grew,' and Now Numbers 100,000 Members." *New York Times* (July 4, 1920): 70.

56 Ibid.

57 The National Farmers Union was organized in 1902 as the Farmers Educational Co-operative Union of America in Point, Texas. It initially advocated "for increased co-operative rights, fair market access for farmers, direct election of senators and voting rights for women." "History," *National Farmers Union*, http://nfu.org/about (accessed 9/6/16).

58 Saloutos and Hicks, 253–258.

59 Ibid., 285.

60 "The Two-Row Cultivator: Readers Say it Saves Labor and Saves Time," *Wallaces' Farmer* 50, no. 35 (August 28, 1925): 1095 (9).

61 Reader, O'Brien County, Iowa, The Voice of the Farm.

62 E.R. Wiggins, "Killing Weeds and Forming a Mulch by Cultivation," *Better Farming* 47, no. 5 (May 1924): 7, 11.

63 Louis W. Arney, "The Tractor Belt," *Better Farming* 46, no. 11 (November 1923): 4.

64 Adams, *The Transformation of Rural Life*, 51.

65 Jay Whitson, "Farmers Whom Agriculture Honors: The First of a Series of Articles on the Iowa Master Farmers of 1927," *Wallaces' Farmer* 53, no. 14 (April 6, 1928): 545 (4); see also "Sixteen Iowa Master Farmers: Leaders in 'Good Farming, Clear Thinking, Right Living' are Chosen," *Wallaces'Farmer* 53, no. 2 (January 13, 1928): 43 (3).

66 "Master Farmers of 1927 Honored: Presented to Corn Belt Over WHO and to Iowa Notables at Banquet," *Wallaces'Farmer* 53, no. 3 (January 20, 1928): 91 (7).

67 "Iowa Farm Businessmen," *Wallaces'Farmer* 53, no. 3 (January 20, 1928): 88 (4).

Bibliography

"255 Acres: 4 Horses; A Tractor; A Truck." *Better Farming* 46, no. 11 (November 1923): All Around the Farm, 9.

Adams, Jane. *The Transformation of Rural Life*. Chapel Hill: University of North Carolina Press, 1994.

"Aged Farmer Makes Over Eight Thousand." *Better Farming* 43, no. 8 (March 1920): 10.

"American Telephone and Telegraph Co. Advertisement." *Farm Journal* 52, no. 1 (January 1928): 53.

Arney, Louis W. "The Tractor Belt." *Better Farming* 46, no. 11 (November 1923): 4.

Beck, Russell H. "The Voice of the Farm: Inoculating Soybeans." *Wallaces' Farmer* 50, no. 9 (February 27, 1925): 298 (10).

Breimyer, Harold F. *Over-Fulfilled Expectations: A Life and an Era in Rural America*. Ames: Iowa State University Press, 1991.

Casey, M.V. "What Kind and Size Truck Shall I Buy? Some Practical Ideas for the Prospective Truck Buyer." *Better Farming* 43, no. 2 (February 1920): 5.

Christensen, Harper. "Electricity on Our Farm." *Farmer's Wife* 33, no. 8 (January 1922): 677.

Clark, Christopher et al. "Household Economy, Market Exchange and the Rise of Capitalism in the Connecticut Valley, 1800–1860." *Journal of Social History* 13, no. 2 (December 1, 1979): 169–189.

———. "The Transition to Capitalism in America: A Panel Discussion." *The History Teacher* 27, no. 3 (May 1, 1994): 263–288.

"Cultivation that Kills Weeds." *Wallaces' Farmer* 50, no. 19 (May 8, 1925): 676 (6).

Duffee, F.W. "Modern Lights Save Farmers Work Hours." *Better Farming* 47, no. 6 (June 1924): 8.

"Farm Federation's Power: Life Topsy, Bureau 'Jes' Grew,' and Now Numbers 100,000 Members." *New York Times*, July 4, 1920, 70.

"Farmer Posing with Farmall F-20 Tractor with a Wind-Powered Water Pump in Background." Author's personal records. Powersville, IA, 1940.

"Farmers Used 78,789 Trucks During 1918." *Better Farming* 43, no. 2 (February 1920): Farm Mechanics, 16.

Gamble, S.E. "Reaping Large Rewards From Good Roads." *Better Farming* 48, no. 1 (January 1925): 12.

"General Motors Delco Light Advertisement." *Farm Journal* 52, no. 3 (March 1928): 67.

———. *Farm Journal* 52, no. 10 (October 1928): 39.

Haecker, A.L. "'Do I Need a Silo?' Half a Million Farmers in the U.S. Have Answered 'Yes.'" *Better Farming* 43, no. 6 (June 1920): 6.

Hahn, Steven and Jonathan Prude, eds. *The Countryside in the Age of Capitalist Transformation: Essays in the Social History of Rural America*. Chapel Hill: University of North Carolina Press, 1985.

Hamilton, David. *Deep River: A Memoir of a Missouri Farm*. Columbia: University of Missouri Press, 2001.

Haney, J.G. "Making Alfalfa and Clover Hay." *Better Farming* 46, no. 7 (July 1923): 3.

Harmon, E.M. "The Cow Testing Association: How it Works for the Herd Owner and the Community." *Better Farming* 47, no. 10 (October 1924): 7.

Hawthorn, Fred W. "More Efficient Tractor Farming." *Farm Journal* 54, no. 3 (March 1930): 41–42.

Henretta, James A. "Families and Farms: Mentalité in Pre-Industrial America." *William and Mary Quarterly* 3rd ser., 35, no. 1 (1978): 6–7.

——. *The Origins of American Capitalism: Collected Essays*. Boston: Northeastern University Press, 1991.

"History." *National Farmers Union*. http://nfu.org/about (accessed 9/6/16).

"Iowa Farm Businessmen." *Wallaces' Farmer* 53, no. 3 (January 20, 1928): 88 (4).

Jones Robert A. to "The Editor." Jones Robert A. to "The Editor." National Association of Farm Manufacturers, Farm Press Release #76, "Increasing the Farm Business," January 2, 1930, Box 24, Record Group Number 503, Accession Number 79–001, Ag Engineering Records, Auburn University Archives, 2.

——. National Association of Farm Manufacturers, Farm Press Release #76, "Care Lengthens Sprayer's Life," January 2, 1930, Box 24, Record Group Number 503, Accession Number 79-001, Ag Engineering Records, Auburn University Archives, 5.

Kirkpatrick, Charles K. "Using Rock Phosphate. How the Walden Farm Has Profited by it." *Wallaces' Farmer* 53, no. 14 (April 6, 1928): 551 (9).

Kulikoff, Allan. *The Agrarian Origins of American Capitalism*. Charlottesville: University Press of Virginia, 1992.

Lucas, Rich. "Radioize the Farm Home." *Better Farming* 48, no. 3 (February, 1925): All Around the Farm, 10.

Lunz, Henry. "Better Farmers See Need of Fanning Mills." *Better Farming* 43, no. 8 (March 1920): 28.

Marchand, Roland. *Advertising the American Dream: Making Way for Modernity, 1920–1940*. Berkeley: University of California Press, 1985.

Marlin, D.F. "Pulling Back to Prosperity." *Wallaces' Farmer* 50, no. 1 (January 2, 1925): 1 (1), 12 (12).

"Master Farmers of 1927 Honored: Presented to Corn Belt Over WHO and to Iowa Notables at Banquet." *Wallaces' Farmer* 53, no. 3 (January 20, 1928): 91 (7).

McCormick-Deering Advertisement." *Farm Journal* 52, no. 1 (January 1928): 25.

——. *Farm Journal* 52, no. 2 (February 1928): 21.

——. *Farm Journal* 52, no. 9 (September 1928): 35.

——. *Farm Journal* 52, no. 12 (December 1928): page unknown.

McKay, Grif. "Good Cows, Well Fed." *Farm Journal* 54, no. 2 (February 1930): 68.

Merrill, Michael. "Cash Is Good to Eat: Self-Sufficiency and Exchange in the Rural Economy of the United States." *Radical History Review* 1977, no. 13 (January 1977): 42–71.

"Motor Skill in Demand: Many Farmers Seek Instruction in Gas Engine Management." *Better Farming* 46, no. 11 (November 1923): All Around the Farm, 9.

"National Electric Light Association advertisement." *Farm Journal* 52, no. 10 (October 1928): 45.

Osterud, Nancy Grey. "Gender and the Transition to Capitalism in Rural America." *Agricultural History* 67, no. 2 (April 1, 1993): 14–29.

Pollay, Richard and Katherine Gallagher. "Advertising and Cultural Values: Reflections in the Distorted Mirror." *International Journal of Advertising* 9 (1990): 359–372.

Price, George. "College on the Farm: 4-H Clubs Provide Opportunity to 'Learn by Doing' at Home.'" *Farm Journal* 52, no. 4 (April 1928): 24.

Purvis, Miller. "New Ideas in Feeding Poultry." *Wallaces' Farmer* 53, no. 11 (March 16, 1928): 431 (7).

Reader, O'Brien, County, Iowa. "Truck and Tractors." *Wallaces' Farmer* 53, no. 12 (March 23, 1928): 480 (10), The Voice of the Farm.

Rudicil, Mr. and Mrs. "Colt Light Advertisement." *Farmers' Wife* 26, no. 12 (December 1926): 611.

Rushing, R.B. "Buying Farm Implements: Good Tools and Implements Pay for Themselves." *Better Farming* 48, no. 1 (January 1925): 13.

Saloutos, Theodore and John D. Hicks. *Agricultural Discontent in the Middle West: 1900–1939.* Madison: University of Wisconsin Press, 1951.

Sellers, Charles. *The Market Revolution: Jacksonian America, 1815–1846.* New York: Oxford University Press, 1991.

"Sixteen Iowa Master Farmers: Leaders in 'Good Farming, Clear Thinking, Right Living' are Chosen." *Wallaces' Farmer* 53, no. 2 (January 13, 1928): 43 (3).

Stephenson, B.N. "A Community Checks up its Corn." *Wallaces' Farmer* 50, no. 1 (January 2, 1925): 8 (8).

"The Two-Row Cultivator: Readers Say it Saves Labor and Saves Time." *Wallaces' Farmer* 50, no. 35 (August 28, 1925): 1095 (9).

"The Voice of the Farm: How Much Progress Can We Stand?" *Wallaces' Farmer* 10 (March 27, 1925): 461.

"Too Busy to Milk." *Farm Journal* 54, no. 1 (January 1930): 40.

"Topics in Season." *Farm Journal* 54, no. 2 (February 1930): 32, 36, 44.

———. *Farm Journal* 54, no. 3 (March 1930): 28.

Wall, Joseph Frazier. *Iowa: A Bicentennial History.* New York: W.W. Norton & Company, 1978. Wallace, Henry A. "The Best Corn in Iowa." *Wallaces' Farmer* 46, no. 2 (January 7, 1921): 43 (1).

———. "Fifth Iowa Corn Yield Contest." *Wallaces' Farmer* 50, no. 7 (February 13, 1925): 1, 18.

Warner, H.W. "Getting More Mileage on the Manure-Spreader." *Farm Journal* 54, no. 2 (February 1930): 38.

Whitson, Jay. "Farmers Whom Agriculture Honors: The First of a Series of Articles on the Iowa Master Farmers of 1927." *Wallaces' Farmer* 53, no. 14 (April 6, 1928): 545 (4).

"Why Some Farms Pay." *Wallaces' Farmer* 50, no. 12 (March 20, 1925): 440 (28).

"Why We Made Our Home Modern." *Farm Journal* 54, no. 2 (February 1930): 32, 89.

Wiggins, E.R. "Comfort, Convenience, and Economy with Electric Power and Light Plants: Thousands of Farmers and Farmers' Wives Point Out Many Advantages of Electricity Both in the Home and the General Farm Activities." *Better Farming* 47, no. 5 (May 1924): 4.

———. "How Tractors Make Farms Pay Profits: Many of the Best Farmers Have Found that Ample Power Provides the Means of Doing Their Work at the Right Time." *Better Farming* 47, no. 3 (March 1924): 6–7, 11.

———. "Killing Weeds and Forming a Mulch by Cultivation." *Better Farming* 47, no. 5 (May 1924): 7, 11.

———. "The Small Grain Thresher-a Profitable Farm Machine: Farmers Save Money Threshing Their Own Grain and Earn Extra Money Threshing." *Better Farming* 47, no. 5 (May 1924): 3.

Wilkins, F.S. "Where Soybeans Replace Oats: Wapello County, Iowa, Community Finds Soy Yields More and Pays Better." *Wallaces' Farmer* 53, no. 12 (March 23, 1928): 477 (7).

Wilson, Ruth. *Buffalo Chips: The History of a Town.* unpublished book, Buffalo Center, IA, year unknown.

Wood, Mabel Travis. "A Pioneer Farm Bureau Celebrates." *Better Farming* 46, no. 4 (August 1923): 3.

4

"MOTHER AND RADIO"

Combatting Urban Gender Stereotypes through Technology Use

There is much being said about systems of marketing, but the main production of the whole country is being left out. It is about the shipping of eggs by farmers' wives.

Mrs. L.R. Marrs, Martinsville, MO, 1921[1]

Women on farms also used technology to perform rural capitalistic modernity. Midwesterners viewed urban discourses not just as positing a rube stereotype, but as an attack on rural conceptions of gender. Indeed, farmers, in part, constructed a capitalistic modern identity through gender. The pattern of audience in the 1920s took on particularly strong gendered elements in which rural people sought to combat urban efforts to frame the "modern woman" as a consumer working only in the home and living in the city. Men and women on Midwest farms exhibited a heightened consciousness of this urban gender conception and exerted great effort to use technology to combat it. Female farmers performed an alternative image of rural women as modern producers outside the home.

This contest over modernity in rural America lends evidence for the broader argument that advanced technology is not inherently male, but that the gendering of technology is, in fact, a social construct. Many science and technology studies (STS) and history of technology scholars have explored the intersection of technology, gender, and identity in ways that do not naturalize the "maleness" of technology. At the same time, some scholars have uncritically adopted the urban notion that technology only shaped male identities without acknowledging the experience of rural women with work and material objects. Ruth Oldenziel, for example, shows how greater access of men to 20th-century institutions such as engineering schools at a time when the middle class viewed the bodies of male workers and

DOI: 10.4324/9781032637952-5

athletes as models of white maleness contributed to the gendering of technology as "male" in the U.S. Oldenziel argues that engineers shaped their identities to conform to this model of maleness to shore up class, race, and gender boundaries and to distinguish themselves from shop floor unionized workers. Technology became "the measure of men and not women as a matter of course," Oldenziel argues.[2] Carroll Pursell similarly contends that the appropriate technology and organic foods movements lost popularity in the 1980s because of their association with "feminine" cultural values such as "Small is Beautiful." Values regarded as feminine conflicted with the culture and politics of the 1980s, which sought to re-masculinize America after its defeat in Vietnam.[3] Some historians have offered similar feminist perspectives on the role of technology in enforcing patriarchies in rural America.[4]

Surprisingly, many feminist-minded scholars simultaneously point out the male gendering of technology as a technique for enforcing social power while also ignoring how women use material objects to construct their own identities. In other words, Oldenziel and Pursell tend to not use *diversity as method* as Fouché does when he examines black technological agency. Existing scholarship implies that the relationship between gender and technology may not conform neatly to the masculine-technology/feminine-nature binary when one focuses on use of material objects.[5] I argue that when one critically examines the use of technology by rural women in America and commits to understanding the social meaning of objects from their perspective, this gender binary around technology weakens.

Rather, I contend that women use technology to perform their identities just as men do. Historians of U.S. agriculture such as Katherine Jellison have offered rich accounts of the use of technology by women on Midwest farms in the early to mid-20th century.[6] I seek to add to this historiography by relating this technology use by women to a larger rural discourse-identity bundle of rural capitalistic modernity. More specifically, in the case of rural Midwesterners, urban discourses framed rural women as bored, isolated in their farm homes, and frustrated by an inability to achieve the material and cultural trappings of modern urban living. A careful reading of primary sources from farmers from the early 20th century reveals that *many rural Americans did not view production and technology as exclusively male spheres and rural women performed their modernity through technology use.* The backlash by rural America to some progressive reform only confirmed the misconception on the part of some urbanites that farmers opposed modernity.[7] For example, stories in the urban-based popular press and reports by progressive reformers following the progressive Country Life Commission report fanned rural resentment. A series of articles in *Harper's Bazaar* in 1912 by Country Life Commissioners Martha Bensley Bruere and Robert Bruere entitled "The Revolt of the Farm Wife!" depicted life for women and children on the farm as unhealthy and monotonous. Similarly, a USDA report in 1919 surveying farm wives framed life for rural women as tedious and lacking the "modern" technologies enjoyed by urban women. One report from Cornell University even claimed a "prevalence of insanity among rural women."[8]

Negative urban stereotypes about rural life following the Country Life Commission Report reached such a feverish level that the USDA dedicated an Associate Economist, Emily Hoag Sawtelle to interview 8,000 female farmers in 1924 to rehabilitate popular images of farm life. Sawtelle's report provides a window both into rural identity and resentment of urban discourses among women on Corn Belt farms in the 1920s. Sawtelle first described the general view of the Country Life Commission Report in rural America: "A rather gloomy picture was painted and put before the people and the impressions it made persists in the minds of city people and writers on rural topics." All the women interviewed by Sawtelle saw the popular conception of farm life appearing in the press as misrepresentative.[9] In addition, the women in the report wanted to "remove the stigma from agriculture," noting that, "The surgeon, the artist, the engineer, are not stigmatized in public thought." Unlike these urban professions where "the glorified part of their calling obliterated the materials with which they work... The soil, the clods-the farmer's medium have been too much stressed." The women Sawtelle interviewed lamented that "the wheat, the cattle, the fruit, his [the farmer's] finished products, have been too little remembered and too little identified with his calling." Thus, Midwestern woman in 1924 saw themselves not as women living in a rural setting but as farmers taking part in agriculture. "It is the problem of the farmer and the farmer's wife," Sawtelle noted,

> To take the stigma from agriculture, so as to elevate it by motive that farming shall cease to take its general reputation from the meaner aspects and begin to assume the character of a lofty calling. Already farmers are objecting to the caricatures in the press which represent them as sorry and disheveled, hay-seedy and dirty.[10]

Throughout the report, Sawtelle reflects a rural obsession with urban stereotypes of farmers and a desire to defy them through rural modernization. The women she interviewed saw negative urban views of farming as not just upsetting but as a threat to a moral way of life from an outside "other." Instead of submitting to urban notions "that American rural life is cracking and bound to break up," by leaving the farm, which Sawtelle characterized as "acknowledgment of defeat," the women considered farming "on no lower round of advancement than merchandizing or any other business."[11] The farm women in the report, therefore, saw the adoption of new technology in terms of a contest over the very survival of what they considered a moral rural identity. Rather than seeing themselves according to the urban conception of women trapped in a backwards place of drudgery and boredom full of social disadvantages, these female farmers performed an identity as "strong, resourceful, capable, and leading personalities in their communities."[12]

Throughout Sawtelle's report, she cites Abigail Adams as a kind of founding heroine who managed the farm in her husband John Adams' absence. Adams is evoked as an archetype of female rural independence, as an example of how women could uphold the family and the farm as a virtuous social system.[13] One can go so far as saying if men had Jeffersonian agrarianism with which to fashion themselves in heroic terms, women had "Abigailian" agrarianism. The report, for example, saw women as the heroic leaders in combating urban stereotypes

> The rural woman through all our agricultural development has held like a creed the determination that while gaining financial advantages, her family should not be needlessly deprived of social privileges. She has endeavored by dint of labor and thought, by substitution and combination, to bring to her family the best that life has to offer. The American farm woman has always been a courageous social pioneer as well as a resourceful frontierswoman.

The woman in the report saw the farm home as preserving democracy:

> The farm home is a home to the whole nation, not merely to those that live within its walls. The nation, in a way, borrows and depends upon the traditions of its farms. This, it seems, is our rock-bed of patriotism, the stabilizer which shall continue to make us a strong and permanent nation. Without it we should be like floating plants without roots.[14]

The women saw it their duty to modernize their farms in order to not only combat false urban stereotype but for the Jeffersonian rationale of preventing the cities from undermining true democracy:

> The farm woman is the strongest supporter of our basic democratic principle – she believes that everyone in America should work, and that everyone in America should share in the higher life. She admits of no second place for American farmers. She makes no apology for her chosen occupation. Rather, she believes its possibilities boundless.[15]

The female farmers also told Sawtelle they saw themselves as *superior* to urban women for several reasons that reveal interesting insights into their performative use of technology. First, several interviewed noted that unlike their urban cousins they worked as partners with their husbands on equal financial footing. "A farm woman," the report noted, "can always start on her way to partnership or economic independence with her garden, her butter, and her hens." Further, "The helping on the farm is not all on one side. The farmer often gives his partner a helping hand with the garden and heavier indoor work." As one Ohio woman wrote, "My husband and I started life on $200, 6 years ago, and today by hard work we are making good, and are looking forward to a home of our own in the near future." As an equal partner in a modernizing business, farmwomen saw themselves as *more modern* than urban women. Sawtelle writes,

Nowhere does a woman have a better chance to be her husband's partner in every sense of the word. The business itself is spread out in front of her door. Its details come to her kitchen. She sees the plans for the work going around her. She hears the talk of the business at her table. The farm papers come into her living room; farm bulletins are on her desk. She has every opportunity for studying the technique of science, and for acquainting herself with the inside workings of a thriving business.[16]

Second, the women in the report saw the farm as a more moral place to raise children than the city. Increased technology use thus became an updated means of enacting an older agrarian role of women heroically preserving a moral family. Sawtelle noted a statement at the 1922 National Agricultural Conference by female farmers, "We stand for the conservation of the American farm home, where husband and wife are partners and where children have the opportunity to develop in wholesome fashion." One Michigan woman expressed a similar identity as a heroine defending her family from less virtuous urban lifestyles,

Our work is harder than the city woman's and there is more of it, but we are free to do as we like in some things. I am raising my family without fear of a landlord ordering me out, because there are too many babies. The little vine-covered cottage is better than an apartment in the city where no children are wanted.

The very landscape of the Corn Belt, the women felt, promised a superior environment for raising children than cities where children "are all too often forced to seek a playground in the street" and where "The city mother is kept in a constant state of anxiety as to the safety of her children." As one woman farming in Wisconsin stated, "In the country there is a lot of room indoors and out for things to grow in. Our families, like the crops, grow up naturally."[17] As partners in a business that preserved more wholesome places for raising children, rural women had constructed an identity as more moral than their urban critics. In contrast, urban women toiled in an inferior position to their husbands and raised their children in a dangerous and unhealthy city, according to these female farmers. That these same urban women looked down on agriculture irked the rural dweller who saw herself as morally superior. A decade later, even President Franklin Delano Roosevelt's Secretary of Agriculture and former chief editor of *Wallaces' Farmer*, Henry Wallace, expressed a similar Jeffersonian view that "The land produced the life-stream of the nation," in the form of "young people bred on the farms."[18]

The farm women interviewed by Sawtelle almost universally viewed the adoption of new technology and accumulation of increased property and wealth as the best ways to challenge the "gloomy picture" of rural life held by urbanites. As Sawtelle writes, "But with the movement for improvement of conditions well under way, protests against calamity stories began to appear and farm people now resent characterization and cartooning as ignorant objects of misguided pity." The rural women objected to the fact that

Casual observers of country life are in the habit of contending that the present day farm woman is more restricted by her household duties than was her Puritan and Revolutionary ancestors: that there has been in the average American farm home, no substantial improvement of conveniences in the past 50 years.[19]

To combat this notion, as one woman in Wisconsin stated, "you may be sure that the farm woman will fall in line with every improvement and wave of progress that is made." Similarly, many of the farm women used Sawtelle's interview as a chance to document a "march of progress" in rural America.[20] One "Iowa farmer's wife"[21] wrote,

Slowly and surely, electricity and gasoline are finding their way into the farm home.... Naturally, if the outbuildings are modernized, it enables the farmer to work faster and realize more capital with which to make further improvements. There's truth in the old saying: 'A barn can build a house sooner than a house can build a barn.'[22]

Other women in the report noted "great improvements that have come about in her lifetime" by listing artifacts they owned in detail and contrasting them with objects possessed by their mothers. These women performed a modern identity for urban observers by creating an inventory of progress. The women sought to present to urban observers a "new farm house" that "does not differ materially from the city home in its opportunities for conveniences." Just because a house stands on a farm "does not mean that it must be merely a crude headquarters for work, completely pervaded by an air of grim business," the report noted.[23] One woman on a farm in Ohio's Miami Valley described her parents as having a 40-acre tract of land. Her mother "had a little 4-hole wood stove, and iron kettle, 2 iron pots, 2 iron skillets, 2 iron griddles, and enameled lined iron kettle and a granite 6-quart preserving kettle, the prize of the neighborhood in canning time." To demonstrate rural progress, the Ohio woman noted,

Now...we have over a hundred acres. I have gas lights in every room, a 3-hole hot plate for hot summer cooking and an extra-large coal range, a washing machine that can be run sitting or standing or by gasoline engine, if I ever am lucky enough to possess one. I can all my vegetables by "cold pack" in a steam cooker. I have all the pans and kettles I can use in aluminum and granite, and I haven't a heavy iron pot in the house. I have a big roomy kitchen with large cupboard cabinet, chest of drawers for towels, 2 large roomy work tables and 2 stoves in it.[24]

Similarly, another woman in Missouri wanted Sawtelle to know she had a gasoline engine to run her water pump and washing machine and that she would soon obtain a car.[25]

Other women interviewed by Sawtelle sought to reframe their technology use within a modern rural discourse to resist efforts by the popular press to evaluate them based on urban conceptions of modernity. One Iowa farm woman, for example, boasted,

> My washing machine has been run with an engine for six years and now I use electricity. I also have a mangle, that is run by electricity that I iron all my clothes with. We farm women never have to watch a meter as we have our own electric plants. Within three miles of my home there are only three out of 14 farmers that haven't electric plants of their own. Eleven of us farm women have the use of electricity, and we don't have someone always sending us a light bill either. There are many things I haven't got, such as an electric vacuum cleaner, but I intend to have one before long to use in place of the hand one I have had for 10 years, and yes, I am going to have a grill to cook my light meals on, as well as other things as soon as I can get them. My sewing machine is run by an electric motor and while I am busy sewing, I have the electric fan to keep me cool.'[26]

The Iowa woman seems to view the lack of electricity from a grid, a condition seen by urban dwellers as a sign of backwardness, as a sense of pride. The fact that she derives her electricity from her own electric plant rather than through a wired meter enhances her identity as an independent Jeffersonian. She sees her electric plant as a subtle act of protest against urban forms of modernization because she does not receive her electricity from an urbanized distant provider but produces it herself on her farm. As a result, she turns urban scorn toward rural people too isolated to connect to a grid into a positive sign of self-reliance. By turning a negative into a positive, the Iowa woman challenges not only rube stereotypes but urban modernity itself. She seeks to reframe modernity into a rural image that contains an element of pre-existing Jeffersonian and German agrarianism: independence. This attitude toward personal electric power plants only makes sense if one blends modernity with rural Jeffersonian or German agrarian identities.

The technologies cited by women in the report fit into a broader rural project of farmers of both sexes constructing a modern self to show, as one Indiana farm woman stated, that rural people "are not behind the times, by any means." The farmer noted that her town had "a fine consolidated school building, modern throughout." A "great many of the people in our community," she noted, "are college-bred men and women and even those who are not, nearly all plan to send their children to college." Similarly, another woman in Indiana, describing herself as "Mrs. Stratton-Porter" presented a sophisticated lifestyle for an urban audience including operating a large flower conservatory. She also claimed to have mastered photography and she declared, "I have written ten books."[27] Surely, no city dweller could classify Ms. Stratton-Porter as a rube.

Further, Sawtelle saw farm journals and other rural publications as instrumental in presenting the rural capitalistic modern rural identity to misinformed urban dwellers. "There is every reason to believe that the courageous spirit with which American farm women are now attacking their problem of social organization will soon be widely reflected in the press," Sawetelle wrote. She continued, "Presently we shall have appearing in our magazines and newspapers strong truthful pictures of farm life. Already occasional glimpses into farm life given in current periodicals show us that we have in the making a new rural literature."[28]

This placement of technology use within the context of constructing a rural modern identity explains why women interviewed by Sawtelle did not speak of technology in economic terms. When the women employed financial justifications for adopting technology, they did so only through vague folksy rhetoric such as "A barn can build a house sooner than a house can build a barn."[29] In other words, the "partners" in the farming enterprise did not conduct a rational economic analysis of each artifact or balance its functions with its costs before making a purchase. Rather, "When the farm women has fully determined for herself that her permanent home is to be upon the farm, she sets about getting all the indoor conveniences as rapidly as practicable," Sawtelle wrote, including "The furnace, the bathroom, electricity, hardwood floors, and other pleasant features."[30] While these items may have made life on the farm more "pleasant," farm wives and husbands ultimately adopted them mainly to convince urban dwellers, themselves, and their neighbors that they were modern capitalistic agrarians rather than backwards rubes. In fact, historian Ronald Kline has shown that studies by home economists as early as the mid-1920s revealed that technologies in the home often did not save rural people time, what historian Ruth Swartz Cohen would later call "the irony of laborsaving household technology."[31] Rural sociologists John Kolb and Edmund S. de Brunner in a 1933 report even cited a Wisconsin study that found farmers "with tractors work longer days on the average than farmers without tractors."[32] In other words, the obvious conveniences of electric lighting or tractors cannot fully explain why famers used electricity or mechanized equipment when they did, nor can it account for how they used these devices. Only an understanding of how urban views of farmers offended rural people, who thought of themselves as moral Jeffersonian and German agrarians, can explain why male and female farmers used technology the way they did in the 1920s.

As I argue throughout this book, resentment and identity, not efficiency or even profit, often drives rural technology use. Farmers, like all people, use technology as a performance. My analysis of Sawtelle's 1924 report suggests that farm-women deeply resented these urban conceptions of the "modern" woman only as a consumer. While the role of women on farms in the U.S. varied from family to family, women in the Corn Belt seemed particularly likely to view themselves as participating in, and sometimes leading, the modernization of both the home and production processes on the farm.

Other women on Midwest farms echoed the view of women in Sawtelle's report that modernity in the 1920s could prevent young people from leaving rural America for urban centers. Female farmer Emma Gary Wallace, for example, told a story of a family able to keep their children from joining the rural-to-urban migration by modernizing their home,

> Eventually the home was entirely remodelled [*sic*] with hardwood floors, electric lights, bath-rooms with plumbing, furnace, and all the modern fittings of any home of refinment [*sic*]. The young people have no desire to go elsewhere.... There is a great deal to be said in favor of living in an atmosphere of prosperity....[33]

Advertisements for technology on farms such as Delco Light power stations reinforced this notion that using the latest technologies would keep young people on the farm.[34]

Similarly, one article describes a school for farm children started by farmer Eugene M. Funk in 1920 that included small experimental plots for each student that "afford good laboratories" as a well as an interior that had "an appearance of a modern country home" as opposed to the old-fashioned one-room schoolhouse. The school also featured instructional fields "to show the value of rotation, fertilizing, proper cultivation and the other things essential to the growing of biggest crops and soil maintenance." Funk started the school because "farming is something like a factory now, and the farmer must be a business man to some extent." Both boys and girls learned in the home and the experimental fields and studied up-to-date business practices. Funk concluded, "It is best for their boys and girls to know something about the business which we hope they will take up. That is farming."[35] Importantly, Funk noted that farming was "like" a factory reflecting an understanding that his work involved preparing farm children for some sort of rural version of modern production without literally copying verbatim the urban industrial model.

Nor did Funk find himself alone in his approach to educating children to become modern capitalists. Harold Breimyer recalled how his high school taught "vocational agriculture" requiring both male and female students to "undertake a livestock production project," experiment with a specified acreage of crops, and keep detailed records of production inputs and yield. Students viewed the agricultural instructor as a "business advisor" and often converted projects into profit.[36] This non-gendering of work in rural schools in which children performed rural capitalistic modernity contrasts sharply with urban models which reserved more advanced technologies and engineering for boys.[37]

Further, this view of childhood, as a training ground for future businesspersons contrasts sharply with memoirs of rural Americans recalling childhood on a farm at the turn of the 20th century where children tended to simply provide labor in production processes, and parents often beat them for not working. This changed view of childhood among farmers as a privileged age for learning and training for the future indicates a shift to a more progressively modern discourse. As Edith Bradley Rendleman recalled of growing up in rural Illinois at the end of the 19th century, "In those days everyone beat their kids something awful." Her memoir recalls many hours of planting beans in fields, picking downed corn, harvesting straw and several other tasks under the constant threat of corporal punishment.[38] Rendleman's account contains nothing about learning accounting techniques or farming methods. Her memoir contains no account of education about plant science, household technologies, or new or future equipment. She certainly never heard the term "experimental plot" or "modern farm home" before and indicates nothing that would suggest she or her parents saw her life in terms of preparing for a better techno-centric future. As Rendleman's memoir title states succinctly *All Anybody Ever Wanted of Me Was to Work*. In fifteen short years between Rendleman's childhood and the Funk school, the image of what a son or daughter on the farm needed to become in the Corn Belt had clearly started to change, at least for some farm families.

Performative Technology Use by Female Farmers in the Midwest

Women participated in production and used technology to perform a rural modern identity as well. The prominent role of women in the business of Midwest farms has origins in German agrarianism in which women brought more land to a marriage than English or early American brides due to a more equitable distribution of inheritance for women in German culture.[39] Thus, this role of women as producers outside the home continued a long rural tradition. Indeed, one often could not easily gender activities of production on farms as both men and women produced many commodities.

When farmers began to sell on capitalist markets and then buy other consumer products, many women saw no reason to abandon the production side of farming simply because urban discourses sought to create "separate spheres."[40] So not only did women seek to combat yokel stereotypes, but they also had to keep urban ideas about gender roles from infiltrating the farm where work processes were organized much differently. In short, rural women faced an extra layer of complexity during the rural-urban conflict. Thus, when Edith Rendleman recalled that her mother in the 1920s "bought a lot of things with that butter, milk, and egg money" on their Illinois farm, including new modern clothes for the family, she did so to perform a modern productive identity that conflicted with perceived urban conceptions.[41] Another woman on a Midwest farm, Grace Gibbard Lentz, wrote a short story for *Better Farming* entitled "Evolution of a Real Farmer's Wife," who "evolved" to manage a large cow operation that eliminated the family's debt. As a result, Lentz wrote "she gained respect for herself. She realized her dream of becoming a really modern farmer's wife."[42] While Lentz presents a fictional woman in her article, the reader gets the impression that if the character is not actually Lentz herself, she at least writes from personal experiences.

The reality of activities on many farms in the Corn Belt reveals that women and men often divided work in and out of the home even though women considered themselves heads of the domestic sphere. Many Corn Belt farm families seemed to view the home and the farm as one hybrid unit of production, which both sexes could occupy and make decisions to modernize. As Jane Adams explains, "a wife could have a commercial strawberry patch… a husband could bake cakes."[43] Ruby Weaver, a woman on one of the Illinois farms interviewed by Adams even recalled how her husband gave her a tractor as a birthday gift early in the 20th century so that they could work in the fields at the same time. Many other women Adams interviewed remembered handling horse-drawn plows and field machinery.[44]

Mildred Armstrong Kalish, who grew up on an Iowa farm in the late 1920s and early 1930s, recalled "though certain work was usually thought of as man's work, on our farm, everyone male, female, and kids, lent a hand to get the job done… The same was true of 'women's work.'" Kalish discusses how women loaded hay in hay mows, shocked wheat, and harvested corn and clover while men often canned meat, washed clothes, or churned butter. Nor did women only participate in "male" activities like harvesting hay casually or do activities that did not require great physical strength. Kalish noted that the women on her family's farm often operated a "murderously heavy hay fork" and "The whole family worked all day in this manner until all of the hay from the fields had been gathered and as much of it possible stored in the mow."[45] Given how women like Kalish's experienced work on a Corn Belt farm early in the 20th century, it should be no surprise that female agrarians did not view using new tractors or other modern artifacts as exclusively "men's work." Nor did women in the Midwest view technology as a "male" threat to a "female" nature or approach farming with a more "nurturing" approach than their husbands, as argued by many contemporary scholars (see Chapter 9).

Given that men participated in work in the home, one should also not be surprised to learn that men did not view rural capitalistic modernity as an exclusively male identity. Even if one takes the male patriarchy as a given, historical evidence show that women also employed the latest technologies and modern business practices to produce commodities. In poultry and egg production, for example, women often published detailed records of "poultry figures" in farm journals such as *Farmer's Wife*.[46] Millions of women invested in new egg machines in the mid-1920s to increase their commercial poultry businesses.[47] Other women on Corn Belt farms in the 1920s, such as L.R. Marrs from Martinsville, Missouri, used middlemen to distribute their eggs as far as Colorado and New York City to take advantage of higher prices outside their local communities.[48] As with a female farmer named only as Mrs. George Penn of Worthington, Ohio, these chicken-raising efforts sometimes even bought entire farms and occupied women for well over eight hours a day. Penn's sale of eggs "bought an 80-acre farm, put up the buildings, put in furnace, lights and bath, bought a car and a cow."[49] Thus, these productive activities went far beyond a casual hobby or simply providing for the home. Rather, women headed aspects of a full-time modernizing business.

Nor did rural women in the 1920s perform a rural modern identity only at the direction of their husbands. Rather, female farmers took initiative on their own to use the latest technologies to perform their identities as modern producers. The historical record suggests that women had technological agency and thought of themselves as business owners. One *Farm Journal* article in 1928, entitled "Mother and the Radio" by self-described "farmer's wife" Ethel Morrison-Marsden, describes a woman on a farm not just using the radio as a domestic consumer but as a savvy businessperson producing a product and manipulating markets. Morrison-Marsden writes,

> Just before dinner the markets come in, with quotations on eggs, poultry and butter. Very often this knowledge means the saving of money to the housewife; she may be shipping eggs to private customers in the city, as I do, and awaiting the day's quotations before setting the price. Or she may be selling poultry and needs the knowledge of the day's poultry market so as not to be at the mercy of the buyer.

The article shows a farmwoman using the radio at a desk seriously jotting down commodity prices (Figure 4.1). The austerity of the photo highlights her business-like persona as it features only a desk, a pen, note cards, and a large black radio speaker. The discourse used by Morrison-Marsden is not one of a bored housewife limited to the domestic sphere, but of technological progress and capitalistic business, as she exclaims, "It [the radio] has made the impossible possible."[50]

Another female poultry producer in Missouri wrote an editorial to *Wallaces' Farmer* sharing modern remedies for sick poultry that increased commercial profit.[51] Similarly, in a 1930 article, Grace Jenney give instructions on how she built a more modern and industrialized poultry house on her farm because "Big business (and that's poultry-raising) demands big methods, and I am not sure if there isn't a decided labor advantage in large individual pens." The article pictures Jenney's "double-decker" poultry house for "200 birds to the floor" as a model of rural capitalistic modernization (Figure 4.2). The photographer took the picture from an upward angle to highlight the impressive size of Jenney's modern structure complete with two levels of large windows. Jenney did not purchase her poultry house or have a builder come to construct one. Rather she designed the entire building herself in the name of "progress." Jenney even subtly, and humorously, remarks on her independence and the relative uselessness of men in the design process stating, "…if you are planning new stairs in any farm building, know your carpenter. In my case, I married him, but I don't advise that as a wholesale program."[52] While Jenny farmed in New England, the journal considered women in the Corn Belt as an audience receptive of her methods.

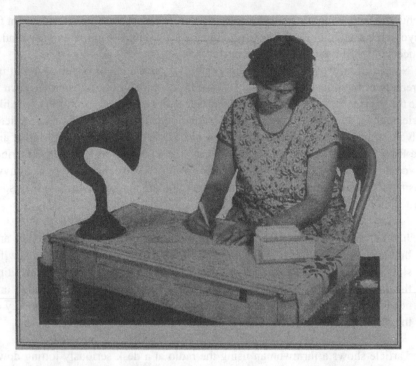

FIGURE 4.1 A female farmer uses radio as a modern producer and seller of commodities, not only as a domestic consumer. Ethel Morrison-Marsden, "Mother and the Radio," *Farm Journal* 52, no. 3 (March 1928): 32.

FIGURE 4.2 Grace Jenny's modern and self-designed double-decker poultry house. Grace Jenney, "Rooftries for our Chickens," *Farm Journal* 54, no. 3 (March 1930): 71.

Women in the Midwest also headed the modernization of large-scale dairy production. S.C. Campbell of Hennepin County, Minnesota, for example, started her own dairy farm. The *Farmer's Wife* described Campbell's farm as,

> Maplewood Farm is a complete, modern dairy plant with well-arranged sanitary barns and milk house, two silos of 190-tons capacity each, and a large, comfortable, modern farm house with huge screened porches and an amazing number of large air windows.

Campbell's farm does not exhibit a small level of milk production simply for family use in the home but a vast network of wholesale distribution to hotels, hospitals, and grocery stores. After describing the details of Campbell's modern farm, the article moralizes the production arrangement by placing it in an idealized family scene that would have looked familiar to a German farmer a century earlier,

> It was a pleasant picture, this comfortable home, with its piano and books and magazines testifying that the values of mental culture had not been ignored, its orderly rooms, its well-kept grounds and neat, substantial buildings, the contented herd of profitable cows in the pasture beyond – and all of it the result of one woman's vision and faith and work.[53]

Another female dairy producer Harper Christensen on the Minnesota-Iowa border extolled the benefits of electricity on her family's farm in a way that showed that she conceived of work in the home and in the barnyard as covered by one umbrella of productive labor

> The greatest asset with an electrically equipped farm is on the labor saving side. We milk with electricity, separate the milk with it, wash and iron with it and in the near future expect to clean the house with electricity. Running the milking machine with a motor is much quicker and handier than a gasoline engine. Sometimes I have cranked a gas engine until I was blue in the face only to have to milk by hand, and the same applies to the cream separator with the motor attached – it runs much more smoothly and more even.

Christensen uses the word "we" implying that more than one family member had responsibility for duties both in and outside of the home. She presents a discourse not as a domestic homemaker but as an owner of a modern dairy business employing the latest technology and electricity to maximize production. *Farmer's Wife* also published regular reports of extensions and cow testing associations entitled "Of Interest to Dairy Women," indicating that women often oversaw "scientific" commercial dairy operations.[54] This journal, describing itself as a "Magazine for Farm Women," published in St. Paul Minnesota, boasted of selling over 800,000 copies a month in 1926.[55]

While one could dismiss the role of women as producers on Corn Belt farms as a subordinate position to farm husbands who operated large machinery in the fields, many people farming in America in the 1920s and 1930s saw the poultry and dairy commodities produced and marketed by women as essential income streams on a successful modern farm. A *Wallaces' Farmer* article in 1925, entitled "Why Some Farms Pay," reported on a Department of Agriculture study of 160 farms in central Indiana that made a profit even during "the depression period." The study concluded that farmers reported "poultry played an important part" on the profitable farms as did less crop diversity and a "scientific layout."[56] In a front-page article in 1925 on "Pulling Back to Prosperity" following the "deflation of the farmer in 1920," Iowa farmer W.J. Breakenridge credited part of his regained prosperity to his wife's modern poultry management techniques and her strict attention to markets. Breakenridge talks about his wife more like a top performing business manager than as a woman whose place belongs in the home stating, "The future of the chicken and egg business looked good so we expanded the farm flock. This was done under the direction of Ms. Breakenridge for it has been her department thuout [*sic*]."[57]

Another article about an event honoring Master Farmers in 1927 noted that,

> All the Master Farmers, in one way or another, gave credit to the folks back home for any success that had come to them. The wives and children were declared to be entitled to a greater percentage of each Master Farmer medal.[58]

Similarly, one article in *Better Farming* urging farmers to increase their poultry production framed women as experts stating, "Thousands of women who see profits in poultry have become expert raisers." The article recognizes the activities of these "expert raisers" not as a hobby or side-job but as an integral part of a successful farm business.[59] Jane Adam's study of women recalling their lives in rural Illinois in the early and mid-20th century found that "no woman I interviewed saw her work as a 'sideline'" and all of them disproved urban conceptions of "separate spheres" for men and women.

Adams also noted that women in interviews evaluated the worth of themselves and others by how hard they worked as commodities producers beyond traditional homemaking duties.[60] Similarly, Louisa Stephens recalls how she began running a poultry business at the age of 13 in Missouri. Her family saw her production activities important enough to send her to courses in poultry management at the Missouri College of Agriculture.[61] Clearly, the prosperous Master Farmers functioned as a dominant image of the modern capitalistic agrarian within a broader conflict between rural and urban economic and social interests. This image of the modern Corn Belt farmer served an important role in rural efforts to combat yokel stereotypes as part of an unspoken cultural dispute over the meaning of modernity at the heart of the rural-urban conflict.

In addition, several articles in farm journals reveal that women's productive work provided the extra income needed to modernize in the first place. Thousands of farming women organized co-operative markets in the Corn Belt selling agricultural products and prepared foods directly to the consumer to buy electric appliances, modern plumbing, college education, or newer farm buildings. One woman farming in Illinois reported $6,700 in profits from these co-operative market sales, allowing her to pay off the mortgage on her family's farm.[62] A North Dakota woman calling herself "Mrs. Albert Limbaugh" discussed how her cake-making business allowed her husband to keep and modernize their farm stating, "My husband and I have always been partners." [63]

Many farm women and men shared tasks on the farm also as a form of insurance so that the wife could keep the farm going in the event of the husband's death. For example, May E. Craw spoke in 1926 about modernizing the farm in Champaign County, Illinois after her husband died,

> we've gone right ahead and prospered. In ten years we've bought the homestead of Mr. Craw's family, remodeled the house, put electricity in house, barnyard and outbuildings and acquired about everything in the way of modern equipment, from a pressure cooker and sweeper to a radio outfit.

Craw attributed her success to the fact that she shared production and business duties with her husband prior to his death and to her careful study of farm journals and farm bureau publications "to keep up with the times." She also regularly listened to the program "the Practical Course in Farming" over her radio. Craw seems to have monitored every aspect of production down to the most mundane detail explaining, "None óf us are afraid to work… At threshing time and corn husking I do every bit of the 'weighing of the grain myself. Then I know what I get.'"[64]

Although farm men and women both sought to construct their identities as modern producers through use of technology, gendering of work processes did occur on American family farms with operation of large field machinery designated as a masculine domain. Evidence suggests that what made fieldwork "men's work" was the large machinery itself, not the act of producing or even of farming crops. Women also wrote into farm journals to discuss how to improve crop production techniques. For example, a woman identified only as Mrs. E. J. Kirk of Ohio wrote to *Wallaces' Farmer* in 1925 on techniques for growing and harvesting barley to supplement the corn crop used to feed hogs. Kirk's editorial used the term "we" suggesting that she participated in the planting and harvesting process as well as the care for large animals.[65] Therefore, work with large field machinery appeared only *weakly* gendered (for more discussion of this topic, see Chapter 10). Many women on farms in the Midwest operated large field machinery and some even produced crops on farms without men. In such cases, women used large field machinery to perform both a rural modern identity and debunk urban notions of a women confined to the home. One female farmer in Grant County, Illinois, for example, posed atop her McCormick-Deering tractor in 1940 to show that "[S]he and her mother run the farm without the assistance of any men." (Figure 4.3).

Female farmers also joined men to organize state farm bureaus and the national American Farm Bureau Federation (AFBF) from 1900 through the 1930s for the purpose of giving farmers more power in modernizing agriculture both in terms of farming methods and political lobbying. Over 300,000 members from 28 states approved the AFBF in March of 1920. Farm bureau meetings in the American Midwest in the 1920s and 1930s demonstrated a rural view that the work processes of women formed *part of* the project of modernization and increased technology use on farms. While such meetings included home economics demonstrations, women also discussed using technology as modern technical producers outside of the home.[66] One *Wallaces' Farmer* article in 1925 reported on a "number of special sessions for women" at the upcoming annual Iowa Farm Bureau Federation

FIGURE 4.3   A female farmer using her tractor to perform rural modernity and production. "Untitled photo, possibly related to: FSA (Farm Security Administration) rehabilitation borrower operating tractor. She and her mother run the farm without the assistance of any men. Grant County, Illinois," *Library of Congress*, https://www.loc.gov/pictures/item/2017810457/ (accessed 8/9/23).

convention "to take up questions especially relating to women's work on farms and the problems of farm betterment." The keynote speaker to the approximately 1,000 women in attendance with their husbands was not an expert in homemaking or domestic work but "L.G. Michael, special investigator of foreign markets for the U.S. Department of Agriculture" to lecture on the "[d]evelopment of foreign markets for American farm products." The article described the other speakers at these sessions, whose lectures they broadcasted over the radio, about "women's work" as "all other important problems now affecting the farmers."[67]

The convention planners literally saw the husbands and wives in attendance as dual productive units seeking to modernize production and expand marketing processes rather than as a male producer and a female domestic consumer. In other words, "women's work" did not seem to have the same pejorative implications within the early 20th-century discourse of rural capitalistic modernity as it may have under our own contemporary rhetoric about gender. Women did attend sessions by home economists, but they framed work processes occurring in the home, such as making clothes, within a broader production context. As one attendee stated, "Everyone seems to have caught the big vision of this work – that it is a means to an end, and that end [is to] better agriculture."[68]

The modernization of activities of women in the home also became linked to a broader effort to bolster the rural cause in the context of the rural-urban conflict. The discourse of rural capitalistic modernity often came tinged with rural resentment and fear of urban political, economic, and cultural power. In the annual convention of the AFBF in December of 1920, for example, male and female speakers urged women to view their work more scientifically and in modern business terms as a means of preventing women form migrating to cities. Not only did the city offer a potential opportunity for a woman to earn an independent income, but some feared women would associate the city with a sophisticated femininity lacking on the farm. Technology and modernization promised not only economic benefits but a means by which rural men and women could reassert their self-worth and their rural identities and protect rural life from a threatening urban world.

Farm journals in the 1920s contain many letters and articles by women urging other women to stay on the farm arguing that they could reach the same wealth and sophistication as their urban cousins by participating in their own modern production processes. In Grace Gibbard Lentz's fictional piece in *Better Farming*, the heroine Marian found herself married to a husband who farmed by himself while she stayed in the home. In this situation, Marian "was a disillusioned farmer's wife. No longer had her form the dainty, graceful curves. She was discouraged farm women." After Marian started using the latest technology to start a successful modern dairy operation, "with her self-respect came back her form, a mature strong woman's form that city women work for in the gymnasiums." Marian's husband, Arthur, had grown apart from his wife, but after she became a success outside of the home, "Arthur realized his dream of a fine herd" and their marriage improved. The picture showed Marian and Arthur in a loving relationship with the caption

"Joy and Pride Swelled in Her Heart When She Caught Arthur's Proud Look and Knew That Together They Had Earned All These Good Things."[69] While such a story strikes the contemporary reader as patriarchal and sexist due to the emphasis on female appearance and the male gaze, it also unexpectedly defends the rural expansion of women's roles as businesswomen and producers. Thus, in Lentz's social context of rural-urban conflict, modern technology use became a way of performing a more empowered, albeit overly sexualized, femininity. Lentz and other rural actors had to confront urban discourses that intertwined modernity with a dominant image of a woman and man who looked and behaved in certain ways.[70] In short, the "rube" that rural Americans sought to counter was not exclusively male. As such, the *Farmer's Wife* and other farm journals also featured articles about the latest fashions for both sexes.

Letters from rural people to farm journals also seek to reinforce Jeffersonian ideas of the morality of rural life.[71] Since rural modernity incorporated Jeffersonian and German notions that the family and productive work moralized rural production and that this virtuous farm supported the rest of the nation, the President of the AFBF, J.R. Howard stated, "The foundation of America, the very heart of America, is the farm home… I am not going to give up until every farm home in the country is as good as any other home."[72] A woman, C.H. Sewall of Oberlin, Indiana expressed similar sentiments as Howard's when serving as the key-note speaker of the Iowa Farm Bureau meeting stating that "twenty-five years ago a farmer's wife would not have been asked to go visiting and talk before an audience of farmers. She was expected to stay at home and do the housework and wipe the noses of the children." Sewall "complimented the men on their graciousness in passing a resolution 'Inviting women into full participation in the task' of advancing the cause of American agriculture."[73] With this greater task of rendering rural America on equal footing with a threatening urban world, women in the Corn Belt also formed hundreds of county "home bureaus" in the 1920s in which attendees would observe home demonstrations and test out new "home equipment."[74]

From the perspective of many rural women, a wide array of urban voices has sought to frame their modern identities as male and immoral at least since the 1920s. The inability of urban observers to grant rural women the status of modern producers despite their performance, in turn, continues to fan rural resentment (see Chapters 9 and 10). Women on farms from the early 20th century onward know that their work as productive units combats the traditional urban view of women reflected in the Census Bureau listing of many female farmers incorrectly as "Women of No Occupation."[75] Jane Adams, writing about her childhood on an Illinois farm in the 1940s, angrily notes the failed attempt by urban "policymakers" to "remove women from farming and make then full-time homemakers," and she laments official narratives contained in literature written by scholars outside the farm that failed to recognize "that wives were as much farmers as husbands."[76]

Thus, women expressed similar concerns over the rural-urban conflict as men, and they similarly viewed technology and modern business practices as means to combat urban industrialism. In a series of articles in *Farmer's Wife* entitled "What are Farm Women Thinking About," women express desires for "more knowledge of business methods," "all the latest machinery," support for cooperative marketing associations, and "less pity and sympathy from city people and greater apprecia- tion on their part of the joys and values of country life." In fact, many saw women as leading the migration to the cities because "the boys go into town because the girls go." Therefore, a central role for women in the modernization of productive processes on the farm promised perhaps the most effective means of reversing rural-to-urban migration.[77]

A culture of the machine had established itself in which rural Midwesterners' identities as modern, progressive, independent capitalists became embedded in the technologies they used as a means of opposing an urbanized industrialism. Both men and women on farms use technology to perform their identities as modern capitalistic producers and content urban conceptions. Technology use functioned as a performance to debunk rube stereotypes, which included the inaccurate im- age of female farmers as bored and unfeminine housewives rather than modern producers and businesswomen. This identity of rural capitalistic modernity would intensify and become more widespread in rural America during the Cold War when the "other" shifted from the threat of urban industrialists and reformers to a col- lectivizing ideology emanating from the Soviet Union.

Notes

1 Mrs. L.R. Marrs, "Shipping Eggs," *Wallaces' Farmer* 46, no. 3 (January 21, 1921): 113 (21).
2 Ruth Oldenziel, *Making Technology Masculine* (Amsterdam: Amsterdam University Press, 1999), Introduction.
3 Carroll Pursell, "The Rise and Fall of the Appropriate Technology Movement in the United States, 1965–1985," *Technology and Culture* 34 (1993): 629–637.
4 See for example Jenny Barker-Devine, *On Behalf of the Family Farm: Iowa Farm Women's Activism Since 1945* (Iowa City: University of Iowa Press, 2013), 201; Barker-Devine, "'The Answer to the Auxiliary Syndrome': Women Involved in Farm Economics (WIFE) and Separate Organizing Strategies for Farm Women, 1976– 1985," *Frontiers: A Journal of Women Studies* 30, no. 3 (February 4, 2010): 117–143; Barker-Devine, "'Hop to the Top with the Iowa Chop': The Iowa Porkettes and Culti- vating Agrarian Feminisms in the Midwest, 1964–1992," *Agricultural History* 83, no. 4 (Fall 2009): 477–502; Nancy Grey Osterud, *Putting the Barn before the House: Women and Family Farming in Early-Twentieth-Century* (New York: Cornell University Press, 2012); Nancy K. Berlage, "The Establishment of an Applied Social Science: Home Economists, Science, and Reform at Cornell University, 1870–1930," in *Gender and American Social Science: The Formative Years*, ed. Helene Silverberg (Princeton, NJ: Princeton University Press, 1998), 185–234.
5 See for example Cowan, *More Work for Mother*.
6 See for example Mary Neth, *Preserving the Family Farm: Women, Community, and the Foundations of Agribusiness in the Midwest, 1900–1940* (Baltimore, MD: Johns Hopkins University Press, 1995); Katherine Jellison, *Entitled to Power: Farm Women and Tech- nology, 1913–1963* (Chapel Hill: University of North Carolina Press, 1993); Jellison, "Women and Technology on the Great Plains, 1910–1940," *Great Plains Quarterly* 8,

no. 3 (July 1, 1988): 145–157; Jellison, "'Let Your Cornstalks Buy a Maytag': Prescriptive Literature and Domestic Consumerism in Rural Iowa, 1929–1939," *Palimpsest* 69 no. 3 (1988): 132–139.

7 Cowan, *More Work for Mother*, 13.

8 Ronald R. Kline, "Ideology and Social Surveys: Reinterpreting the Effect of 'Laborsaving' Technology on American Farm Women," *Technology and Culture* 38, no. 2 (April 1997): 355–385.

9 Emily Hoag Sawtelle, "The Advantages of Farm Life: A Study by Correspondence and Interviews with Eight Thousand Farm Women," unpublished manuscript, U.S. Department of Agriculture, March 1924, 1, https://archive.org/stream/CAT31046460#page/n4/mode/1up (accessed 9/29/16).

10 Ibid., 12.

11 Ibid., 19.

12 Ibid., 2.

13 Ibid., 23.

14 Ibid., 14, 29.

15 Ibid., 4.

16 Ibid., 4–5.

17 Ibid., 21–22, 24.

18 Isenberg, *White Trash*, 216–217. Isenberg also makes the point that Wallace argued that children born on farms produced "the life-stream of the nation" to combat eugenic notions at the time regarding the rural poor during the Great Depression, particularly in the South, as genetically inferior.

19 Sawtelle, "The Advantages of Farm Life," 1, 8–9.

20 Ibid., 4, 8–9.

21 I should note that whenever I use the terms "farm woman" or "farmwife," I do so because the subject referred to herself that way at the time, whereas I use the less pejorative and more accurate descriptions "female farmer" or "farmer."

22 Sawtelle, 5.

23 Ibid., 24.

24 Ibid., 10.

25 Ibid., 11–12.

26 Ibid., 10.

27 Ibid., 19, 28.

28 Ibid., 21.

29 Ibid., 5.

30 Ibid., 25.

31 Ronald R. Kline, "Ideology and Social Surveys," 355–385.

32 John H. Kolb and Edmund S. de Brunner, *Rural Social Trends* (New York: McGraw-Hill, 1933), 65–66, quoted in Kline, "Ideology and Social Surveys," 379.

33 Emma Gary Wallace, "Making the Young People Contented," *Better Farming* 46, no. 1 (January 1923): 4.

34 "Delco-Light: Keeps the Young Folks on the Farm," *Indiana Farmer's Guide* (June 22, 1918): 21, American Periodicals database.

35 Thomas J. Delohery, "Teaching Practical Farming in Funk Grade School: Junior Agricultural High School Recognized as Best by State Authorities," *Better Farming* 43, no. 1 (January 1920): 5, 10.

36 Breimyer, *Over-Fulfilled Expectations*, 67.

37 See for example, Lerman, "Categories of Difference."

38 Jane Adams, ed., *All Anyone Ever Wanted Me to Do Is Work: The Memoirs of Edith Bradley Rendleman* (Carbondale: Southern Illinois University Press, 1996), 83–85.

39 Historian Sonya Salamon has confirmed that two different husband-wife relationships existed on farms according to ethnicity. Families with German heritage, according to

Salamon, featured women more prominently in work processes outside the home than did "Yankee" families without a German lineage. Salmon contributes this greater involvement of German-American women in farming activities to the fact that in those families, the wife also brought more land into the marriage. Sonya Salamon, *Prairie Patrimony: Family, Farming and Community in the Midwest* (Chapel Hill: University of North Carolina Press, 1992).

40 Ibid., 49–50.
41 Ibid., 91.
42 Grace Gibbard Lentz, "The Evolution of a Real Farmer's Wife," *Better Farming* 47, no. 6 (June 1924): 4.
43 Adams, *The Transformation of Rural Life*, 85.
44 Ibid., 96–98.
45 Mildred Armstrong Kalish, *Little Heathens: Hard Times and High Spirits on an Iowa Farm During the Great Depression* (New York: Random House, 2007), 108–111.
46 F.A. Millard, "My Poultry Figures," *Farmer's Wife* 23, no. 8 (January 1921): 312.
47 "The Farm Women's Poultry Business," *Farmers' Wife* 26, no. 12 (December 1926): 604–605.
48 Marrs, "Shipping Eggs," 113 (21).
49 Clara M. Sutter, "The Farm Woman's Poultry Business: These Chickens Buy a Farm," *Farmer's Wife* 31, no. 6 (June 1928): 36.
50 Ethel Morrison-Marsden, "Mother and the Radio," *Farm Journal* 52, no. 3 (March 1928): 32, 65.
51 "How to Get More Eggs," *Wallaces' Farmer* 51, no. 1 (January 1, 1926): 22 (22).
52 Grace Jenney, "Rooftries for Our Chickens," *Farm Journal* 54, no. 3 (March 1930): 70–71.
53 Bernice H. Irwin, "A Modern Dairy Farm: Minnesota Woman Demonstrates the Profitableness of Milk Production," *Farmer's Wife* 23, no. 8 (January 1921): 312–313.
54 "Of Interest to Dairy Women: The Cow Keeps Her Place as Man's Best Friend, Nutritionally," *Farmer's Wife* 33, no. 8 (January 1922): 696.
55 *The Farmer's Wife* 29, no. 12 (December 1926): Front Cover.
56 "Why Some Farms Pay," *Wallaces' Farmer* 50, no. 12 (March 20, 1925): 440 (28).
57 D.F. Malin, "Pulling Back to Prosperity," *Wallaces' Farmer* 50 (January 2, 1925): 1 (1), 12 (12).
58 "Master Farmers of 1927 Honored," 91 (7).
59 E.R. Wiggins, "How to Make Poultry Produce Profits," *Better Farming* 48, no. 1 (January, 1925): 6, 19.
60 Adams, *The Transformation of Rural Life*, 2–4.
61 Louisa Stephens, "A Farm Girl's Poultry: How She Grew Up in the Business and Has Made it a Success," *Farmer's Wife* 33, no. 8 (January 1922): 693.
62 Carroll Streeter, "Building Markets by Cooperation," *Farmer's Wife* 31, no. 6 (June 1928): 9, 32.
63 "How Some Women Succeed: True Stories About Real Farm Women," *Farmer's Wife* 27, no. 5 (October 1924): 126–127.
64 "How Some Women Succeed," *Farmer's Wife* 29, no. 5 (May 1926): 270, 301.
65 E.J. Kirk, "Barley for Hogs Next Summer," *Wallaces' Farmer* 50, no. 7 (February 13, 1925): 222 (18).
66 Gilbert C. Fite, *American Farmers, the New Minority* (Bloomington: Indiana University Press, 1981), 39; "June Meeting of Farm Bureau Federations," *Indiana Farmer's Guide* (July 10, 1920): 32; "Farm Federation's Power," 70.
67 "Iowa Farm Bureau Convention," *Wallaces' Farmer* 50, no. 2 (January 9, 1925): 43 (11).
68 "With the Women of the Farm Bureau," *Wallaces' Farmer* 50, no. 4 (January 23, 1925): 113 (15).
69 Lentz, 4.

70 See for example "Hearts and Homes: A Review of Spring Styles," *Wallaces' Farmer* 53, no. 15 (April 13, 1928): 604 (22).
71 See for example Mrs. C.A.B., "Letters From our Farm Women: Not Sorry I Stuck," *Farmer's Wife* 31, no. 6 (June 1928): 8.
72 Bess M. Rowe, "American Farm Bureau Convenes," *Farmer's Wife* 23, no. 8 (January 1921): 289.
73 "Keeping Up With Father and the Boys," *Wallaces' Farmer* 46, no. 3 (January 21, 1921): 114 (22).
74 "Home Demonstration Agents: News of Important Results Being Accomplished in Various States," *Farmer's Wife* 23, no. 8 (January 1921): 290.
75 "Dignifying the Work of Farm Women," *Farmer's Wife* 23, no. 8 (January 1921): 1.
76 Adams, *The Transformation of Rural Life*, 2.
77 Bess M. Rowe, "What Are Farm Women Thinking About?" *Farmer's Wife* 29, no. 5 (May 1926): 268–269.

Bibliography

Adams, Jane. *The Transformation of Rural Life*. Chapel Hill: University of North Carolina Press, 1994.

Adams, Jane, ed. *All Anyone Ever Wanted Me to Do Is Work: The Memoirs of Edith Bradley Rendleman*. Carbondale: Southern Illinois University Press, 1996.

Barker-Devine, Jenny. "'Hop to the Top with the Iowa Chop': The Iowa Porkettes and Cultivating Agrarian Feminisms in the Midwest, 1964–1992." *Agricultural History* 83, no. 4 (Fall 2009): 477–502.

———. *On Behalf of the Family Farm: Iowa Farm Women's Activism Since 1945*. Iowa City: University of Iowa Press, 2013.

———. "'The Answer to the Auxiliary Syndrome': Women Involved in Farm Economics (WIFE) and Separate Organizing Strategies for Farm Women, 1976–1985." *Frontiers: A Journal of Women Studies* 30, no. 3 (February 4, 2010): 117–143.

Berlage, Nancy K. "The Establishment of an Applied Social Science: Home Economists, Science, and Reform at Cornell University, 1870–1930." In *Gender and American Social Science: The Formative Years,* edited by Helene Silverberg, 185–234. Princeton, NJ: Princeton University Press, 1998.

Breimyer, Harold F. *Over-Fulfilled Expectations: A Life and an Era in Rural America*. Ames: Iowa State University Press, 1991.

Cowan, Ruth Schwartz. *More Work for Mother: The Ironies of Household Technology from the Open Hearth to the Microwave*. New York: Basic Books, 1983.

"Delco-Light: Keeps the Young Folks on the Farm." *Indiana Farmer's Guide,* June 22, 1918, 21, American Periodicals database.

Delohery, Thomas J. "Teaching Practical Farming in Funk Grade School: Junior Agricultural High School Recognized as Best by State Authorities." *Better Farming* 43, no. 1 (January 1920): 5, 10.

"Dignifying the Work of Farm Women." *Farmer's Wife* 23, no. 8 (January 1921): 1.

"Farm Federation's Power: Life Topsy, Bureau 'Jes' Grew,' and Now Numbers 100,000 Members." *New York Times*, July 4, 1920, 70.

Fite, Gilbert C. *American Farmers, the New Minority*. Bloomington: Indiana University Press, 1981.

"Hearts and Homes: A Review of Spring Styles." *Wallaces' Farmer* 53, no. 15 (April 13, 1928): 604 (22).

"Home Demonstration Agents: News of Important Results Being Accomplished in Various States." *Farmer's Wife* 23, no. 8 (January 1921): 290.

"How to Get More Eggs." *Wallaces' Farmer* 51, no. 1 (January 1, 1926): 22 (22).

"How Some Women Succeed." *Farmer's Wife* 29, no. 5 (May 1926): 270, 301.

"How Some Women Succeed: True Stories About Real Farm Women." *Farmer's Wife* 27, no. 5 (October 1924): 126–127.

"Iowa Farm Bureau Convention." *Wallaces' Farmer* 50, no. 2 (January 9, 1925): 43 (11).

Irwin, Bernice H. "A Modern Dairy Farm: Minnesota Woman Demonstrates the Profitableness of Milk Production." *Farmer's Wife* 23, no. 8 (January 1921): 312–313.

Isenberg, Nancy. *White Trash: The 400-Year Untold History of Class in America.* New York: Viking, 2016.

Jellison, Katherine. "'Let Your Cornstalks Buy a Maytag': Prescriptive Literature and Domestic Consumerism in Rural Iowa, 1929–1939." *Palimpsest* 69 no. 3 (1988a): 132–139.

———. "Women and Technology on the Great Plains, 1910–40." *Great Plains Quarterly* 8, no. 3 (July 1, 1988b): 145–157.

———. *Entitled to Power: Farm Women and Technology, 1913–1963.* Chapel Hill: University of North Carolina Press, 1993.

Jenney, Grace. "Learning too Fast?" *Farm Journal* 54, no. 3 (March 1930a): 71.

———. "Rooftries for Our Chickens." *Farm Journal* 54, no. 3 (March 1930b): 70–71.

"June Meeting of Farm Bureau Federations." *Indiana Farmer's Guide,* July 10, 1920, 32.

Kalish, Mildred Armstrong. *Little Heathens: Hard Times and High Spirits on an Iowa Farm During the Great Depression.* New York: Random House, 2007.

"Keeping Up With Father and the Boys." *Wallaces' Farmer* 46, no. 3 (January 21, 1921): 114 (22).

Kirk, E.J. "Barley for Hogs Next Summer." *Wallaces' Farmer* 50, no. 7 (February 13, 1925): 222 (18).

Kline, Ronald R. "Ideology and Social Surveys: Reinterpreting the Effect of 'Laborsaving' Technology on American Farm Women." *Technology and Culture* 38, no. 2 (April 1997): 355–385.

Kolb, John H. and Edmund S. de Brunner. *Rural Social Trends.* New York, 1933, 65–66. Quoted in Ronald R. Kline, "Ideology and Social Surveys: Reinterpreting the Effect of 'Laborsaving' Technology on American Farm Women." *Technology and Culture* 38, no. 2 (April 1997): 379.

Lentz, Grace Gibbard. "The Evolution of a Real Farmer's Wife." *Better Farming* 47, no. 6 (June 1924): 4.

Lerman, Nina E. "Categories of Difference, Categories of Power Bringing Gender and Race to the History of Technology." *Technology and Culture* 51, no. 4 (October 2010): 893–918.

Malin, D.F. "Pulling Back to Prosperity." *Wallaces' Farmer* 50, no. 1 (January 2, 1925): 1 (1), 12 (12).

Marrs, Miss. L.R. "Shipping Eggs." *Wallaces' Farmer* 46, no. 3 (January 21, 1921): 113 (21).

"Master Farmers of 1927 Honored: Presented to Corn Belt Over WHO and to Iowa Notables at Banquet." *Wallaces' Farmer* 53, no. 3 (January 20, 1928): 91 (7).

Millard, F.A. "My Poultry Figures." *Farmer's Wife* 23, no. 8 (January 1921): 312.

Morrison-Marsden, Ethel. "Mother and the Radio." *Farm Journal* 52, no. 3 (March 1928): 32, 65.

Mrs. C.A.B. "Letters From our Farm Women: Not Sorry I Stuck." *Farmer's Wife* 31, no. 6 (June 1928): 8.

Neth, Mary. *Preserving the Family Farm: Women, Community, and the Foundations of Agribusiness in the Midwest, 1900–1940*. Baltimore, MD: Johns Hopkins Press, 1995.

"Of Interest to Dairy Women: The Cow Keeps Her Place as Man's Best Friend, Nutritionally." *Farmer's Wife* 33, no. 8 (January 1922): 696.

Oldenziel, Ruth. *Making Technology Masculine*. Amsterdam: Amsterdam University Press, 1999.

Osterud, Nancy Grey. *Putting the Barn before the House: Women and Family Farming in Early-Twentieth-Century*. New York: Cornell University Press, 2012.

Pursell, Carroll. "The Rise and Fall of the Appropriate Technology Movement in the United States, 1965–1985." *Technology and Culture* 34, no. 3 (1993): 629–637.

Rowe, Bess M. "American Farm Bureau Convenes." *Farmer's Wife* 23, no. 8 (January 1921): 289.

———. "What Are Farm Women Thinking About?" *Farmer's Wife* 29, no. 5 (May 1926): 268–269.

Salamon, Sonya. *Prairie Patrimony: Family, Farming and Community in the Midwest*. Chapel Hill: University of North Carolina Press, 1992.

Sawtelle, Emily Hoag. "The Advantages of Farm Life: A Study by Correspondence and Interviews with Eight Thousand Farm Women." Unpublished manuscript, U.S. Department of Agriculture, March 1924, 1. https://archive.org/stream/CAT31046460#page/n4/mode/1up (accessed 9/29/16).

Stephens, Louisa. "A Farm Girl's Poultry: How She Grew Up in the Business and Has Made it a Success." *Farmer's Wife* 33, no. 8 (January 1922): 693.

Streeter, Carroll. "Building Markets by Cooperation." *Farmer's Wife* 31, no. 6 (June 1928): 9, 32.

Sutter, Clara M. "The Farm Woman's Poultry Business: These Chickens Buy a Farm." *Farmer's Wife* 31, no. 6 (June 1928): 36.

The Farmer's Wife 29, no. 12 (December 1926): Front Cover.

"The Farm Women's Poultry Business." *Farmers' Wife* 26, no. 12 (December 1926): 604–605. "Untitled photo, possibly related to: FSA (Farm Security Administration) rehabilitation borrower operating tractor. She and her mother run the farm without the assistance of any men. Grant County, Illinois." *Library of Congress*. https://www.loc.gov/pictures/item/2017810457/ (accessed 8/9/23).

Wallace, Emma Gary. "Making the Young People Contented." *Better Farming* 46, no. 1 (January 1923): 4.

"Why Some Farms Pay." *Wallaces' Farmer* 50, no. 12 (March 20, 1925): 440 (28).

Wiggins, E.R. "How to Make Poultry Produce Profits." *Better Farming* 48, no. 1 (January 1925): 6, 19.

"With the Women of the Farm Bureau." *Wallaces' Farmer* 50, no. 4 (January 23, 1925): 113 (15).

5

RUMBLING DOWN MAIN STREET

Cold War Ideology and the "American Way"
Encouraging Rural Capitalistic Modernity[1]

> Khrushchev has said to us- "We will bury you-your grandchildren will be
> Communist." The best weapon we have pointed against him is our overabun-
> dance of food-this is even more important than any missile in the long run.
>   J.K. Stern, President-American Institute of Cooperation, March 30, 1961[2]

Fourth of July or state fair parades in small towns throughout the Corn Belt, in
places like Charles City and Des Moines, Iowa (Figure 5.1) rehearse a familiar
performance. Expressions of patriotism involve more than simply nighttime fireworks. Rather, residents sit along main streets in lawn chairs while a steady column
of huge pieces of farm machinery that seem to dwarf the nearby brick buildings
rumble by, driven by farmers and their families waving American flags. Often older
tractors accompany newer models, all of which sparkle in pristine condition. As
in Figure 5.1 showing a parade in Des Moines, Iowa in 2016, old tractors like
the Case International 1466 pictured here or newer equipment often have flags attached to them with the driver wearing work clothes, like the farmer in this picture,
to highlight his or her rural heritage.[3] The celebration masks deep and unwavering
disagreements among the spectators over which company produces the best farm
machinery. For my family, Case International makes the only proper tractor and no
Brinkman would ever farm in something painted green or blue. Of course, the more
interesting question as I grew older was the meaning of this annual performance.
Why, after all, did large farm equipment, American flags, fireworks, and bands
playing the "Star Spangled Banner" seem to "go together" on the main streets of
the Corn Belt? Why do people on these huge rumbling machines wave flags and
why do people on the street smile and wave flags back when they see an artifact that
seems so out of place to an outside observer on a paved street? In the language of

DOI: 10.4324/9781032637952-6

FIGURE 5.1 Carol M. Highsmith, Not surprisingly in the largely rural Midwest state, tractors play a prominent role in the downtown Des Moines, Iowa, parade marking the opening of the annual Iowa State Fair, 2016, photograph, Carol M. Highsmith Archive, Library of Congress, Prints and Photographs Division, at https://www.loc.gov/pictures/item/2016630279/ (accessed 11/2/23).

Irving Goffman and Stephen Hilgartner, what exactly *is* the performance?[4] Surely, the hundreds of people lining a Main Street in the Corn Belt see the latest combine as more than a way to increase the efficiency of soybean harvesting.

At the dawn of the Cold War, a significant number of Midwest farmers had developed a rural modern identity as a means of combating the threats posed by urban industrialism and rube stereotypes. This new sense of self tended to adopt the modern sensibilities of urban dwellers but decoupled it from the image of the factory. In place of this urban notion of efficiency and progress, farmers strategically preserved notions borrowed from German and Jeffersonian agrarianism such as the importance of a farmer's control of technology within a family-based production process. Rural dwellers would continue the German practice of performing identity through using material objects and urban dwellers still observed and evaluated these displays. Hence, farmers and their urban observers would debate the new meaning of modernity through a familiar pattern of audience.

This environment describes the stage in the Corn Belt when American farmers began to turn their attention not just to the threat of urban reforms, but to counter the Soviet menace. Soviet collectivization simply seemed less of a threat in the Corn Belt prior to 1945 with some American farmers even participating in delegations to teach Soviet peasants how to use American-made tractors.[5] The Cold War spurred the development of the rural capitalistic view of technology as associated with modernity because it highlighted the dichotomy between American capitalism and Soviet/Red Chinese collectivism. By projecting America's technological superiority in agriculture promising abundant food production, the U.S. could attract allies and stop the spread of communism. The central role of the farmer in the contest between communism and capitalism would leave a lasting effect on the rural mind. Nationalistic concerns would strengthen rural capitalistic modernity to such an extent that a new identity of rural ultramodernity would emerge as the Cold War progressed. Farmers faced the injustice of a persistent urban rube stereotype (or it's more contemporary "hick" moniker) even after having "won" the Cold War though modern capitalistic practices. Corn Belt agrarians responded by performing an identity that framed the farmer as even more modern than her unappreciative urban cousins.

Tractors and milking machines took on symbolic significance as soon as World War II ended. When Soviet Premier Joseph Stalin forced millions of Russian and Ukrainian peasants off their land and into huge government-operated "agro-towns," the successful American family farmer became a symbol of the superiority of the free market respecting private property over collectivized workers.[6] Every local newspaper in the American Corn Belt from Waukesha, Wisconsin, to Cedar Rapids, Iowa, to Chicago, Illinois highlighted this ideological struggle around agriculture and reported with subtle pride on the inability of the Soviet Union to feed its own people.

Agricultural technology became embedded with the ideology that market incentives win over government control. Such notions also preserved Jeffersonian tendencies to view farmers as heroically overcoming obstacles as well as German agrarian ideas of independence and distrust of an outside other after the New Deal era in which federal government aid to farmers dealt a blow to both. The Cold War allowed Corn Belt farmers to fall back into the familiar identity of rural capitalistic modernity following what they saw as the temporary blip where the Great Depression and wartime production represented extraordinary circumstances calling for greater federal government involvement (again, whether farmers in fact were independent following the New Deal era is not my question as I am more interested in the way farmers viewed themselves and their technologies). For example, a note from the editor of *Electricity on the Farm Magazine*, W.J. Ridout, Jr., in a May 1953 issue entitled "A Defeat for Communism" echoed the sentiments of many farmers in the Corn Belt who saw private technological production among farmers in any country as a victory for rural capitalistic modernity. In celebrating reports that farmers in Yugoslavia and other Eastern Bloc countries had resisted collectivization, Ridout, whose magazine targeted farmers who wished to modernize through greater use of electrification, dramatically wrote:

> In Yugoslavia, defeat has been conceded in the effort of the government to collectivize farming. From what we hear agriculture is well along the road back to private ownership in that country. Stubborn resistance is also taking place in the

Russian satellite countries. The farmers in these countries will certainly draw new strength from the victory scored in Yugoslavia. Without freedom, incentive is lost, with[out] incentive, hope is lost.[7]

Political satirists and commentators at the time expressed a sentiment similar to that expressed by Ridout. A 1949 cartoon by the famous *Washington Post* cartoonist Herb Block, a Soviet government official has a plow shaped as the Soviet hammer and sickle symbol harnessed to a Russian peasant who sweats with exhaustion as he pulls the plow by a sign that reads "Marshall Stalin Plan." Above this scene of oppression on a piece of higher ground a Western European farmer with a pipe in his mouth drives a modern tractor labeled "Marshall Plan" with a large grin on his face. In the background, another farmer drives a tractor in front of a modern and prosperous barn and silo. Below, on the Soviet side, another Soviet official explains to another Russian peasant dejectedly watching his comrade pulling the plow "It's the same thing without mechanical problems." Another oppressed Russian peasant looks up at the modern tractor with jealousy.[8] Even at the dawn of the Cold War, Block saw rural capitalistic modernity as an important weapon in attracting allies and combating the spread of communism, hence the "Marshall Plan" label on the tractor.

The centrality of agriculture in Cold War diplomacy meant that domestically, the modern farmer became a Cold War warrior who almost had a duty to buy the latest machinery, educate himself, and increase production. At that time, when an American farmer purchased the latest tractor or the newest combine and raised output, he did so not just to assert his own version of modernity over visions of urban reformers, but to show his patriotism. The rural "modern way" became the "American way" in which discourses of rural-urban conflict transformed into a more inclusive nationalistic discourse (Figure 5.2).

In Figure 5.2, a cartoon published in the *Farmer's Weekly Review* in rural Illinois in 1950, one man labeled "Big Business" and another named "Small Business" help each other pull on a rope hoisting a large block named "America's Industrial Expansion." The two men exclaim "All together now!" underneath a title of "The American Way." Most farmers in the Corn Belt seeing this cartoon would have identified with the man representing "Small Business."[9] This inclusive message contrasts sharply with the political cartoons in the 1920s in which industrial and other urban interests attacked or threatened farmers (recall Figure 2.1 in Chapter 2, for example). Those arguing against domestic price controls for agricultural commodities and land limits further promoted the identity of the American farmer as an independent capitalist employing technology as opposed to the collectivized and government-controlled Russian peasant using old-fashioned farming techniques. Politicians arguing in favor of ending New Deal-era price controls accused the opposition of attempts to socialize agriculture.[10] Such strategy clearly represents the ambiguous relationship most farmers felt toward New Deal programs due to their incompatibility with their rural modern identities. After all, viewing policies

FIGURE 5.2 *Farmer's Weekly Review* editorial cartoon, 1950 showing a shift in discourse from urban-rural conflict to a more inclusive nationalism. The *Farmer's Weekly Review* was published in Will County, Illinois. "The American Way," *Farmer's Weekly Review*, cartoon, October 24, 1950, 1.

by urban reformers and a socialite president from New York City as saviors of the family farm would not have affirmed either the German or the Jeffersonian ethos already existing in the Corn Belt. Rather, the view of urbanites and a distant government as a threat fit the Jeffersonian notion of farmers heroically overcoming obstacles and the German distrust of an outside other much better. On the other hand, farmers formed part of Roosevelt's New Deal coalition and understood that they benefitted from programs such as the Agricultural Adjustment Act.[11] Opposing a foreign Soviet menace that threatened farmer's control of work allowed rural people to resolve this paradox.

Agricultural technology became an even more important Cold War weapon from 1954 to 1955 when it took center stage in the struggle for power between Georgi Malenkov and Nikita Khrushchev.[12] As Stalin's secretary of the Central Committee, Khrushchev proposed in 1949 the idea of the *agrogorod*, or agrarian city, which extended the policy of forced collectivization.[13] In fact, Khrushchev owed his eventual victory in gaining the premiership to Malenkov's inability to respond to an agricultural crisis and keep pace with American production.[14] As a result, Khrushchev, regarded by many Midwest newspapers as the most dangerous world leader since Hitler, sought to vastly improve Soviet agricultural production to show the World that collectivization could out-produce privatized free market capitalism.[15] The *Chicago Daily Tribune* even reported in 1959 that as "communist chief of the Ukraine" after 1938, Khrushchev had helped to "direct the program of planned starvation to force collectivization of the farms. Five million Ukrainians died."[16]

Increased mechanization formed the center of Khrushchev's Seven Year Plan, but he also socialized Soviet peasants even further by shifting from collectivized farm towns to large state-operated compounds.[17] One Iowa newspaper editor likened Khrushchev's state-operated facilities to "huge Roman slave estates, state owned, state managed and worked by laborers having no personal stake in the land."[18] Similarly, the *Chicago Daily Tribune* referred to collectivized Chinese farmers as "dehumanized guinea pigs" and Soviet-style farms as "coercive communal living."[19]

While Stalin's successor may have despised the West, he also admired the American farmer's ability to use technology to boost production. As a result, the premier opened the Soviet Union to tours by Midwest farmers and politicians as a "cultural exchange" and allowed high ranking Soviet officials to visit the U.S.[20] The American corn-hog economy fascinated Khrushchev, and he even proposed to introduce corn to Soviet farming in "the American style" in 1955.[21] Khrushchev famously became the first Soviet premier to tour Iowa in 1959, an event that did not go unnoticed by Corn Belt farmers.[22]

Meanwhile, the U.S. State Department and the Central Intelligence Agency recognized agricultural technology as a Cold War weapon. In classified documents from 1955, these agencies recommended organizing cultural exchanges as military campaigns with the goal of strategically gaining more useful information than the Soviet delegates could acquire from tours of the U.S.[23] Farmers and politicians returned from these trips in the 1950s and 1960s and debunked the notion of an "agricultural crisis in the Soviet Union" while at the same time reporting on inefficient farming methods.[24] These farmer delegates often attributed the failures of Soviet farming to the lack of individualized incentive caused by too much government control.[25] Almost all of the farmer delegates and government officials who toured the Soviet Union noted the use of manual labor and old fashioned tools by collectivized Russian peasants rather than the up-to-date hardware operated by American farmers.[26] Several travelers derided the Soviets, in a blend of faith in technology with 1950s male chauvinism, for using women as manual labor in the fields declaring "Most of these women – in the late teens, twenties and early thirties – showed little of the femininity of American women."[27] In contrast, "feminine" American women such as those described by Grace Gibbard Lentz in Chapter 4 used the latest labor-saving technologies to produce as modern capitalists.[28]

In speeches at town meetings and in articles filling newspapers of the upper Midwest, rural residents received (and resonated with) one clear message: the capitalist American farmer who employed modern technology and business practices

FIGURE 5.3   *Waukesha Daily Freeman* cartoon, 1955, mocking failures in Soviet agriculture and highlighting the victory of the American farmer. D.W. McDowell, "State Agriculture Alert to Responsibilities," *Waukesha Daily Freeman*, July 29, 1955, Editorial Page.

was winning the Cold War (Figure 5.3).[29] In Figure 5.3, for example, a cartoon published in rural Wisconsin in 1955, a "Soviet Farm Guest" sits devouring a large pile of sweet corn served on a plate labeled "Good Will and Friendliness." In the background one American woman brings even more corn while a man holds a plate of butter asking, "More Butter Ivan?" The carton, labeled "And there's a lot more where this came from" conveys that the American farmer's use of modern technology has defeated Soviet farming efforts so thoroughly that even Russian farm representatives arrive to the U.S. hungry and enjoy the abundance of American commodities.[30]

In many farmers' minds, reports in Corn Belt newspapers of the failure of Khrushchev's agricultural policies to increase food production by the 1960s confirmed the superiority of privatized capitalist agrarians using modern technology.[31] Midwest newspapers also reveled in the failure of communal agriculture in Communist China by the late 1950s.[32] The *Chicago Tribune* mocked "Professor" Khrushchev for lecturing Communist China about the advantages of collectivism while his own collective farms failed exclaiming, "If Prof. Khrushchev while in America had talked less and studied more, he might have learned how to increase production."[33]

Roscoe Drummond's editorials in the *Cedar Rapids Gazette* serve as an illustrative example of this nationalistic modern discourse. In one article from 1962, he boasted, "Soviet agriculture is faltering, failing, and falling behind. Today it is in a colossal mess for two reasons. Communist farming through collectivization doesn't work."[34] In 1963 when the U.S. Secretary of Agriculture announced that Moscow would soon purchase a billion tons of wheat from the U.S., Drummond asked "Why? Primarily because its collectivized, state-managed agriculture is inefficient and unproductive. Soviet agriculture is exactly the way communism says it should be."[35] (Figure 5.4).

In one cartoon in Figure 5.4, published in Cedar Rapids, Iowa in 1962, "Soviet collective farm officials" exit an airplane onto an American runway. In front of the plane, Uncle Sam mocks the officials by asking an American government bureaucrat, "Maybe they can solve our surplus problem?" The bureaucrat answers, "Let 'em mess around a bit." The cartoon accompanied a Roscoe Drummond editorial asking rhetorically "Why can't Communist Countries Feed Their People?"[36] In a 1963 cartoon entitled "Small Problem," a tombstone at the head of an open grave

FIGURE 5.4 Cartoons in the *Cedar Rapids Gazette* accompanying 1962 and 1963 Drummond editorials. The message is that the American farmer is winning the agricultural Cold War. Drummond, "Why Can't Communist Countries Feed Their People?" *Cedar Rapids Gazette*, March 12, 1962, Editorial Page; Drummond, "Communism's Many Ailments Add Up to One Big Flop." *Cedar Rapids Gazette*, October 21, 1963, Editorial Page.

paraphrases Khrushchev's famous declaration to the U.S., "We'll bury the U.S. here."[37] Inside the grave, a shovel sticks out labeled "K." for Khrushchev stating, "Help! I'm stuck." Beside the grave, Uncle Sam approaches with a rescue ladder with a flag inscribed with the words "Wheat Deal." The cartoon, which accompanied another editorial by Drummond entitled "Communism's Many Ailments Add Up to One Big Flop," saw the Soviet need for U.S. wheat as the ultimate sign that the American farmer had thoroughly won his or her part of the Cold War.[38]

Reinforcing this message of agricultural superiority, the American Feed Market Association published a booklet entitled "Mightier than Missiles" in 1961 declaring, "The greatest of all weapons is America's production of food and fiber...its agriculture and agribusiness."[39] The idea that agricultural technologies served a greater role in combating communism than missiles also appeared at agricultural fairs throughout the U.S by the early 1960s.[40] By 1960, the Department of Agriculture and the President had issued several reports characterizing the failure of Soviet agriculture as an important Cold War victory.[41] An editorial by William Ryan in the *Cedar Rapids Gazette* noted in 1962,

> Persistent failure of communist agriculture is regarded in Washington as one of the most important aspects of the Cold War. President Kennedy is known to feel that if the USSR, while it still is led by those obsessed with the notion of world domination, should ever out-produce the United States in food and consumer goods, the cause of the whole free world would be in danger.[42]

The apparent victory of American agriculture had a clear effect on farmers' identities as users of technologies. Rather than a thoughtless worker using a manager's machine in a collective Soviet system, the machine itself reinforced the farmer's identity as an independent, hard-working, capitalist. As the *Buffalo* (Iowa) *Center Tribune* declared, "It's likely that the toughest barrier to the solution of the Soviet farm problem, as with other economic problems, will prove to be a system where the few give the orders and the many meekly obey."[43] Just as when rural Americans encountered the idea of industrialization in the 1910s and 1920s, when farmers themselves participated in the development of rural modernity as a means of resistance against urban interests, they later, in the 1950s and 1960s, bolstered the same rural capitalistic identity against a collectivized "other" in the communist world. The fact that the Soviet peasant had been stripped of his agency and starved by the centralized government served as proof, from the farmer's perspective, of the validity of the capitalist ideology of the independent American agrarian whose production capabilities, in contrast, grew greater with the continual adoption of advanced hardware. In Figure 5.5, for example, an Iowa farmer in 1963 has taken a photo of three generations of Farmall tractors from the 1930s to the mid-1960s, the M (1939–1953), Super C (1951–1954), and the 460 (1958–1963) in front of the family farm home with a family member on the latest 460 to show the family's progress from older to new technology. The 460 featured the latest technology in 1963

FIGURE 5.5 "Three Farmall Tractors on an Iowa Farm." Author's personal records. Charles City, Iowa, 1963.

including a 6-cylinder engine, a hydraulic hitch, and a torque amplifier allowing a gear shift without a clutch.[44] Thus, the 460 serves as a performance of the farm family's rural modernity and the chronological placement of the machines shows that the farmer has also achieved progress.

The establishment of the modern rural capitalist identity in the Midwest Corn Belt had a reciprocal effect of spurring growth in farmers' land holdings and their use of increasingly mechanized equipment. As a result, technology, ideas of modernity, and capitalist Cold War ideologies became co-constructed. The discourse of rural capitalistic modernity alleviated social unrest over the grow-or-die phenomenon after World War II, in which some family farms failed while others grew, by framing the process as a natural consequence of competitive agricultural capitalism. It gave an ideological justification for the massive rural-to-urban migration spurred by the economic pressures of thin profit margins and economies of scale. Residents of rural America accepted the growth of the successful farmers at the expense of their defeated counterparts because (at least in part) rural capitalistic modernity suggested that superiority derived from the effective use of modern business principles and the best new technology.

The Persistence of the Yokel Myth in American Popular Culture at the Height of the Cold War

Despite the strengthening of rural capitalistic modernity during the Cold War among farmers in the Corn Belt, the myth of the backward Jeffersonian resistant to technological change persisted among urban observers of agriculture. Farmers from the 1950s onward still faced urban yokel stereotypes notwithstanding their performative use of the latest technologies. The resentment in the Corn Belt over what, to farmers, seems like an urban refusal to acknowledge their true modernity still forms an important feature of modern rural identities. Men and women in the Corn Belt have maintained the same unspoken rural obsession with how urban observers view them, a pattern of audience that intensified during the rural-urban conflict of the 1920s. As a result, farmers use the latest technologies to perform a modern identity to combat urban hick stereotypes that seem to remain remarkably difficult to dislodge in the broader American imagination.

The dominant image of the Jeffersonian agrarian resistant to social and technological change has pervaded American popular culture since at least the 1920s. Television shows such as the three CBS sitcoms airing in the 1960s and early 1970s, *Green Acres*, *The Beverly Hillbillies*, and *Petticoat Junction*, portrayed rural Americans exclusively from the perspective of urban dwellers; by doing so, they reinforce rube stereotypes. In *Green Acres*, for example, a sophisticated New York lawyer, Oliver Wendell Douglas, and his glamorous Hungarian wife, Lisa Douglas, abandon their Manhattan penthouse for a dilapidated farm in the rural town of Hooterville.[45] The entire comedic vehicle of *Green Acres* revolves around this reasonable urbane straight man encountering a backward and bizarre rural "other."

In a 1965 article previewing the show for urban viewers, the *New York Times* even characterized Douglas as "a kind of reverse rube."[46] The Douglases repeatedly fail to bring urban modernity to rural yeomen, who include Mr. Haney, a yokel con man, Eb Dawson, a hillbilly farmhand, and Fred and Doris Ziffel, the Douglas' elderly neighbors whose old-fashioned ways include repeated attempts to pass off a talking pig as their son, Arnold.[47]

Throughout the sitcom, which aired from 1965 to 1971 in the midst of the Cold War, technology plays a key role in demonstrating Douglas' modernity as well as his neighbors' backwardness.[48] None of the farms in Hooterville appear prosperous, and none of the farmers around Douglas use modern mechanized equipment. One running joke throughout the show occurs when Mr. Haney cons Douglas into purchasing poorly made or badly conceived material objects, often from an old-fashioned truck or a junk shop, suggesting that Hooterville lacks respectable retail options. Mr. Haney accentuates the comedic effect of this repeated scene with a noticeable high-pitched country accent and new ways to try to fool Mr. Douglas.

One can view the rube salesman's presentation of these poorly designed technologies as a symbolic ritual that reinforces urban identity by "othering" rural dwellers. Namely, Mr. Haney's hucksterism stands in for the urban suspicion that the rural performance of modernity through technology use is also a scam. One can view Mr. Haney's hocking of old-fashioned junk as a fictitious pattern of audience through a lens of urban stereotypes about rural life as a means of blocking or derailing real-life rural performances. Mr. Haney, for example, sells Douglas a tractor because Douglas, the urbanite, insists on modernizing farming in Hooterville. A Corn Belt farmer in 1965 would view scenes with this tractor as ridiculous, and perhaps even insulting for several reasons. First, the tractor requested by Douglas is an old model with metal wheels that farmers would have used prior to World War II, not in the 1960s. For example, in 1965, the first year *Green Acres* aired, International Harvester released the famous Farmall 1206 featuring large severe-duty 18.4x38" rubber tires and a revolutionary DT-361 turbocharged engine that delivered over 100 horsepower. The Farmall 1206 delivered an impressive 12,000 pounds of pull and 16 forward speeds. In just two years, International Harvester sold over 10,000 of these machines and even more farmers bought the upgraded models in the late 1960s.[49] So in attempting to show an urban dweller's view of a "modern" tractor, the writers of *Green Acres* did not even present a contemporary machine. While the supposedly more sophisticated city dweller Douglas bought an outdated machine, real farmers in the Midwest operated the latest in mechanical engineering.

Further, when Douglas presents his tractor to his farm hand, Eb Dawson, Dawson has no idea how to use it. Rather, the farm hand attempts to hook a moldboard plow designed for horses to the back of the tractor. The scene symbolically places Douglas on the "modern" tractor seat and Dawson on the ground behind the plow as if he had never used a tractor before. As with all the technologies sold by Mr. Haney, the gag concludes with the tractor falling apart with neither Douglas nor Dawson knowing how to fix it.[50] With the collapse of the tractor, any notion of a modern rural identity falls apart as well. Rural technologies, and hence rural modernity, are hoaxes when explored further by the urbane Douglas who exposes rural America as anti-modern and defunct. Mr. Haney embodies this urban notion that rural performative use hides an underlying anti-modern yeomanism existing backstage.

When rural Americans act in *Green Acres*, they do so repeatedly as rubes. In one scene, Mr. Haney and a group of yokels forming the Chamber of Commerce plan to attract economic development by holding the Olympics in Hooterville. The fact that the board does not understand Douglas' argument that the small town cannot attract such an important event highlights the unworldliness of Mr. Haney and his provincial neighbors.[51] In yet another scene, Fred Ziffel, clad in old-fashioned overalls, attempts to open a bank account for Arnold, and becomes offended when a sophisticated banker tells him that he cannot open an account for a pig.[52]

While one cannot definitively determine whether farmers in the Corn Belt viewed *Green Acres* and other sitcoms set in rural settings as insulting or simply humorous, CBS did decide to remove it along with its other rural-based comedies from the air in 1971 because of low ratings. The *Atlanta Constitution* noted that *Green Acres* found its greatest appeal in the Southeast, not the Midwest despite its focus on rural characters, implying that its lack of popularity outside of the South may have led to its demise.[53] No Midwest state even rated *Green Acres,* or any of the other CBS country-based comedies, among the top 25 of television shows in 1971. The article also notes that only local stations in the South chose to continue *Green Acres* in syndication after CBS discontinued it.[54] Thus, regardless of whether Midwest farmers expressed objections to the show, one can hardly view it as popular in the Corn Belt by the early 1970s. Indeed, any viewer can recognize that the ridiculousness of the displays of rural backwardness in the show reinforced the urban project of "othering" the American farmer, which accelerated in the 1920s, well before *Green Acres.* Arnold and other rural characters in *Green Acres* serve as urban ways to prevent a rural capitalistic modern identity from becoming mainstream in the American consciousness. Such rube stereotypes maintain notions of urban superiority and the morality of urban or suburban lifestyles.

Interestingly, Eddie Albert, the actor who played Oliver Wendell Douglas, viewed *Green Acres* through the lens of some early organic food advocates who associated rural lifestyles with the unaltered Jeffersonian ideal. While Albert declared in an interview in 1965 that, "*Green Acres* is still part of the American dream" of "getting back to the soil," he himself lived on an estate in the plush coastal neighborhood of Pacific Palisades, California where he planted an organic garden. One article highlighting Albert's "grounds" pictures him sitting under an arbor taking a break from gardening and enjoying a cup of tea. He wears a pair of khaki pants and a polo shirt with dress shoes and black socks. His hair appears perfectly styled as if on the set of *Green Acres.* Another photo shows Albert dressed in the same manner in his garden wielding a small gardening trowel. The author of the article seems surprised that the actor plants most of the trees and flowers himself. In another interview from a suite in a downtown hotel, Albert noted how he also formed Eddie Albert Productions "which will aid in national and international campaigns against environmental pollution."[55]

Albert, who described himself as "a conservationist," also traveled the country to give speeches about conserving wildlife and worked with universities to develop "model organic farms which would grow only organic food." The author of one article, Norma Lee Browning, noted that, "The organic garden at his Pacific Palisades home had helped make him an expert." Albert described his estate as, "We grow corn, beets, carrots, lettuce, and other vegetables. We can't grow enough to feed the whole family so we buy more at health food stores." Albert went onto note, "I own several farms which I visit regularly and I'm training people to run and organize them. They're learning modern agricultural techniques and, at the same time, providing more organic foods which are free of DDT and other pesticides."

Thus, for Albert, the dream of "all men" of leaving the city for the country depicted in *Green Acres* meant reoccupying an idealized Jeffersonian pastoral space where one would practice non-technological organic production rather than rural capitalistic modernity. Even though he uses the term "modern agriculture," one should not view Albert's version of modernity as the same as that held by most Midwest farmers.

Ironically, in the same article, Albert discusses his "agricultural" activities (recall that he did not even feed his own family with his garden), the actor immediately turns to describing his work with the "glamorous" and urbane Eva Gabor who played his wife on *Green Acres*.[56] On the show, Gabor famously wore "a $150,000 square-cut necklace" as well as negligees designed by the acclaimed fashion designer, Jean Louis.[57] Gabor even insisted that CBS hire "one of Hollywood's most expensive and expert" hairdressers, Peggy Shannon, and makeup artists, Gene Hibbs, as part of her contract. One author writing for urban readers in Los Angeles in an article entitled "Hungary Meets Hillbilly, U.S.A." thought it amusing that Gabor had trouble understanding the country accents when visiting the set of *Petticoat Junction* and declared "What's a chic Hungarian like her doing in a barnyard?"[58] Prior to *Green Acres*, Albert acted alongside the equally glamorous Audrey Hepburn for the movie "Roman Holiday" for which he received an Academy Award nomination.[59] Indeed, the Corn Belt farmers of the late 1960s would likely have not related to Albert's Hollywood lifestyle working with Gabor or to his organic garden on the Pacific Coast between Brentwood and Malibu, California any more than they would have seen the outdated tractor driven by Oliver Wendell Douglas as representative of their modern capitalistic identities. To the contrary, one can view Albert as the archetypal urban outsider who reinforced farmers' resentment to the discourse of reform advocated by city dwellers. From the farmer's perspective, Albert is the kind of outsider that needs to be resisted much as Samuel Insull threatened rural identities in the 1920s, in a familiar repeat of this pattern of audience. This book will show that this pattern would continue with the further rise of the organic and sustainable foods movement after the 1970s.

To conclude, the Cold War strengthened the identity of rural capitalistic modernity that had formed in rural America in the 1920s. This buttressing of rural modernity occurred during the 1950s and 1960s because rather than fully conflicting with competing urban identities, as it had during the rural-urban conflict of the 1920s, it now partially aligned with nationalistic ideologies and political discourses opposed to the perceived threat of communism.

Importantly, rural capitalistic modernity did not abandon its distrust of urban industrialism. Rather, the Cold War resulted in a paradox in rural performative use. On the one hand, it encouraged farmers to use the newest technologies as a means of carrying out the patriotic aims of urban politicians and government officials to contest the Soviet Union. Farmers willingly viewed their technology use as fighting the Cold War because they saw parallels between the collectivization of the Soviet peasant and the urban industrialization of the American farmer at places such as Hawthorn Farm in the 1920s. Both systems violated Jeffersonian and German agrarianism, which served as foundations for rural capitalistic modernity, by denying farmers control over work processes and engage in family-based production on their own land. Rural capitalistic modernity retained its resentment of urban dwellers and its distrust of urban industrialization.

Even though rural modernity strengthened in the Corn Belt because of its usefulness in promoting of Cold War objectives, rural resentment against urban actors may have even increased during this period due to the stubborn persistence of images of the farmer as a "hick" resistant to change. Farmers failed to debunk urban rube stereotypes even through their use of the most contemporary technologies outproduced the Soviet *agrogorod*. Urban television producers fanned rural resentment already present in the Corn Belt coming out of the rural-urban conflict in the 1920s. I will identify other urban voices perpetuating the hick stereotype of Midwest farmers from the 1950s to the present in Chapter 7. Thus, farmers during the Cold War saw themselves as both supporting and opposing urban discourses and ideologies. As a result, a more inclusive discourse ironically *increased* farmers' resentment of the urban "other" and an acute consciousness of rube stereotypes that would drive the formation of an ultramodern rural identity.

Notes

1 Part of this chapter is published in the article, Brinkman and Hirsh, "Welcoming Wind Turbines and the PIMBY ('Please in My Backyard') Phenomenon."
2 J.K. Stern, "The Weapon Khrushchev Can't Match: American Agriculture," speech delivered at Farmer's Night Program, Roaring Spring, Pennsylvania, March 30, 1961, in *Vital Speeches of the Day* 27, no. 15 (1961): 469–471.
3 Carol M. Highsmith, *Not Surprisingly in the Largely Rural Midwest State, Tractors Play a Prominent Role in the Downtown Des Moines, Iowa, Parade Marking the Opening of the Annual Iowa State Fair,* 2016, photograph, *Carol M. Highsmith Archive, Library of Congress, Prints and Photographs Division,* at https://www.loc.gov/pictures/item/2016630279/ (accessed 11/2/23).
4 Stephen Hilgartner, *Science on Stage: Expert Advice as Public Drama* (Stanford, CA: Stanford University Press, 2000). See also Lynn Hunt, "The American Parade: Representations of the Nineteenth-Century Social Order," in *The New Cultural History*, ed. Lynn Hunt (Berkley: University of California Press, 1989), 131–153. Hunt argues that urban parades in the U.S. from 1825 to 1880 constituted a "public, ceremonial language," whereby Americans developed and performed ethnic, class, and gender identities.
5 Thomas Hughes, *American Genesis: A Century of Invention and Technological Enthusiasm* (New York: Viking, 1989), 171–175.
6 Henry Shapiro, "Malenkov Regime Acted to Stop Soviet Food Crisis," *Waukesha Daily Freeman* (January 21, 1954), 9.
7 W.J. Ridout Jr., "A Defeat for Communism," *Electricity on the Farm Magazine*, May 1953, Editor's Note, http://www.papergreat.com/2014/09/checking-out-1953-issue-of-electricity.html (accessed 9/16/16).
8 Herbert Block, "It's the Same Thing Without the Mechanical Problems," *Washington Post*, cartoon, January 26, 1949, 21, https://www.loc.gov/exhibits/herblocks-history/ticktock.html (accessed 6/16/15).
9 "The American Way," *Farmer's Weekly Review*, cartoon (October 24, 1950), 1.
10 "Farm Bureau Opposes Socialized Agriculture," *Waukesha Daily Freeman* (September 9, 1953), Editorials; "What Makes Communists Tick?" *Chicago Daily Tribune* (September 8, 1950), 20; "The Family Farm," *Chicago Daily Tribune* (September 20, 1952), 10; "Pilgrim's Progress," *Chicago Daily Tribune* (August 11, 1950), 14.
11 James L. Roark, et al., *Understanding the American Promise: A History of the United States Vol. 2,* 3rd ed. (Boston, MA: Bedford's/St. Martin's, 2017), 678–685.

12 Shapiro, "Malenkov Regime Acted to Stop Soviet Food Crisis," 9.

13 "Biographies: Nikita Khrushchev and Frol Kozlov," *Communist Affairs* 1, no. 1 (1962): 13–17.

14 Harold Milks, "Khrushchev Rips Malenkov Group on Farm Issue," *Cedar Rapids Gazette* (December 16 1958), 16; "The Two Faces," *Chicago Daily Tribune* (August 30, 1959), 24; Lazar Volin, "Khrushchev and the Soviet Agricultural Scene," in *Soviet and East European Agriculture*, ed. Jerzy F. Karcz (Berkeley: University of California Press, 1967), 9–10.

15 William R. Ryan, "Nikita Khrushchev: Bold Gambler, He Has More Potential for Mischief than Hitler Before Munich," *Cedar Rapids Gazette* (November 24, 1957), 16; Jake Booth, "Russia Faces Big Odds in Drive to Improve Farming," *Cedar Rapids Gazette* (April 25, 1955), 15; "Khrushchev, Heretic," *Chicago Daily Tribune* (December 19, 1958), 12.

16 "The Two Faces," 24.

17 "The State of Soviet Agriculture," *Buffalo Center Tribune* (December 3, 1959): page unknown.

18 Joseph Alsop, "Khrushchev Gambles on Farms," *Cedar Rapids Gazette* (January 28, 1958), Editorial Page, 6.

19 "Red China's Communes," *Chicago Daily Tribune* (December 27, 1958), 8.

20 "Russ Ag Men Eye Specialties," *Waukesha Daily Freeman* (July 22, 1955), 4.

21 "Seek to Boost Corn Output," *Waukesha Daily Freeman* (February 25, 1955), 3.

22 "Nikita Khrushchev's Visit to Iowa," film, 1959, a Special Pool Telecast on WHO-TV (available at Iowa State University Library University Archives, Film K-3056), https:// www.youtube.com/watch?v=LZ8WZB0sWwU (accessed 5/27/15).

23 U.S. Department of State, *Intelligence Report: U.S. and Soviet Gains from Agricultural Exchange* (Washington, DC: GPO, 1955), Office Memorandum, 1–7, at http://www. scribd.com/doc/138124499/1955-Report-on-US-USSR-Agricultural-Exchange-Visits (accessed 3/27/15).

24 Peggy Brown, "Diplomatic Farmers: Iowans and the 1955 Iowa Delegation to the Soviet Union," *The Annals of Iowa* 72, no. 1 (2013): 50, 53.

25 "Farmers End Russian Tour," *Waukesha Daily Freeman* (August 22, 1955), 8.

26 Dick Hanson, "State Farms Are Becoming Dominant in Soviet Union," *Cedar Rapids Gazette* (December 13, 1959), 11; Rex Conn, "Soviet Agriculture Still Lags, Says Minn. Farmer," *Cedar Rapids Gazette* (July 29, 1958), 15; Ovid A. Martin, "Russia Pins Ag Hopes on Work of Peasant Women," *The Cedar Rapids Gazette* (October 20, 1959), 24.

27 Martin, "Russia Pins Ag Hopes on Work of Peasant Women," 24.

28 Lentz, "The Evolution of a Real Farmer's Wife," 4.

29 D.W. McDowell, "State Agriculture Alert to Responsibilities," *Waukesha Daily Freeman* (July 29, 1955), Editorial Page.

30 Ibid.

31 William L. Ryan, "Early Farm Plan Almost Spelled Nikita's Downfall," *Cedar Rapids Gazette* (April 1, 1959), 4A; Phil Newsom, "Agriculture Is Soviets' Big Problem," *Cedar Rapids Gazette* (January 27, 1961), 14; Henry S. Bradsher, "Mismanagement, Weather Plague Russian Agriculture," *Cedar Rapids Gazette* (April 27, 1964), 13; "Why the Slavs Hunger," *Chicago Daily Tribune* (November 20, 1950), 18; "Production for Use," *Chicago Daily Tribune* (April 13, 1950), 16.

32 "Red China's Flop," *Chicago Daily Tribune* (August 29, 1959), 12.

33 "Chinese Seminar by Prof. Khrushchev," *Chicago Daily Tribune* (October 3, 1959), 12.

34 Roscoe Drummond, "Look at World Through K's Eyes; He's Got Troubles," *Cedar Rapids Gazette* (July 27, 1962), Editorial Page.

35 Drummond, "Communism's Many Ailments Add Up to One Big Flop," *Cedar Rapids Gazette* (October 21, 1963), Editorial Page; for similar opinions among farmers regarding collectivization, see also "Farmers Want to Own Their Land," *Wallaces' Farmer* (November 4, 1950), 6.

36 Drummond, "Why Can't Communist Countries Feed Their People?" *Cedar Rapids Gazette* (March 12, 1962), Editorial Page
37 For an example of Khrushchev's use of the word "bury" to threaten the U.S. in terms of agricultural production see J.K. Stern, "The Weapon Khrushchev Can't Match: American Agriculture."
38 Drummond, "Communism's Many Ailments Add Up to One Big Flop."
39 *Mightier than Missiles*. Chicago: American Feed Manufacturer Association, 1961, 2–3; "Food Is Stronger Weapon Not Generally Recognized," *Quad City He*rald (November 2, 1961), 6; Barker-Devine, "'Mightier than Missiles:' The Rhetoric of Civil Defense for Rural American Families, 1950–1970," *Agricultural History* 80, no.4 (2006): 415–435.
40 "Food Mightier than Missiles," *St. Petersburg Times*, February 8, 1962, 3–B. news. google.com/newspapers (accessed 5/21/15).
41 Gaylord P. Goodwin, "Study Compares American and Soviet Agriculture," *Cedar Rapids Gazette* (September 29, 1962), 11B; "U.S., Soviet Agriculture Compared," *Cedar Rapids Gazette* (July 20, 1965), 18; "Russia Beset by Problems in Agriculture," *Cedar Rapids Gazette* (January 10, 1964), 17.
42 William L. Ryan, "Khrushchev Risks Hersey to Solve Food Troubles," *Cedar Rapids Gazette* (April 13, 1962), 22.
43 "The State of Soviet Agriculture," page unknown.
44 Robert N. Pripps, *The Complete Book of Farmall Tractors: Every Model 1923–1973* (Minneapolis, MN: Motorbooks, 2020).
45 It is difficult when viewing the show to surmise the exact geographical location of Hooterville.
46 Val Adams, "Eddie Albert Due in Comedy Series," *New York Times* (August 17, 1965), 67.
47 Ed Stephan et al., "Green Acres," *IMDB*, http://www.imdb.com/title/tt0058808/ (accessed 5/26/16).
48 Ibid.
49 Lee Klancher and Kenneth Updike, *Red Tractors, 1957–2022: The Authoritative Guide to International Harvester and Case IH Tractors 3rd ed.* (Austin, TX: Octane Press, 2022).
50 *Green Acres-A Few Scenes with Mr. Haney (3)* (December 25, 2013: MyyyClips), https://www.youtube.com/watch?v=5KaAO56WTe8 (accessed 5/26/16).
51 Ibid.
52 *Arnold Ziffel Tests his Civil Rights - Green Acres - 1967 & 1968* (September 2, 2014: Shatner Method, MGM). https://www.youtube.com/watch?v=fTEcL7bw6U4 (accessed 5/26/16).
53 "CBS Kills Comedies," *Atlanta Constitution* (March 18, 1971), 8B.
54 Ibid.
55 "Eddie Albert: A Green Thumb," *Washington Post* (November 15, 1970), 34; Marion Purcelli, "Green Acres, an Electronic Shangri-La," *Chicago Tribune* (December 5, 1965), N10.
56 Norma Lee Browning, "Eddie Albert Just Loves Green Acres on and Off the Screen," *Chicago Tribune* (May 24, 1970), S2.
57 Adams, "Eddie Albert Due in Comedy Series," 67.
58 Hal Humphrey, "Hungary Meets Hillbilly, U.S.A.," *Los Angeles Times* (August 9, 1965), C20.
59 Browning, "Eddie Albert Just Loves Green Acres on and Off the Screen," 52.

Bibliography

Adams, Val. "Eddie Albert Due in Comedy Series." *New York Times*, August 17, 1965, 67.
Alsop, Joseph. "Khrushchev Gambles on Farms." *Cedar Rapids Gazette*, January 28, 1958, Editorial Page, 6.

Arnold Ziffel Tests his Civil Rights - Green Acres - 1967 & 1968. September 2, 2014: Shatner Method, MGM. https://www.youtube.com/watch?v=fTEcL7bw6U4 (accessed 5/26/16).

Barker-Devine, Jenny. "'Mightier than Missiles:' The Rhetoric of Civil Defense for Rural American Families, 1950–1970." *Agricultural History* 80, no. 4 (2006): 415–435.

Block, Herbert. "It's the Same Thing Without Mechanical Problems." Cartoon. *Washington Post*, January 26, 1949, 21. http://www.loc.gov/exhibits/herblocks-history/ticktock.html. (accessed 6/16/15).

Booth, Jake. "Russia Faces Big Odds in Drive to Improve Farming." *Cedar Rapids Gazette*, April 25, 1955, 15.

Bradsher, Henry S. "Mismanagement, Weather Plague Russian Agriculture." *Cedar Rapids Gazette*, April 27, 1964, 13.

Brinkman, Joshua T. and Richard F. Hirsh. "Welcoming Wind Turbines and the PIMBY ('Please in My Backyard') Phenomenon: The Culture of the Machine in the Rural American Midwest." *Technology and Culture* 58, no. 2 (2017): 335–367.

Brown, Peggy. "Diplomatic Farmers: Iowans and the 1955 Iowa Delegation to the Soviet Union." *The Annals of Iowa* 72, no. 1 (2013): 31–62.

Browning, Norma Lee. "Eddie Albert Just Loves Green Acres on and Off the Screen." *Chicago Tribune*, May 24, 1970, S2.

"CBS Kills Comedies." *Atlanta Constitution,* March 18, 1971, 8B.

"Chinese Seminar by Prof. Khrushchev." *Chicago Daily Tribune,* October 3, 1959, 12.

Conn, Rex. "Soviet Agriculture Still Lags, Says Minn. Farmer." *Cedar Rapids Gazette*, July 29, 1958, 15.

Drummond, Roscoe. "Look at World Through K's Eyes; He's Got Troubles." *Cedar Rapids Gazette*, July 27, 1962, Editorial Page.

———. "Why Can't Communist Countries Feed Their People?" *Cedar Rapids Gazette*, March 12, 1962, Editorial Page.

———. "Communism's Many Ailments Add Up to One Big Flop." *Cedar Rapids Gazette*, October 21, 1963, Editorial Page.

"Eddie Albert: A Green Thumb." *Washington Post*, November 15, 1970, 34.

"Farm Bureau Opposes Socialized Agriculture." *Waukesha Daily Freeman*, September 9, 1953, Editorials.

"Farmers End Russian Tour." *Waukesha Daily Freeman*, August 22, 1955, 8.

"Farmers Want to Own Their Land." *Wallaces' Farmer,* November 8, 1950, 6.

"Food Is Stronger Weapon Not Generally Recognized." *Quad City Herald*, November 2, 1961, 6.

"Food Mightier than Missiles." *St. Petersburg Times*, February 8, 1962, 3–B. news.google.com/newspapers (accessed 5/21/15).

Goodwin, Gaylord P. "Study Compares American and Soviet Agriculture." *Cedar Rapids Gazette*, September 29, 1962, 11B.

Green Acres-A Few Scenes with Mr. Haney (3), December 25, 2013: MyyyClips. https://www.youtube.com/watch?v=5KaAO56WTe8 (accessed 5/26/16).

Hanson, Dick. "State Farms Are Becoming Dominant in Soviet Union." *Cedar Rapids Gazette*, December 13, 1959, 11.

Highsmith, Carol M. *Not Surprisingly in the Largely Rural Midwest State, Tractors Play a Prominent Role in the Downtown Des Moines, Iowa, Parade Marking the Opening of the Annual Iowa State Fair.* 2016. Photograph. *Carol M. Highsmith Archive, Library of Congress, Prints and Photographs Division.* https://www.loc.gov/pictures/item/2016630279/ (accessed on 11/2/23).

Hilgartner, Stephen. *Science on Stage: Expert Advice as Public Drama*. Stanford, CA: Stanford University Press, 2000.

Hughes, Thomas P. *American Genesis: A Century of Invention and Technological Enthusiasm*. New York: Viking, 1989.

Humphrey, Hal. "Hungary Meets Hillbilly, U.S.A." *Los Angeles Times*, August 9, 1965, C20.

Hunt, Lynn. "The American Parade: Representations of the Nineteenth-Century Social Order." In *The New Cultural History*, edited by Lynn Hunt, 131–153. Berkley: University of California Press, 1989.

"Khrushchev, Heretic." *Chicago Daily Tribune*, December 19, 1958, 12.

Klancher, Lee, and Kenneth Updike. *Red Tractors, 1957–2022: The Authoritative Guide to International Harvester and Case IH Tractors 3rd ed.* Austin, TX: Octane Press, 2022.

Lentz, Grace Gibbard. "The Evolution of a Real Farmer's Wife." *Better Farming* 47, no. 6 (June 1924): 4.

Martin, Ovid A. "Russia Pins Ag Hopes on Work of Peasant Women." *The Cedar Rapids Gazette*, October 20, 1959, 24.

McDowell, D.W. "State Agriculture Alert to Responsibilities." *Waukesha Daily Freeman*, July 29, 1955, Editorial Page.

Mightier than Missiles. Chicago, IL: American Feed Manufacturer Association, 1961.

Milks, Harold. "Khrushchev Rips Malenkov Group on Farm Issue." *Cedar Rapids Gazette*, December 16, 1958, 16.

Newsom, Phil. "Agriculture Is Soviets' Big Problem." *Cedar Rapids Gazette*, January 27, 1961, 14.

"Nikita Khrushchev's Visit to Iowa." Film, 1959. A Special Pool Telecast on WHO-TV (available at Iowa State University Library University Archives, Film K-3056), https://www.youtube.com/watch?v=L78WZB0sWwU (accessed 5/27/15).

"Pilgrim's Progress." *Chicago Daily Tribune*, August 11, 1950, 14.

Pripps, Robert N. *The Complete Book of Farmall Tractors: Every Model 1923–1973*. Minneapolis, MN: Motorbooks, 2020.

"Production for Use." *Chicago Daily Tribune*, April 13, 1950, 16.

"Red China's Communes." *Chicago Daily Tribune*, December 27, 1958, 8.

"Red China's Flop." *Chicago Daily Tribune*, August 29, 1959, 12.

Ridout Jr., W.J. "A Defeat for Communism." *Electricity on the Farm Magazine*, May 1953, Editor's Note. http://www.papergreat.com/2014/09/checking-out-1953-issue-of-electricity.html (accessed 9/16/16).

Roark, James L., et al. *Understanding the American Promise: A History of the United States Vol. 2*, 3rd ed. Boston, MA: Bedford's/St. Martin's, 2017, 678–685.

"Russ Ag Men Eye Specialties." *Waukesha Daily Freeman*, July 22, 1955, 4.

"Russia Beset by Problems in Agriculture." *Cedar Rapids Gazette*, January 10, 1964, 17.

Ryan, William L. "Early Farm Plan Almost Spelled Nikita's Downfall." *Cedar Rapids Gazette*, April 1, 1959, 4A.

———. "Khrushchev Risks Heresy to Solve Food Troubles." *Cedar Rapids Gazette*, April 13, 1962, 22.

———. "Nikita Khrushchev: Bold Gambler, He Has More Potential for Mischief than Hitler Before Munich." *Cedar Rapids Gazette*, November 24, 1957, 16.

"Seek to Boost Corn Output." *Waukesha Daily Freeman*, February 25, 1955, 3.

Shapiro, Henry. "Malenkov Regime Acted to Stop Soviet Food Crisis." *Waukesha Daily Freeman*, January 21, 1954, 9.

Stephan, Ed, Brian Washington, Jerry Roberts, Kenneth Chisholm. "Green Acres." *IMBd*. http://www.imdb.com/title/tt0058808/ (accessed 5/26/16).

Stern, J.K. "The Weapon Khrushchev Can't Match: American Agriculture." Speech delivered at Farmer's Night Program, Roaring Spring, Pennsylvania, March 30, 1961. In *Vital Speeches of the Day* 27, no. 15 (1961): 469–471.

"The American Way." Cartoon. *Farmer's Weekly Review,* October 24, 1956, 1.

"The Family Farm." *Chicago Daily Tribune*, September 20, 1952, 10.

"The State of Soviet Agriculture." *Buffalo Center* (Iowa) *Tribune,* December 3, 1959, page unknown.

"The Two Faces." *Chicago Daily Tribune*, August 30, 1959, 24.

"Three Farmall Tractors on an Iowa Farm." Author's personal records. Charles City, Iowa, 1963.

U.S. Department of State. *Intelligence Report: U.S. and Soviet Gains from Agricultural Exchange.* Washington, DC: GPO, 1955, Office Memorandum. http://www.scribd.com/doc/138124499/1955-Report-on-US-USSR-Agricultural-Exchange-Visits. (accessed 3/27/15).

"U.S., Soviet Agriculture Compared." *Cedar Rapids Gazette,* July 20, 1965, 18.

Volin, Lazar. "Khrushchev and the Soviet Agricultural Scene." In *Soviet and East European Agriculture*, edited by Jerzy F. Karcz. Berkeley: University of California Press, 1967, 9–10.

"What Makes Communists Tick?" *Chicago Daily Tribune,* September 8, 1950, 20.

"Why the Slavs Hunger." *Chicago Daily Tribune*, November 20, 1950, 18.

6

"WE FEED THE WORLD"

Rural Globalized Ultramodernity[1]

> We need your angus cattle to satisfy the customers at our 200 restaurants in
> Japan. My customers like juicy, well-marbled, tender and delicious beef. They
> won't settle for anything less. That's why we selected Certified Angus Beef when
> we were looking for American beef to use in our 200 new Tokyo restaurants.
>
> Yashiro Honma, Tokyo restaurant executive, 1992[2]

While conducting research for this book, my uncle traveled from his farm in Iowa
to visit me in Virginia and to meet my young daughter. Walking through my house,
he immediately noticed a bound copy of old farm journals from the 1920s sitting on
my desk. The thick book had a leather cover, and the pages of the journals had dis-
colored over many years, giving the volume an impressive appearance that made
the reader feel as if he or she discovered a rare and ancient document. The content
of the journals also contained many different points of possible interest, making
them fun to read. One could find advertisements for old medicinal remedies, classic
automobiles, and fashions, along with editorials about historical events and jokes
and folk stories that have fallen out of popular taste. The journals even contained
recipes showing what people on Midwest farms ate in the 1920s. But my uncle
gleefully flipped through the pages trying to answer just one question: what corn
yield did farmers obtain in 1925? For me, the journal provided a window into a
way of life in 1920s America. For my uncle, on the other hand, it spoke chiefly to
production. Predictably, the farmers in the journal reported much lower corn yields
in 1925 compared to today's standards. As such, my uncle saw the farm journals
as evidence of the progress achieved by current farmers. In other words, the historian (me) and the farmer (my uncle) literally "saw" two completely different
documents.

DOI: 10.4324/9781032637952-7

The farm journal's testament to the advancements achieved by the contemporary farmer reminded my uncle to show me a YouTube video that one of his neighbors, Jared Schrage, had recently posted. The video features Schrage's 2014 corn harvest filmed from above with the latest drone technology. Set to music, one sees an impressive combine shooting corn into a grain wagon pulled by an equally imposing John Deere tractor. Next, the camera runs along Schrage's newest semi-truck such that the lights of the adjacent tractor illuminate the metallic side of the huge vehicle. The video highlights the majesty of Schrage's farm equipment both during the day and at night.

The drone makes the technology appear fast and efficient, as does the increased recording speed. The tractor and wagon race across the landscape more rapidly than they would in real time with freshly harvested corn and pull the empty bin adjacent to the moving combine for the next round of bushels. The whole process of the combine, wagon, and tractor working together presents a carefully choreographed circle with little wasted movement.

Later, the camera features the farmer driving the tractor, which appears to almost fly down the harvested portion of the cornfield. One also notices how remarkably clean the farm equipment appears and how ordered Schrage's corn rows look. The viewer cannot find a single weed in the five-minute video or cite an instance in which dirt or dust appears on Schrage's machinery. The yellow corn falls into the pristine trailer bed so evenly that it almost looks like yellow water. The video concludes with the farmer standing triumphantly on top of the semi-truck bin watching hundreds of bushels of corn cascade into the spotless metallic surface.[3]

My uncle proudly held up a tablet so I could see this video in its entirely, giving me a look as if to ask, "Well, are you impressed?" I then realized that Schrage did not stand alone in his drone filming of his harvest. The side bar of YouTube recommends a multitude of similar videos loaded by Midwest farmers showing their harvesting and planting operations and many have well over one thousand views. Several of these videos show the latest farming equipment accompanied by either inspiring music or country songs celebrating the virtues of farm work or hardships heroically overcome by generations of Corn Belt farmers.[4]

A non-farmer, like me, closes YouTube wondering why so many videos of the harvest exist and why they have thousands of views. How many times, after all, can one watch combines and tractors driving in a cornfield without becoming bored? Does a drone actually make a monotonous Iowa landscape appear more exciting? Of course, the answer to these questions speaks to the heart of this book's main argument. The filming and viewing of these videos represent instances of performative use. Each video posted by farmers on YouTube serves the same function as the family farm photos with members posing with technology: to perform an ultramodern identity showing the use of the latest technologies.

The drones in these cases do nothing to enhance farming profits and, in fact, impose an added cost. The videos exist solely to enhance the virtuous view farmers have of themselves and to present their ultramodernity to outside urban observers and to each other. Indeed, the farmers in making and watching these videos engage in performative use. Just as my uncle did not realize he read the 1920s farm journals in a culturally specific way, the farmers making and watching the drone movies do not conceive of their actions as forming and reinforcing a complex bundle of ultramodern discourses and identities. People use technologies to perform their embodied identities in ways determined by historical and social contexts. By showing the drone video on the tablet, my uncle acted out an ultramodern identity because he sought to display modernity greater than an outsider, like me, had achieved. Indeed, my uncle and his farming colleagues in Iowa operating large GPS-guided combines and analyzing "big data" use a greater number of advanced technologies than the average city dweller.

Additionally, the farmers filming the drone videos perform their identities as ultramodern users. None of the videos feature morose music or songs with lyrics that disparage the globalized food system because the farmers used the drones and their farm equipment to reinforce their virtuous sense of self and to dispel yokel stereotypes. The lyrics in the songs extoling the morality of work or celebrating the success of farmers in the face of environmental and financial obstacles exist as part of a long genealogy of rural discourses and identities that advance an agrarian hero myth.

Further, I have argued that people perform their identities not only in the case of blatantly theatrical uses of technology, such as filming YouTube videos, but through the everyday mundane employment of artifacts. As with the drone, the farmer does not use the equipment he films simply for rational economic reasons, but because it helps him to form and reinforce his unspoken sense of self, which he regards as moral. Nor have I contended that farmers exhibit unique performativity. Rather, all people use technology performatively in unarticulated ways determined by a socially determined practice resulting from historical factors. As a result, the relationship we have with material objects is not entirely rational or voluntary, but historically and culturally contingent.

By the 1980s, the Cold War discourse of rural capitalist modernity had morphed into an ultramodern discourse in which many farmers thought of themselves as surpassing their urban cousins in terms of technological savvy. From the 1950s to the present, Corn Belt agrarians developed self-images as heads of sophisticated technoscientific systems through computerized networks, remote sensing, and site-specific technologies. Farmers have gone beyond using technology as a means of claiming their own version of modernity. Instead, they use it as a means of constructing an identity of *greater* modernity. Hence, this discourse frames the farmer not as a user of science and engineering, but as someone who remains on par with scientists and engineers – as an expert who combines technical and scientific knowledge with practical experience. Farmers' use of practical knowledge confers on them some kind of enhanced moral standing deriving from Jeffersonian agrarianism. An emphasis on the practical played an important role in American rural identity as early as the 18th century.[5] The discourse of *rural globalized ultramodernity* continues this practice-based ethic, but in a way that reinforces an updated identity. As Joseph Frazier Wall explains in his 1978 history of Iowa, the farmer thinks of himself as "no longer the simple tiller of the soil; he had become a remarkable hybrid himself-part geneticist, part chemist, part mechanic, part processor."[6]

Just as rural capitalistic modernity built on and added to the existing identities of traditional Jeffersonian and German agrarianisms, ultramodernity incorporated aspects of rural capitalistic modernity including the goal of eliminating urban stereotypes of the "bumpkin" farmer. Despite the farmer's self-image as having technical knowledge superior to urban expertise, the rural-urban conflict remains an unspoken ingredient within the ultramodern rural self-image. Farmers still use technology to perform their identities and combat urban rube stereotypes. When discussing letters written by Iowans on farms in the 1850s and 1930s, for example, Wall notes a common thread among farmers in these eras and agrarians in 1978 notwithstanding drastic technological changes. Wall claims that farmers in all three eras "were highly suspicious of town folk with their superior ways, their lack of understanding of the farmer's problems."[7]

Similarly, ultramodernity retains traditional German agrarian rural values. Farmers still imbue the dominant image of an independent producer on a "family farm" with morality and retain the cultural practice of displaying wealth through productive artifacts. As Wall recognized, this ultramodernity has not created an identity that completely abandons prior rural identities when he writes, "Yet, in spite of the revolutionary changes that have occurred in agriculture during the last century and a half, ... the farmer himself has remained a remarkably consistent feature in our society."[8]

The retention of elements of Jeffersonian and German agrarianism within the rural globalized ultramodern identity explains why newspaper articles about the financial crisis in the Corn Belt in the 1980s quote farmers as expressing shame at seeking government aid. As one social worker in Kansas stated, "Kansas farmers always felt if you're poor and working hard, then you're not working as hard as you should." She added, "Here people think it's immoral to be poor."[9] One Iowa farmer even related going home and vomiting after she stood in line to receive government assistance because such aid threatened her notion of independence.[10] Many Kansas farmers avoided going to government offices to seek help even though their children were malnourished because of fear that their neighbors would know of their poverty. In both Iowa and Kansas, farmers were moved to tears by the need to receive food stamps.[11] More contemporary observers have noted higher suicide rates among Midwest farmers than the general population attributed to reluctance to seek counseling when faced with financial stresses caused by weather and global commodity price changes. As one psychologist practicing in Iowa stated in 2022, "the things that make a good farmer – independence and a willingness to take risks – can work against them when they need help... They're reluctant to reveal what they perceive as weaknesses."[12] Therefore, such remnants of German agrarianism – the association of prosperity with morality; the desire to perform a moral living for yourself and others; and a fierce desire to feel independent – still shape rural identities. In addition, according to the farmer in the Corn Belt, he still has not become an urban industrialist. Rather, he or she uses technology to form and perform an updated version of *rural* modernity.

Unlike identities and discourses prevalent through the Cold War, farmers guided by ultramodernity no longer view themselves in nationalistic terms, or as using technology as proof of the superiority of American capitalism over Russian socialism, but as globalized businessmen using a high level of business and technical expertise to compete in an ultra-competitive global market. The farmer no longer views him- or herself as a single modern user of technology. Rather, Corn Belt agrarians regard themselves as the head of a sophisticated technoscientific network of scientists, agricultural engineers, crop and soil analysts, commodities traders, and accountants all connected to the farmer's field through computerized networks, remote sensing, and site-specific technologies. This shift in the farmer's identity from an American farmer to a world farmer parallels in many ways the change in the early 20th century from a collective to an individual producer. In both instances, the farmer not only operates within different agricultural networks of production, but also sees himself or herself in those terms and moralizes the new self-image by combining it with prior identities. In both cases, the agrarian uses technology to develop and maintain this moral identity. In the most recent version of this cultural practice of performativity, farmers go beyond using the technology as a means of reclaiming their own version of modernity from urban efforts to define it but use it as a means of constructing an identity of *greater* modernity.

Further, rural denizens view this ultramodern identity as an inborn or inherited trait arising out of an underlying view of history as a progressive march towards better material objects with greater productive capacity. This idea of progress arises not just from many years of using technology to reinforce an identity of rural capitalistic modernity, but through active use of artifacts, old and new, which highlights how both the machine and the farmer have modernized.

Hence, this ultramodern discourse frames the farmer not as the one expert combining technical and scientific knowledge with practical experience and use of artifacts to produce.[13] For example, when farmer Rodney J. Fee rejoined *Successful Farming* as their Senior Livestock Editor in 1992, he found it necessary to create trust with his readers by opening his debut article entitled "Production" with "After 12 years of what my dad would have called 'good practical experience' down on the farm, I'm back again at *Successful Farming*, behind a computer screen (an editor's modern-day typewriter)." If Fee's practical experience as a farmer did not gain his reader's trust, he then recalled how he attended a presentation at the "'92 farm outlook session" in Kansas City, Missouri by economist Bill Hemming about ending "the USDA's long-established beef cattle quality grading system." While Fee ultimately agreed with Hemming's argument, he instead highlighted in large bold letters in the middle of his text "All we need is another economist telling us how to make money on cattle!"[14] Fee's article is clearly a performance of identity. The new livestock editor sought to establish trust by aligning himself according to an ultramodern identity among his readers that saw the practical modern farmer as the superior expert with higher moral standing than the urbanized "other" represented by the economist. In bringing up his father's valuing of practical farm experience, Fee not only positioned himself as an ultramodern "us," but also subtly implies an inborn modernity and expertise that he shares with his intended audience.

JoAnn Wilcox, who served as one of the chief technology editors of the farm journal *Successful Farming* in the 1990s, also sought to establish the farmer as the ultimate expert by emphasizing practical uses for advanced technologies. She titled her monthly column "Production," as a way of appealing to her reader's production ethos.[15] In one article, she launched "Yield Monitor Watch 2000," in which she supplied a Minnesota farmer with a yield monitor system, a global positioning system, and yield map software and documented how she and the farmer used the technology on his combine during a year.

Wilcox proudly begins her article by stating

Every high-tech thing I know, I learned in farmer kindergarten, so to speak. By that I mean I started at ground zero and learned by doing. I've gone to countless meetings on precision agriculture. I've ridden in many combines with yield monitors. And I've climbed on dozens of tractors, sprayers, fertilizer applicators, and ATV equipped with high-tech gadgets.[16]

The article, clearly written to demonstrate the farmer's ultramodern technical know-how, showed a picture in which Wilcox and the ultramodern farmer stood in front of his largest and most impressive combine. The huge machine sits in the farmer's massive shed with a metal ceiling and features another grain chute in the background. Wilcox and the farmer prepare to render the combine even more modern by installing the new yield monitoring system. The farmer admitted he feared looking foolish during the experiment in front of other technologically savvy neighbors.[17] The article pictures Wilcox handling the technology with the male farmer as equals in the project of ongoing ultramodernization (Figure 6.1).

Ironically, by adopting an ultramodern identity, the farmer relinquishes agency as a member of an ultramodern technical network. The farmer adopts technologies as an embodied practice. Simply put, the farmer must adopt the newest, most sophisticated, technology without question because it represents ultramodernity and practically, but also because he or she will fail financially by resisting the imperatives of the system.[18] The farmer finds himself ensnared in a technological

FIGURE 6.1   Wilcox and farmer Brent Olsen using a yield-monitoring device. JoAnn
Wilcox, "Every High-Tech Thing I Know I Learned in Farmer Kindergarten," *Successful Farming* (December 2000): 24.

momentum that he himself helped to create and, from some farmer's point of view,
he still controls.[19]

On the other hand, the real financial pressures exerted by the system do not
render the farmer's actions less performative or strategic. Advertisements and editorials in farm journals published in the 1990s and 2000s reflect farmers' use of
technology to re-enforce a rural ultramodern globalized identity. Journals such as
Successful Farming, a popular farm journal printed since 1902 in Des Moines,
Iowa, and its online version, *Agriculture.com*, presented articles by technology
and machinery editors to show that farmers had reached a level of technology use
that made them more modern than the average American. For example, machinery
editor Dave Mowitz in a January 1992 edition of *Successful Farming* praised the
"wonder of a farmer's mechanical genius" and exclaimed, "There isn't a staff of
engineers in this world that can top the inventiveness of farmers." Mowitz boasted
that without engineering degrees, farmers had invented the combine, the corn
picker, the four-wheel drive tractor, the pivot sprinkler, and the pneumatic planter.
Mowitz then announced that the journal planned to team up with Conoco to spon-
sor a National Farmer Inventors Conference, solely to highlight "success stories"
of ultramodernity.[20]

In another editorial in *Successful Farming* in September of 2000, machinery editor Larry Reichenberger expressed amazement that his tractor in Kansas used remote radio sensor fuel monitors connected to his fuel company that then used a global positioning satellite to rout a gasoline truck directly to his tractor in real time. While Reichenberger projected the same ultramodernity as Mowitz, he also wrote "Despite our reluctance, we were surprised this spring to realize that site-specific farming technology found us anyway."[21] Reichenberger's fuel supplier simply incorporated him, perhaps involuntarily, into a completely new technological system of fuel, satellites, and remote sensing.

While Reichenberger reveals recognition by some farmers of their ironic lack of agency in forming ultramodern identities, they tend to overlook such self-reflexivity when interacting with urban dwellers. When under the observation of city folk, the farmer once again becomes conscious of "yokel" stereotypes and normally presents himself or herself as an ultramodern businessperson who willingly, and skillfully, uses technology. In such cases, the rural modern discourse driven by rural-urban contestations over the meaning of modernity re-asserts itself. For example, a February 1992 article about a farm family from Missouri participating in a Smithsonian Festival of American Folklife, told the story about how they resisted efforts of the festival to frame their lifestyle as an example of old-fashioned "folk" culture, to show the urban viewers the true ultramodern character of Midwest farming. The article stated proudly,

> With Washington's Smithsonian Institution for a stage and a million eastern urbanites for an audience, the Harlan Borman family taught visitors more than just cow-milking and pie baking. They brought modern agriculture to Capitol Hill.[22]

In a ritual that may have looked quite familiar to German farmers observed by Rush in the 18th century, the article celebrated how the farm family used technology and success in producing to debunk urban "hick" notions of farm life. The author continued to write "Their mission: To help educate the more than one million urban visitors to the Institution during those two weeks about life down on the farm as it really functions today." The article described the Borman's farm in Missouri to highlight its ultramodernity and impressive productive capacities:

> The Borman's farm 575 acres just off Interstate 70 east of Columbia. With a milking herd of about 100 head, their rolling herd average is just over 20,000 pounds of milk and 750 pounds of fat. Their registered Holstein herd started from 4-H projects and has been shown successfully at several state and national shows.
>
> They take pride in forages. Last year's sixth cutting off one 15-acre field yielded nearly 13 tons. Part of the farm is irrigated with a center-pivot system. They plant 150 acres of corn, 100 acres of double-cropped wheat and no-till beans.

FIGURE 6.2   Images constituting a performance of an ultramodern rural identity. Rodney J. Fee, "Successful Family Farm: Capitol Hill's Country Crusaders," *Successful Farming* (February 1992): 42–43.

A DeLaval Computerized Feeding System meters concentrates to high producers. Programmable transponders program each cow's feed needs.[23]

Above this description of ultramodernity, the author included two revealing photos. (Figure 6.2). One photo shows the family around the dinner table to highlight several features of an older Jeffersonian and German rural morality such as family, production, and independence. The second image pictures two women in a highly technical milking station with the caption "Judy and daughter Kate take an active role in milking when the men are gone in the fields. The clean double six parlor speeds milking chores." The photos pair the latest hardware with a family-based production process.

A few more ads and articles (of literally hundreds) from *Successful Farming* demonstrate the character of this ultramodern discourse. One category of articles draws on Mowitz's imagery of the farmer as an ultramodern inventor of a completely new technology. For example, an article from June 2000 told the story of David Herbst who invented a way to use solar powered controllers to open and close valves on an irrigation system as part of "an ambitious land-leveling program to make the operation more efficient" allowing him to expand a farm "his grandfather had founded as a teenager on 10 acres" to a 3,800-acre farm. Herbst won the American Farm Bureau Federation's Young Farmer and Ranchers Award contest because his use of irrigation technologies offered "unique approaches to caring for the environment and managing in an era of rented land and labor shortages." (Figure 6.3).[24] Figure 6.3 shows Herbst in work clothes but he does not perform grueling manual labor. Rather, the photos show him programming his complex computer systems both on the main irrigation console mounted on a large green trailer with wheels and on an irrigation pipe.

FIGURE 6.3 *Successful Farming*, 2000, showing a farmer-inventor with his computer powered irrigation equipment. Bill Eftink, "Missouri Family Bets on Irrigation: Irrigation Boosts Corn Yields by 30 to 40 Bu an Acre," *Successful Farming* 98, no. 7 (May-June 2000): 62–63, 65.

Another article in January of 1992 related how farmers modified their corn and soybean drills to create a "super drill" that could accurately plant in different row sizes and configurations (Figure 6.4). The article characterized Paul Beckman and Marian Calmer as the "Sperry, Iowa innovator[s]" and described the device as mounting "an old IH 400 Cyclo blower" on "a John Deere 750 no-till drill." The rest of the article described how the farmers designed and built the drill in great technical detail. Beckman and Calmer's invention allowed the farmers to plant rows with no seeds directly across from each other. Calmer claimed, "Equidistant spacing provides each plant more growing space, thereby boosting yields." In Beckman's words, he created "the ultimate do-it-all planter with tremendous down pressure to cut through the residue." The picture accompanying the article, Figure 6.4, shows Calmer with other male family members kneeling in a perfectly cultivated field with their impressive super drill in the background connected to a tractor complete with a complex mechanism of gears, drums, wheels, and pistons. The title of the article proudly stated, "Farmer ingenuity creates new generation of high yield planters."[25]

Another category of *Successful Farming* articles highlighted farmers creatively using advanced technologies in new and novel ways.[26] Articles from 2000, for example, demonstrated how Midwest agrarians take it upon themselves to employ satellite data to manage herbicide application and track cattle grazing requiring ultramodern interpretation of graphing data (Figure 6.5).[27] In one photo from a 2000 article in *Successful Farming* entitled "Precision Pastures," a cow is fitted with a GPS unit from the collar so that farmers can map their herds' grazing patterns.[28]

FIGURE 6.4   *Successful Farming*, 1992, showing farmer-inventors with their modified row planter. Dave Mowitz, "Super Drills: Farmer Ingenuity Creates New Generation of High Yield Planters," *Successful Farming* (January 1992): 42.

FIGURE 6.5   *Successful Farming*, 2000, cover story shows ultra-modern farmers using sophisticated satellite monitoring technology and knowledge of plant biology to monitor herbicide application. Mike Holmberg, "Sky-High Scouting: From Ultralights to Satellites, you have More Opportunities to Get a Bird's Eye View of Your Crops," *Successful Farming* (May-June 2000): 34–36, 38.

In another article from the same year, a photo, Figure 6.5, shows one farmer holding a satellite-mapping device in a field to identify the location of different species of weeds with more precision. The article describes the two ultramodern farmers as engaging in "sky-high scouting" using ultralights and satellites.[29]

In a special edition of *Successful Farming* entitled *Agriculture.com* (which was printed in the early 1990s), the magazine dedicated a whole issue to articles showing how farmers surpassed urban dwellers in their use of computers to purchase goods, trade commodities online, manage crops and cattle in various ways, and share knowledge (Figure 6.6).[30] In an article from a 2000 issue of *Agriculture.com* entitled "Milking the Net: dairy producer bases business decisions on Internet info," Ohio farmer Jay Herron poses with his newest John Deer tractor. He has one foot up on the front wheel well with one hand in his pocket. Tools purchased on the Internet sit on the ground next to the tractor as well as on the huge machine to displayed to the camera. One sees Herron's prosperous farm buildings, silos, and grain bins in the background to indicate his success. The photo draws the viewer's eyes upward as four large Harvester silos in the background rise into the sky.

Jay Herron used the Internet to buy or do research on all these farm items, including the tractor.

FIGURE 6.6 *Agriculture.com*, 2000, shows a farmer posing with his online purchases with machinery and silos in the background to indicate his success. Chester Peterson, Jr., "Milking the Net: Dairy Producer Bases Business Decisions on Internet Info," *Agriculture.com* (Fall 2000): 36.

The silo directly behind the machine in the center of the picture displays an American flag that seems to overlook Herron and his ultramodern display of technology. The caption to the picture states "Jay Herron used the Internet to buy or do research on all these farm items, including the tractor."[31] The article invites the reader to count the number of new technologies purchased or researched on the Internet as if such a tally proves the ultramodern identity of not just Herron but of all farmers in the Corn Belt. Herron's tally of ultramodernity mirrors how farmwomen in the 1924 USDA report in Chapter 4 presented a similar inventory of artifacts to demonstrate their rural capitalistic modernity. Thus, the article shows an updated version of a familiar rural performance.

Another article from 2000 in *Successful Farming* discusses use of computers and sophisticated analysis of milk production. It pictures a young farmer at a computer managing each cow from his office. He appropriately wears a "Pioneer" seed shirt with the older farmer symbolically positioned behind him dressed in old-fashioned overalls (Figure 6.7). The older farmer has his right hand on the young farmer's shoulder as if to proudly encourage his son to carry on an even more modern legacy. The image is clearly one of ultramodern progress from an old farmer in overalls to a young computer-using "pioneer," the dominant image that rural ultramodern agrarians would like urban dwellers to see. The article, written by the farmers in the photo, Jim and Joe Stewart, boasts, "We're totally computerized. We even record what time, which stall, pounds of milk, and who milked a

Joe Stewart (sitting) and his father, Jim, say their computers help them meet the needs of each cow as an individual.

FIGURE 6.7 *Successful Farming*, 2000. Chester Peterson Jr., "Simply incredible!: Cows Accept the Challenge, Milk Goes Up 8,000 Pounds," *Successful Farming* (April 2000): 26–27.

certain cow at each milking." Stewart went onto explain, "If one of us is away from the farm, through the use of a modem and a laptop computer, we can call up information to keep abreast of what's happening." Such a detail-oriented use of technology, according to Stewart, allowed him to see "when a cow isn't meeting our bottom line" or correct for human error.[32]

An article in 1992 featured two farmers in Kansas who modified a Caterpillar tractor to run between narrow rows of corn thereby increasing yields (Figure 6.8). The article reinforces the view farmers have of themselves as the true modern technology users ahead of both urban businesses and university scientists and engineers. The author states, "The Larned, Kansas, innovators spent 9 months on and off on making the modifications to mount the narrow tracks on a Caterpillar AG4 (an experimental tractor Caterpillar never brought to market)." The article happily noted that the two farming brothers' theory that Caterpillar tracks increased yields by avoiding soil compaction by standard tractor wheels was "confirmed by a 3-year Iowa State University-USDA study which found a 14% increase in corn yields when track rather than wheel tractors were used."[33] As with Jay Herron's photo with his machine, the two brothers pose with their modified tractor with the camera capturing the full length of the added caterpillar tracks. One of the brothers poses

FIGURE 6.8 *Successful Farming*, 1992, shows farmers posing with their modified Caterpillar tractor. Chester Peterson, Jr., "Cats in Corn: The Crane Brothers Tapped the Advantages of Cat Tractors by Converting One to Run Between Rows," *Successful Farming* (January 1992): page 40 in unnumbered pages in between numbered pages 38 and 40.

with his leg up on the implement attached to the front of the tractor. By using and modifying technologies, "the Crane boys" and other farmers reinforce an identity that frames him or her as more modern than even the most cutting-edge tractor companies or innovative university scientists or engineers.

One further article from *Agriculture.com* presented a short biography of a farmer, Gene McCool, who traveled the world designing computer software for companies and used his computer knowledge on his western Iowa farm. The farmer hoped to trade in his "exotic" life as a computer consultant to farm with his wife and son full time in Iowa. When asked "What's ahead for cows and computers?" McCool's answer exhibited an ultramodern vision of highly technical progress

Visualize this: sick cow; it's 2 a.m. and storming. You take a video of the cow, download it to your computer, e-mail to the vet, who reviews the video and re-turn e-mails a treatment until he can arrive. Computer chips will soon track my cattle all the way through the packer. In the future a computer chip implanted in a cow will be able to tell me her temperature, metabolic rate, whether she is sick, and so on. I envision an implanted chip will tell me the perfect time to AI, or tell me when labor has started. Until now, computers in ag, for the most part, have been used as record keepers. In the future, computers will become as much a tool as a tractor or pickup.

The accompanying photographs show McCool posing in both traditional rural clothes and sophisticated business attire at his farm with a computer perched on a fence.[34] In each picture, one sees McCool's cows and barn in the background to challenge urban notions that maintain a dichotomy between rural and sophisti-cated modern imagery. The article and the photo convey that farmers participate intimately in a computer revolution as ultramodern developers and designers to combat urban stereotypes of the yokel farmer (Figure 6.9). The computer, like the tractor and pickup truck noted by McCool, becomes another device farmers can use to perform their ultramodern identities.

Finally, advertising in *Successful Farming* from 1990 to 2000 appealed directly to an ultramodern rural identity. For example, an ad for *Agriclick.com*, pictures a computer mouse divided into sections with functions for the ultramodern farmer. The list of ultramodern items encompassing the purview of the agrarian include "Product innovations, crop report, market forecast, cash markets, futures analyst, events calendar, weather vane, technology advisor, government correspondent, and business news." The list of functions appeals to the farmer's view of himself or her-self as an ultramodern globalized businessperson using technology to wield both economic and political power. The caption states, "Long before the internet, farm-ers learned from other farmers. Their knowledge was a collection of experiences, each generation teaching the next. And each generation understanding farming as only farmers can."[35]

FIGURE 6.9   *Agriculture.com*, Fall 2000 shows the farmer as an ultra-modern software developer. John Walter, "People Pages: He Blends Computers and Cows," *Agriculture.com* (Fall 2000): 16.

An ad for Aventis Crops proclaimed, "America's farmers thrive on innovative thinking" and then elaborated for two pages,

> The American farmer is a born innovator. To be successful, he had to know how to fix things. Or learn fast. When an idea has merit, he'll try it out. Then if it works to his satisfaction, he'll show you a better way to do it than you ever imagined.[36]

In another ad, the Council for Biotechnology Information framed the farmer as the expert on using biotechnology to grow and manage soybeans.[37] This image of the farmer as the ultimate expert appeared again in an ad for Pioneer Forage seeds in 1992. The photo shows the farmer surrounded by a team of smiling experts including an agronomist, microbial products specialist, corn breeder, pathologist, nutritionist, and microbiologist.[38] The photo appeals to the farmer's self-image as the head of an ultra-modern technoscientific network.

Rural globalized ultramodernity became such a ubiquitous rural identity in the Corn Belt by the late 1990s that it inspired a widely successful television show called *AgDay* based in South Bend, Indiana. The anchorman, Al Pell, farmed in

Indiana but he projected a "bankerly image" consisting of "dark suits and conservative ties" and the president of the show, Jeff Pence, came from a local farming family. When discussing his motivation for producing the show, Pence explained, "America's image of the average farmer is a country bumpkin. But a successful farmer has more capital under his control than most other businessmen, often in the range of $500,000 to $1 million in land and equipment."

To dispel this "bumpkin" misconception, *AgDay* presented "segments on global positioning in marketing farm products, the role of technology in pest control, and an interview with a stock market-expert." One urban reporter surprisingly observed, "*AgDay* is not 'Hee Haw.' The show is taped on a chrome-and-glass set that would do any Washington-politics show proud. Computer graphics spin across the screen to introduce segments, and weather maps benefit from all the high-tech wizardry viewers expect." From 1982 to 1995, the show aired on 150 stations in 130 markets blanketing the Midwest and stretching as far west as Seattle, Washington and as far east as West Palm Beach Florida. *AgDay*'s popularity also attracted many national advertisers such as DuPont Ag Chemical and Ford New Holland Tractors, companies benefitting from the ultramodern rural identity the show projected. In an indication that Midwestern women still perform rural modernity, "the show says its viewers are split almost evenly between men and women."[39]

This ultramodern rural identity still exists as the dominant discourse identity bundle in the Corn Belt. The *Wall Street Journal* in 2016, for example, reported how farmers in Iowa and Manitoba, Canada, just north of the American Corn Belt, altered software and hardware to produce their own devices. Jim Poyzer of Boone, Iowa redesigned a microprocessor to monitor and adjust seed placement on his planter. Poyzer then moved onto developing a solar-powered sensor to monitor soil temperature. Another farmer in Colby, Kansas, Lon Frahm, attributed his expansion to achieve a 30,600-acre farm by 2017 to "constant analysis of efficiency and scale-cents per pound or units per acre." In an interview, Frahm highlighted his use of mobile apps to monitor moisture levels in fields and reducing seed and chemical costs by using Internet deals to bypass local co-ops. The article described the work processes of the Kansas farmer, who holds three degrees in business and agricultural economics, as the following:

> His nine-person team totes tablet computers that control sprinkler systems miles away, switching them off when sensors report adequate moisture. When his tractors are rolling, automated systems monitor the number and type of seeds being sown in each row, to maximize planting on fertile ground and avoid wasting seed on poor soil.
>
> The equipment beams the data to remote servers, where Mr. Frahm and his team analyze it to determine how machinery can be run more efficiently, or where they can spray fewer chemicals. This year Mr. Frahm has been testing automated insect traps that deliver updates on the number and type of bugs killed to help time pesticide spraying.

While one may regard Frahm's farm as violating rural globalized ultramodernity since his employees do not own the equipment or property, Frahm is not an outsider seeking to control production and he makes sure to cater to his employee's needs to use technology performatively. Frahm gives the farmers who work for him, many of whom have stayed for decades, more than enough benefits to make up for their lack of ownership of the equipment. The article states, "His employees hold year-round, salaried positions with health care and four weeks of vacation. He brings them on ocean cruises as bonuses and outfits them with matching white Ford pick-ups." In addition, the article suggests Frahm allows his employees to use and exercise control over work processes. The employees on Frahm's farm are more than simply mindless factory workers because they are tasked with interpreting data and working with the most advanced technology. Frahm goes as far as giving his employees $1000 extra per year just to donate to charity, declaring that "These jobs offer a higher quality of life than if you were out trying to [farm] on your own." In addition, Frahm's operation differs significantly from plans of urban industrialism such as Hawthorn Farms in that the farmers live in their own homes, rather than barracks, and farm for the benefit of their own families. Yet, some farmers in Colby, Kansas accuse Frahm of being a corporate manager rather than a farmer and often ask, "haven't you got enough by now?"[40] Therefore, large technological farms at a certain point create ambivalence even among the most ultramodern agrarians connected to the view that they consolidate production too much and threaten to destroy already struggling rural communities.

Two other farmers in Elbow Lake, Minnesota worked with software engineers to develop mobile applications to map soil fertility and rocks in fields. While the author of the article, Jacob Bunge, accurately documented practices reflecting an ultramodern identity, he incorrectly (in my view) characterized innovation among farmers as a recent "technology revolution sweeping North America's breadbasket" resulting from a fall in commodity prices. Predictably, the farmers presented to Bunge, a *Wall Street Journal* reporter, the self-image of rational businessmen by proffering economic explanations for their "high-tech tinkering." The farmers did not voice motivations such as their identities or sense of morality as arising from a German agrarian production ethos as motivating factors because they took such notions for granted. Rural ultramodern identity, as I have argued throughout this book, is unarticulated.

Further, the idea that farmers use their creative energies to overcome a fall in farm income lends itself to the Jeffersonian myth of the agrarian as a frontier hero overcoming hardship through grit and determination.[41] Poyzer, for instance, stated, "poverty is the mother of invention," even though he admitted his planting mechanism saved him only about $1,000 per year on seeds and cost him $750. Something other than $250 in savings in the first year of the device, which also depreciates over time, must explain why Poyzer spent many hours developing the complex planting mechanism. Poyzer's explanation for this small economic gain, that "farmers are trying to optimize everything," reflects a deep-seated adoption of the myth of the Jeffersonian frontier hero stubbornly resisting hardship. Additionally, Bunge roots his uncritical reporting in the common assumption that technological change is solely driven by rational economic or functional motives. Yet, Poyzer gives no indication that his solar-powered soil sensor will save him any money, only that it "could help him get a jump on planting."[42]

In addition, Bunge's uncritical assumption that farmers' innovative use of technology results wholly from a decrease in farm incomes implies that farmers do not engage in such behavior when commodity prices rise. Farmers in the Corn Belt have modified and re-designed material objects on their farms at least since the 1920s in innovative ways regardless of the levels of commodity prices.[43] The actions observed by Bunge do not, therefore, constitute a recent "technology revolution" among farmers but a deeply embedded and historically formed cultural practice. While a drop in commodity prices may occur, the income farmers receive from corn does not really explain why this group of agrarians reacts to such financial pressure by choosing to innovate and become even more high tech. Such a reaction is not, as Bunge and most observers of agriculture assume, inevitably directed by material conditions farmers face. This reaction to prices only makes sense within a globalized ultramodern identity that farmers have formed over many years of performative use.

Other contemporary reports have documented technology use consistent with a rural globalized ultramodern identity. Many of these pieces also show an urban misunderstanding of the ultramodern way farmers think of themselves reminiscent of Bunge's *Wall Street Journal* article. Reports regarding the unveiling of the prototype of the Case IH self-driving tractors at the 2016 Farm Progress Show in Boone, Iowa, varied considerably between publications intended for urban and rural audiences. The *Wall Street Journal*, for instance, described the only advantage of the machine as, "An autonomous tractor could theoretically run around the clock." Rather than outlining other features of the machine, the article concentrated on the threat it posed to jobs and the dangers of running into objects such as houses or dogs.[44]

Kelly McSweeney, robotics reporter for the global technology magazine *ZDNet*, with its U.S. office in San Francisco, California, lamented that "autonomous tractors could turn farming into a desk job" and highlighted the potential of the tractor to "steal jobs from human workers."[45] Chicago-based reporter Mario Parker, writing for *Bloomberg* headquartered in New York City, also critiqued that the Autonomous Concept Vehicle "features everything but the farmer."[46] In addition to their urban locations, none of these publications consider farmers their primary audience or even a significant portion of their readership. *ZDNet*, for example targets "IT professionals and decision makers" in Silicon Valley and similar information technology hubs while *Bloomberg* and the *Wall Street Journal* write for financial industry professionals based in New York City and other financial market centers.[47]

In contrast, reports of the self-driving tractor prototype from farm journals or other publications based in the Corn Belt reflect a rural ultramodern discourse. More specifically, these rural reporters viewed the autonomous tractor more positively as bringing progress to the farm and described its functions in much greater technical depth. Further, these rural reports acknowledged how the self-driving tractor prototype would fit into a larger system of autonomous and precision technologies employed by the ultramodern farmer, such as remote sensing satellites. For example, an article posted by the *Nebraska Rural Radio Association* described the possible uses for the autonomous tractor as "Better use of labor, integration into current machinery fleets, plus the flexibility to work unmanned around the clock with real time data monitoring – and, in the future, the ability to automatically respond to weather events." Rather than characterizing the prototype as a threat to labor, the article contended that it would "help farmers and agribusinesses sustainably boost production and productivity at these times, through the ability to make the most of ideal soil and weather conditions, as well as available labor." The author embraced the coming of "a future autonomous era" because technologies like the self-driving tractor left the farmer in control of production and devices remotely. The article went onto describe the device in great technical detail, including the interface and how the farmer would program and operate the tractor. The ultramodern agrarian would farm by controlling a multitude of automated machines and precision technologies through several computer screens displaying the devices from multiple angles as well as "big data" tracking atmospheric or soil conditions.

The devices address the safety concerns of the tractors running into buildings or animals through "a complete sensing and perception package, which includes radar, LiDAR (range finding lasers) and video cameras to ensure obstacles or obstructions in the tractor's path or that of the implement are detected and avoided." The farmer would read the images from the cameras and monitor the laser data remotely through a tablet interface. In many cases, the farmer would program the tractors to operate with existing satellites to operate automatically, without human input, to take advantage of ideal conditions. For example, "on private roads, they can be sent to another field destination where conditions are better – soils are lighter or there has been no rain."[48]

Other farm journals and rural-based publications reported the autonomous tractor prototype in similar technical detail and in discourse that embraced the association of the device with progress. *CropLife*, headquartered in rural Ohio, for instance, reported excitement that the machines left "the Farm Progress Show crowd buzzing."[49] *Farms.com*, based in Ames, Iowa, stated that farmers "flocked" to see the "futuristic" new machines described as "An interesting new concept that illustrates some of the new technologies being incorporated into the farm machinery of the future."[50] In *Farmersadvance.com*, published in rural Michigan, the author described the unveiling of the autonomous tractor in Boone, Iowa,

Once the first images of the video dedicated to the tractor (https://youtu.be/boOzbF6pkQ8) and its incredible operational capacities appeared on the screen, everyone understood that they were witnessing a glimpse into the future of farming, one that could feature fully autonomous machinery: something which could redefine the agriculture of tomorrow.[51]

A similar article in *Farm and Diary*, published in rural Ohio, stated that the autonomous tractor

Enables farmers to access tractor and implement data, wherever they are, from different locations, while checking fields from the comfort of their pick-up, while tending livestock or from home, and always whenever they need. This facilitates right-time decision making to enhance operational efficiency and productivity.

Again, the article emphasized that farmers maintained control and ownership of their data. While describing the tractor in great technical detail, the article noted that it

Follows optimized in-field paths, which are automatically generated by the software... In the future this concept will be able to utilise [sic] previously collected yield data for the variable application of inputs and to carry out operations with maximum precision, year after year.

Finally, the *Farm and Dairy* author noted that the self-driving tractor "is able to work alongside other autonomous machines and can also work in tandem with machines driven by an operator."[52] All of the farm journals and rural reports maintained that the autonomous tractor ensured "maximum flexibility, efficiency and sustainability" as well as to increase "the ability to make the most of short operating windows."

None of these rural articles expressed concern with the displacement of the farmer or rural labor from the actual driving of the tractor. Rather, rural authors saw the idea of the Corn Belt agrarian in control of an entire system of advanced automated and precision technologies from remote interfaces as enhancing the ultramodern identity that rural residents have developed. The concern of urban reporters that eliminating the farmer from sitting on the tractor somehow makes the task "fake" farming fails to appreciate the ultramodern self-image farmers have of themselves. As with urban portrayals of farm machinery in *Green Acres* in the 1960s as old fashioned, these urban reporters tend to bring their own ideas of Jeffersonian agrarianism to their views of the self-driving Case IH. Urban observers, not Corn Belt agrarians, hold a dominant image of farming as involving a farmer sitting on a tractor.

Rural Midwesterners themselves embrace the autonomous tractor prototype much more because using it allows them to perform their identities as ultramodern. Farmers do not define themselves just according to whether they drive a tractor but, rather, as users of the most advanced and newest technologies to increase the productivity of their private property. The farmer has exceeded the modernity of his urban cousins, few of whom program or control complex systems of precision and automated technologies or monitor and manipulate big data. Further, the authors writing for rural audiences view the self-driving tractor as a moral device because it promises to increase production and allows the heroic farmer to meet the growing threat of global food shortages. Rather than seeing the automated tractor as leading to moral dilemmas such as the displacement of labor and turning farming into "a desk job," the ultramodern agrarian views the use of the machine as promoting virtuous results such as increased productivity and greater control over property and material objects.

British global business correspondent, Nathan King displayed a similar misunderstanding of rural globalized ultramodernity when reported from a 2016 farm equipment show in Des Moines, Iowa. King declares with surprise, "If it goes in the ground and grows, there's now an App for that." The Englishman interviewed Scott Meldrum, a sales representative and designer with John Deere from the inside of the cab of the tractor which King mocked as a machine that looked more like something from a racetrack than a corn field. Meldrum explained, pointing to a computer mounted on the tractor console, "All the tractor health information will come through here, as well as auto-steer functions, documentation functions, and product application functions." King then discussed drone technology with a representative from Crop Copter used to spot crop wind damage from the air and employing infrared technology to determine plant health. Again, King interviewed the sales representative rather than any farmers at the convention.

Throughout the report, King seems to have a condescending tone as if he wants the viewer to ask, "Is this all necessary?" a sarcasm lost on the sales representatives at the farm show who clearly have, from King's perspective, an impure financial interest. Of course, King is ultimately right in noticing that the technology seems excessive to an outside observer, but he interprets the significance of this hunch the wrong way. Namely, his own analysis concluded that the "small family farmer will find it tough to keep up with bigger, more industrial farms, when it comes to equipment like this" without interviewing a single farmer at the show. King even went as far as reading Meldrum the new tractor's price tag of $641, 240.85 asking, "That's a lot of money?!" Meldrum responded, "Yes it is a pretty steep number. But when you look at it it's gonna be there to get your job done." Despite their ultramodern performance, an outsider like King still viewed the tractor companies as pushing small family farmers unwillingly towards industrial agriculture. It never occurred to King that farmers may have already established the kind of large ultramodern businesses that could allow them to use such machines.

On the other hand, it rarely occurs to Meldrum that the family farmer *will not* use his product. As such, Meldrum and King talk past one another throughout the piece. As with other urban observers of rural life a century ago, King made no effort to understand the production ethos or rural discourses or identities of the farmers themselves. King makes the erroneous assumptions that "all this data, when used properly, is designed to do one thing: boost yields of Iowa's two most important crops, corn and soybeans." Once again, King's report reflects an assumption that technological change is solely driven by economic concerns. Further King commits the common mistake of focusing only on the articulated benefits of technology use (a significant theme of this book). Surely, from King's perspective, country yokels could have no voice in the process of modernization regardless of how many apps or computer functions they have in their tractors or combines. In contrast, all that Meldrum had to say from his ultramodern rural perspective to moralize the steep price tag on the new tractor was "It will get your job done," an agrarian virtue lost on King who sarcastically concludes "Farms of the future, large and small, will help feed a growing planet, one App at a time."[53] In addition to "feeding a growing world," the farmer clearly has even more work to do to present a convincing performance through technological use, at least from a rural perspective.

King's view of technology use by farmers in the Corn Belt contrasts sharply with a statement released by the Iowa Pork Producers Association (the "Association") in 2016 encouraging the public to celebrate "National Agriculture Day... to reflect and be grateful for those who grow and raise our abundant and safe food supply."[54] The Association's announcement of the event included a "manifesto" of rural globalized ultramodernity. First, the declaration framed technology in heroic terms as allowing farmers to feed a growing world population. This practice (of viewing the farmer as a hero overcoming difficulty) represents a persistent feature of Jeffersonian agrarianism.[55] The rural hero now takes on an updated form with globalized and technological dimensions. For example, the statement urged the public to "thank a farmer" because

> Thousands of hardworking Iowa farm families work diligently every day to bring you the safest, most wholesome and affordable food found anywhere in the world. Each American farmer now feeds nearly 150 people in the U.S. and abroad, up from 25 people in the 1960s.

This increased productivity renders the farm heroic because "U.S. agriculture has evolved and is embracing new and emerging technologies in an effort to be as productive and efficient as possible to meet the nation's and world's growing food demands."

Second, the Association viewed the farmers' use of the most contemporary technologies within vast networks as a point of pride still unappreciated by the urban "other." The statement sought to define the farmer as a modern user of technology rather than a small yeoman rube by extoling farmers as bringing about progress. In a paragraph that summarizes the ultramodern view of progress, the Association concluded the piece,

> Farming has come a long ways since fields were plowed with a couple of horses and livestock suffered through the extreme elements of Mother Nature. Your

food still comes from the farm, but it's raised by hardworking men and women with greater efficiency thanks to decades of technological improvements. Quite simply, Iowa farmers are doing more – and doing it better!

One senses that the authors of the announcement still suspect that the urban public views the farmer as Eb Dawson on *Green Acres* instead of as a "born innovator." The Association emphasized that

Many pork producers and other farmers today have college degrees, they seek out and receive continuing education, they maintain business plans, keep elaborate records, and use the Internet and smartphones to help run their farming enterprises. This is today's farmer!

Third, the Association moralized Corn Belt farmers' use of large and expensive technologies within socio-technical networks by associating it with the family farm and productive work, two virtues prized by both Jeffersonian and German agrarianism. Not only does the statement aim to distinguish the ultramodern farmer from the anti-modern bumpkin; it seeks to separate the moral family farm from the category of immoral industrialized agriculture:

You may have heard some people refer to it as "big ag," "factory farms" or "industrialized farming." These are some of the favorite phrases people opposed to modern farming like to use.

A family farmer who raises 10,000 pigs and 500 beef cows a year, harvests 3,000 acres of corn and soybeans, uses computer and satellite technology, maintains a grain mill, owns a million dollars' worth of modern farm implements and has his own trucks to haul livestock and grain may sound like "big ag," but it's simply life on the farm today. Farming today is not done by huge corporations. It's done by real people, families who have deep roots in agriculture and wouldn't do anything else for a living.[56]

The Association's declaration thus articulates the rural globalized ultramodern identity to promote it to an urban public. This performance of an ultramodern identity both to convince an outside "other" of farmers' morality and to reinforce a rural sense of self repeats itself as a common cultural practice in the Corn Belt. For example, Missouri farmer and *Wallaces Farmer* journalist Mindy Ward reported on a Dow Chemical display at the Iowa State Fair which asked fairgoers "how do you measure perfect?" Apparently, visitors in the Dow tent encountered buttons that they would press to vote for the answer to this question among three candidates for "How do you measure perfect?: 1. Working with family. 2. Caring for the land. 3. Feeding the world."

Of course, Ward concluded that what made the Dow display so appealing was that it led the person taking the "test" to conclude that all three items "make farming the perfect vocation." Ward's description of the importance of these three items reiterates the Associations statement defining an ultramodern identity. Ward recalled,

> My eldest daughter would say that 'working with family,' is what makes being involved in the agriculture industry perfect. From our hot summers of putting up fence to cold winters in the lambing barn, it was all about being around family…. For her, agriculture is family.

Ward then stated,

> For my husband perfection comes from "caring for the land." Our operation may be small, but he is constantly looking to improve it. Whether it is rotationally grazing our sheep flock so not to overgraze our pastures or planting cover crops to improve soil health, he is always looking for ways to have our livestock work in harmony with our land. After all, like even the largest farmer, there is a desire to leave the land better than you started for the next generation.

While Ward's discourse on the land exhibited a greater concern with sustainability than the statement by the Association, they both shared the basic sentiment in the valuing of personal property and viewing progress, or "improving the land," as a virtue representing an updated version of German agrarianism. Further, Ward regards her husband's sustainable practices as fitting into rural ultramodernity (unlike the discourse of outside sustainable advocates discussed in Chapter 8) because he independently chose to implement these work and management practices on his own for the perceived benefit of the family (i.e. "the next generation"). Finally, Ward presented an updated Jeffersonian notion of the farmer as a hero feeding a growing world population by using advanced technologies in ways that debunk urban rube stereotypes about rural America,

> The youngest member of the family would definitely say that achieving perfection in the agriculture industry means, 'feeding the world.' With her mind set on breeding the next generation of seed and a college bill to back that up, our daughter has pursued her only dream- -helping to feed a hungry world through plant biotechnology.

Ward concluded "the measure of a perfect livelihood comes from working with family, caring for the land and feeding the world. Spot on, Dow. Spot on."[57]

Importantly, Ward's article demonstrates that this rural globalized ultramodern identity resides not only on the institutional level among "agribusiness" companies or organizations but also as a deeply imbedded rural consciousness among farmers and their families. Both Dow Chemical, through their sign, and Ward's family, through using technology, intended to signal this ultramodernity and its morality to other farmers and, more significantly, to an outside observer.

In addition, Ward had another more subtle intention in writing her article. Namely, by personalizing the Dow sign, she intended to place "the family" back at the center of ultramodernity rather than Dow Chemical, an undeniable symbol of a less moral industrialized agriculture. While Ward agreed with Dow's assessment of what makes farming the "perfect profession," she aimed to present the company as only the spokesman for a moral identity originating within farm families. As with farm journal advertisers in the 1920s, Dow Chemical knows that by reinforcing farmers' identities as ultramodern and moral, they promote the use of Dow's chemicals and other technologies without even mentioning the company's available products. The rural practice of performative use completes Dow's sales job for them.

The Association's statement and Ward's article reflect how farmers use technology to form a new sense of self that incorporates elements of prior rural identities and frames the farmer as virtuous. To more fully understand why the Association and Ward exert such great effort to market farmers' ultramodern identities and why part of the Association's sales pitch involves drawing a distinction between the family farm and industrialized agriculture, one must explore how their discourse functions in a broader social context. Both the past rural-urban conflict and the current perceived struggle between traditional farmers and the organic and sustainable foods movement call on farmers to defend the morality of mainstream agriculture. This clash over rural modern and organic discourses, according to Corn Belt agrarians, also presents an updated urban effort to portray farming as unfeminine much as urban industrialism did in the 1920s. In both eras, rural women resented perceived urban efforts to characterize them as yokels and construct rural production as a purely masculine domain. In addition, Corn Belt farmers react to new technologies both positively and negatively according to how devices and systems fit into rural identities. The following chapter will discuss persistent rural resentment of yokel stereotypes from the Cold War to the present and how this pattern of audience continues to drive performative use of technology.

Notes

1 Part of this chapter is published in the article, Brinkman and Hirsh, "Welcoming Wind Turbines and the PIMBY ('Please in my backyard') Phenomenon."
2 "Angus: the Business Breed," *American Angus Association* advertisement, *Successful Farming* (January 1992): 36–37.
3 Jared Schrage, *Schrage Corn Harvest (11–9–2014),* https://www.youtube.com/watch?v=J3yZzlpRGFM (accessed 6/20/16).
4 See for example T.K. Farms, *T K FARMS DARKE CO. OHIO ,2014 Corn Harvest, Job of the Grain Cart Operator,* November 1, 2014, https://www.youtube.com/watch?v=8c6bng93j2U (accessed 6/20/16); Devon Murray, 2015 *Iowa Farming-Murray Farms, Inc.,* January 29, 2016, https://www.youtube.com/watch?v=lxJAUhfn3Ew (accessed 6/20/16); The Burbank Blues, *Final Corn Harvest Aerials,* November 15, 2014, https://www.youtube.com/watch?v=givo2lUxBaw (accessed 6/20/16); Schrage, Jared, *Jensen Grain Farms Harvesting Corn in Early Nov. Snow,* November 16, 2014, https://www.youtube.com/watch?v=-PN5E9zygEM (accessed 6/20/16); Agrimap

Services, *Boyd Grain Farms Fall 2014*, October 2, 2014, https://www.youtube.com/watch?v=0pXG3jfXy0U (accessed 6/20/16).

5 Benjamin R. Cohen, *Notes from the Ground* (New Haven, CT: Yale University Press, 2009), 34–35.

6 Wall, *Iowa*, 133–134.

7 Ibid., 135.

8 Ibid., 134.

9 Keith Schneider, "New Product on Farms in Midwest: Hunger," B24.

10 Andrew H. Malcolm, "Problems on Farms Take Their Toll on Family Life," A17.

11 Schneider, B24; Malcolm, A17.

12 Kendall Crawford, "Suicide Rates Are Higher Among Farmers. Some Midwest States Are Teaching Communities How to Help," *Iowa Public Radio*, September 22, 2022, https://www.iowapublicradio.org/health/2022-09-23/suicide-rates-are-higher-among-farmers-some-midwest-states-are-teaching-communities-how-to-help (accessed 10/10/23).

13 See also Brinkman and Hirsh, "The Effect of Unarticulated Identities and Values on Energy Policy."

14 Rodney J. Fee, "Production," *Successful Farming* (February 1992): 23.

15 See for example JoAnn Wilcox, "If There's a Rattle and a Sputter, Aren't You Going to Fix It?" *Successful Farming* (May–June 2000): 33.

16 Wilcox, "Every High-Tech Thing I Know I Learned in Farmer Kindergarten," *Successful Farming* (December 2000): 24.

17 Ibid.

18 The view that the pressures of a rational and reductionist system compel farmers to adopt increased mechanization and scientific practices is discussed extensively by Deborah Fitzgerald. Fitzgerald, *Every Farm a Factory*, Introduction.

19 Thomas Hughes, "Technological Momentum," 101–113; Hughes, "The Evolution of Large Technological Systems," in *The Social Construction of Technological Systems: New Directions in the Sociology and History of Technology*, ed. Wiebe E. Bijker, Thomas P. Hughes, and Trevor J. Pinch (Cambridge, MA: MIT Press, 1994), 51–82.

20 Dave Mowitz, "Production," *Successful Farming* (January 1992): 27.

21 Larry Reichenberger, "Technology Is Trickling Down Into Every Aspect of Agriculture," *Successful Farming* (September 2000): 39.

22 Rodney J. Fee, "Successful Family Farm: Capitol Hill's Country Crusaders." *Successful Farming* (February 1992): 42–44.

23 Ibid., 43.

24 Bill Eftink, "Missouri Family Bets on Irrigation: Irrigation Boosts Corn Yields by 30 to 40 Bu an Acre," *Successful Farming* 98, no. 7 (May–June 2000): 62–63, 65.

25 Dave Mowitz, "Super Drills: Farmer Ingenuity Creates New Generation of High Yield Planters," *Successful Farming* (January 1992): 42–43. See also Rich Fee, "Long-Armed Sprayers: With a Little Ingenuity and a Lot of Built-In Strength, these Sprayers Reach Out and Touch a Bunch of Acres," *Successful Farming* 90, no. 2 (February 1992): 26–27.

26 See for example "Sensor on Center Pivots Offer Site Specific Irrigation," *Successful Farming* (December 2000): Production, 36; Lisa Foust Prater, "Pioneers Blaze Trail Through Digital Divide: South Dakotans Go On-Line to Learn, Do Business, Keep in Touch," *Agricultre.com* (Fall 2000): 18–21; Dan Looker, "Click and Contract: Online Marketing gets Serious," *Agriculture.com* (Fall 2000): 10–11.

27 Larry Reichenberger, "Precision Pastures: GPS-Equipped Cattle Map Grazing Patterns," *Successful Farming* (May-June 2000): Special Bonus Page, unnumbered page 1 in between numbered pages 44 and 45; Mike Holmberg, "Sky-High Scouting: From Ultralights to Satellites, You have More Opportunities to Get a Bird's Eye View of Your Crops," *Successful Farming* (May–June 2000): 34–36, 38.

28 Larry Reichenberger, "Precision Pastures," Special Bonus Page, unnumbered page 1 in between numbered pages 44 and 45.

29 Holmberg, "Sky-High Scouting."

30 Chester Peterson, Jr., "Milking the Net: Dairy Producer Bases Business Decisions on Internet Info," *Agriculture.com* (Fall 2000): 36; Peterson, Jr., "Simply Incredible!: Cows Accept the Challenge, Milk Goes Up 8,000 Pounds," *Successful Farming* (April 2000): 26–27.

31 Peterson, Jr., "Milking the Net."

32 Peterson Jr., "Simply incredible!" 26–29.

33 Peterson, Jr., "Cats in Corn: The Crane Brothers Tapped the Advantages of Cat Tractors by Converting One to Run Between Rows," *Successful Farming* (January 1992): page 40 in unnumbered pages in between numbered pages 38 and 40.

34 John Walter, "People Pages: He Blends Computers and Cows," *Agriculture.com* (Fall 2000): 16.

35 "Agriclick.com advertisement," *Successful Farming* (April 2000): 53.

36 "Aventis Corps advertisement," *Successful Farming* (April 2000): 12–13.

37 "Council for Biotechnology advertisement," *Successful Farming* (May–June 2000): 5.

38 "Pioneer advertisement," *Successful Farming* (January 1992): 59.

39 Marc Spiegler, "Hot Media Buy: The Farm Report," *American Demographics* (October, 1995): 18–19.

40 Jacob Bunge, "Plowed Under: Supersized Family Farms are Gobbling Up American Agriculture," *The Wall Street Journal* (October 23, 2017).

41 Recall Peterson's description of this frontier myth feature of Jeffersonian agrarianism in Peterson, "Jefferson's Yeoman Farmer as Frontier Hero," 9–19.

42 Jacob Bunge, "Farmers Reap New Tools From Their Own High-Tech Tinkering," *The Wall Street Journal*, April 18, 2016, TECH, http://www.wsj.com/articles/farmers-reap-new-tools-from-high-tech-tinkering-1461004688 (accessed 6/1/16).

43 See for example Kline, *Consumers in the Country.*

44 Andrew Tangel, "Farm Show Visitors Marvel, Scoff at Self-Driving Tractor," *Wall Street Journal* (September 1, 2016), Business.

45 Kelly McSweeney, "Autonomous Tractors Could Turn Farming Into a Desk Job: CNH Industrial Revealed its Concept for a Self-Driving Tractor that Farmers Control Via Tablet or Computer. Naturally, We Had to Ask Whether this Robotic Farmer would Steal Jobs from Human Workers," *ZDNet*, September 2, 2016, Robotics, http://www.zdnet.com/article/autonomous-tractors-could-turn-farming-into-a-desk-job (accessed 9/14/16).

46 Mario Parker, "Tractor for Modern Farm Features Everything But the Farmer," *Bloomberg*, September 1, 2016, Technology, http://www.bloomberg.com/news/articles/2016-09-01/robot-tractor-draws-crowds-on-debut-at-iowa-farm-industry-show (accessed 9/14/16).

47 "About Us," *ZDNet*, http://www.zdnet.com/about (accessed 9/15/16).

48 "CNH Industrial Brands Reveal Concept Autonomous Tractor," *Nebraska Rural Radio Association-KTIC Radio*, August 30, 2016, News-Agricultural News, http://kticradio.com/agricultural/cnh-industrial-brands-reveal-concept-autonomous-tractor (accessed 9/15/16).

49 Matthew J. Grassi, "New Case IH Autonomous Tractor Concept Leaves Farm Progress Show Crowd Buzzing," *CropLife*, September 1, 2016, http://www.croplife.com/equipment/new-case-ih-autonomous-tractor-concept-leaves-farm-progress-show-buzzing (accessed 9/15/16).

50 Joe Dales, "Farm Progress Show 2016 Highlights and New Products," *Farms.com*, September 4, 2016, News, http://www.farms.com/ag-industry-news/farm-progress-show-2016-highlights-and-new-products-375.aspx (accessed 9/15/16).

51 "Autonomous Concept Tractor Shows a Vision into the Future of Ag," *Farmersadvance.com*, September 6, 2016, http://www.farmersadvance.com/story/news/2016/09/06/autonomous-concept-tractor-shows-vision-into-future-ag/89916340 (accessed 9/15/16).

52 "Driverless Tractors Unveiled at Farm Progress Show," *Farm and Dairy*, August 30, 2016, Other News, http://www.farmanddairy.com/news/driverless-tractors-unveiled-at-farm-progress-show/359165.html (accessed 9/15/16).
53 *Iowa Farms Use Drones and Data to Improve Crop Yields*, Nathan King (2016; Des Moines, IA: CCTV America's, 2016). https://www.youtube.com/watch?v=F2LIKr96pF4 (accessed 10/23/23).
54 "National Agriculture Week 2016: Thank a Farmer," *Iowa Pork Producers Association*, March 14, 2016, http://www.iowapork.org/national-agriculture-week-2016-thank-farmer/ (accessed 5/31/16).
55 Peterson, "Jefferson's Yeoman Farmer as Frontier Hero," 9–19.
56 "National Agriculture Week 2016."
57 Mindy Ward, "How Do You Measure Perfect?" *Wallaces Farmer*, September 9, 2016, Show-Me Life, http://farmprogress.com/blogs-how-measure-perfect-11319#authorBio (accessed 9/21/16).

Bibliography

"About Us." *ZDNet*. http://www.zdnet.com/about (accessed 9/15/16).
"Agriclick.com advertisement." *Successful Farming*, April 2000, 53.
Agrimap Services. *Boyd Grain Farms Fall 2014*, October 2, 2014. https://www.youtube.com/watch?v=0pXG3jfXy0U (accessed 6/20/16).
"Angus: The Business Breed." *American Angus Association* advertisement. *Successful Farming*, January 1992, 36–37.
"Autonomous Concept Tractor Shows a Vision into the Future of Ag." *Farmersadvance.com*, September 6, 2016. http://www.farmersadvance.com/story/news/2016/09/06/autonomous-concept-tractor-shows-vision-into-future-ag/89916340 (accessed 9/15/16).
"Aventis Crops advertisement." *Successful Farming*, April 2000, 12–13.
Brinkman, Joshua T. and Richard F. Hirsh. "Welcoming Wind Turbines and the PIMBY ('Please in My Backyard') Phenomenon: The Culture of the Machine in the Rural American Midwest." *Technology and Culture* 58, no. 2 (2017): 335–367.
———. "The Effect of Unarticulated Identities and Values on Energy Policy." In *The Handbook of Energy Transitions*, edited by Katherine Araújo, 71–85. London: Routledge, 2022.
Bunge, Jacob. "Farmers Reap New Tools From Their Own High-Tech Tinkering." *The Wall Street Journal*, April 18, 2016, TECH. http://www.wsj.com/articles/farmers-reap-new-tools-from-high-tech-tinkering-1461004688 (accessed 6/1/16).
———. "Plowed Under: Supersized Family Farms are Gobbling Up American Agriculture." *The Wall Street Journal*, October 23, 2017.
"CNH Industrial Brands Reveal Concept Autonomous Tractor." *Nebraska Rural Radio Association-KTIC Radio*, August 30, 2016, News-Agricultural News. http://kticradio.com/agricultural/cnh-industrial-brands-reveal-concept-autonomous-tractor (accessed 9/15/16).
Cohen, Benjamin R. *Notes from the Ground*. New Haven, CT: Yale University Press, 2009.
"Council for Biotechnology Advertisement." *Successful Farming*, May–June 2000, 5.
Crawford, Kendall. "Suicide Rates Are Higher Among Farmers. Some Midwest States Are Teaching Communities How to Help." *Iowa Public Radio*, September 22, 2022. https://www.iowapublicradio.org/health/2022-09-23/suicide-rates-are-higher-among-farmers-some-midwest-states-are-teaching-communities-how-to-help (accessed 10/10/23).

Dales, Joe. "Farm Progress Show 2016 Highlights and New Products." *Farms.com*, September 4, 2016, News. http://www.farms.com/ag-industry-news/farm-progress-show-2016-highlights-and-new-products-375.aspx (accessed 9/15/16).

"Driverless Tractors Unveiled at Farm Progress Show." *Farm and Dairy*, August 30, 2016, Other News. http://www.farmanddairy.com/news/driverless-tractors-unveiled-at-farm-progress-show/359165.html (accessed 9/15/16).

Eftink, Bill. "Missouri Family Bets on Irrigation: Irrigation Boosts Corn Yields by 30 to 40 Bushels an Acre." *Successful Farming* 98, no. 7 (May-June 2000): 62–63, 65.

Fee, Rich. "Long-Armed Sprayers: With a Little Ingenuity and a Lot of Built-In Strength, These Sprayers Reach Out and Touch a Bunch of Acres." *Successful Farming* 90, no. 2 (February 1992): 26–27.

Fee, Rodney J. "Production." *Successful Farming* (February 1992a): 23.

———. "Successful Family Farm: Capitol Hill's Country Crusaders." *Successful Farming* (February 1992b): 42–44.

Fitzgerald, Deborah. *Every Farm a Factory: The Industrial Era in American Agriculture*. New Haven, CT: Yale University Press, 2003.

Grassi, Matthew J. "New Case IH Autonomous Tractor Concept Leaves Farm Progress Show Crowd Buzzing." *CropLife*, September 1, 2016. http://www.croplife.com/equipment/new-case-ih-autonomous-tractor-concept-leaves-farm-progress-show-buzzing (accessed 9/15/16).

Holmberg, Mike. "Sky-High Scouting: From Ultralights to Satellites, You Have More Opportunities to Get a Bird's Eye View of Your Crops." *Successful Farming* (May–June 2000): 34–36, 38.

Hughes, Thomas P. "Technological Momentum." In *Does Technology Drive History? The Dilemma of Technological Determinism*, edited by Merritt Roe Smith and Leo Marx, 101–113. Cambridge, MA: MIT Press, 1994.

Iowa Farms Use Drones and Data to Improve Crop Yields. Nathan King. 2016; Des Moines, IA: CCTV America's, 2016. https://www.youtube.com/watch?v=F2LIKr96pF4 (accessed 10/23/23).

Kline, Ronald R. *Consumers in the Country: Technology and Social Change in Rural America*. Baltimore, MD: Johns Hopkins University Press, 2000.

Looker, Dan. "Click and Contract: Online Marketing gets Serious." *Agriculture.com*, Fall 2000, 10–11.

Malcolm, Andrew H. "Problems on Farms Take Toll on Family Life." *New York Times*, November 20, 1984, A1, A17.

McSweeney, Kelly "Autonomous Tractors Could Turn Farming into a Desk Job: CNH Industrial Revealed its Concept for a Self-Driving Tractor that Farmers Control via Tablet or Computer. Naturally, We Had to Ask Whether this Robotic Farmer Would Steal Jobs from Human Workers." *ZDNet*, September 2, 2016, Robotics. http://www.zdnet.com/article/autonomous-tractors-could-turn-farming-into-a-desk-job (accessed 9/14/16).

Mowitz, Dave. "Production." *Successful Farming* (January 1992a): 27.

———. "Super Drills: Farmer Ingenuity Creates New Generation of High Yield Planters." *Successful Farming* (January 1992b): 42–43.

Murray, Devon. 2015 *Iowa Farming-Murray Farms, Inc.*, January 29, 2016. https://www.youtube.com/watch?v=lxJAUhfn3Ew (accessed 6/20/16).

"National Agriculture Week 2016: Thank a Farmer." *Iowa Pork Producers Association*, March 14, 2016. http://www.iowapork.org/national-agriculture-week-2016-thank-farmer/ (accessed 5/31/16).

Parker, Mario. "Tractor for Modern Farm Features Everything But the Farmer." *Bloomberg*, September 1, 2016, Technology. http://www.bloomberg.com/news/articles/2016-09-01/robot-tractor-draws-crowds-on-debut-at-iowa-farm-industry-show (accessed 9/14/16).

Peterson, Tarla Rai. "Jefferson's Yeoman Farmer as Frontier Hero: A Self Defeating Mythic Structure." *Agriculture and Human Values* 7, no. 1 (1999): 9–19.

Peterson, Jr., Chester. "Cats in Corn: The Crane Brothers Tapped the Advantages of Cat Tractors by Converting One to Run Between Rows." *Successful Farming* (January 1992): page 40 in unnumbered pages in between numbered pages 38 and 40.

———. "Milking the Net: Dairy Producer Bases Business Decisions on Internet Info." *Agriculture.com* (Fall 2000): 36.

———. "Simply Incredible!: Cows Accept the Challenge, Milk Goes Up 8,000 Pounds." *Successful Farming* (April 2000): 26–29.

Pioneer advertisement. *Successful Farming,* January 1992, 59.

Prater, Lisa Foust. "Pioneers Blaze Trail Through Digital Divide: South Dakotans Go On-Line to Learn, do Business, Keep in Touch." *Agricultre.com* (Fall 2000): 18–21.

Reichenberger, Larry. "Precision Pastures: GPS-Equipped Cattle Map Grazing Patterns." *Successful Farming* (May-June 2000): Special Bonus Page, unnumbered page 1 in between numbered pages 44 and 45.

———. "Technology Is Trickling Down Into Every Aspect of Agriculture." *Successful Farming,* (September 2000): 39.

Schneider, Keith. "New Product on Farms in Midwest: Hunger." *New York Times*, September 29, 1987, A1, B24.

Schrage Corn Harvest (11–9–2014). https://www.youtube.com/watch?v=J3yZzlpRGFM (accessed 6/20/16).

Schrage, Jared. *Jensen Grain Farms Harvesting Corn in Early Nov. Snow,* November 16, 2014. https://www.youtube.com/watch?v=-PN5E9zygEM (accessed 6/20/16).

"Sensor on Center Pivots Offer Site Specific Irrigation." *Successful Farming,* December 2000, Production, 36.

Spiegler, Marc. "Hot Media Buy: The Farm Report." *American Demographics*, October, 1995, 18–19.

Tangel, Andrew. "Farm Show Visitors Marvel, Scoff at Self-Driving Tractor." *Wall Street Journal*, September 1, 2016, Business.

The Burbank Blues. *Final Corn Harvest Aerials,* November 15, 2014. https://www.youtube.com/watch?v=givo2lUxBaw (accessed 6/20/16).

T K Farms. *T K FARMS DARKE CO. OHIO ,2014 Corn Harvest, Job of the Grain Cart Operator,* November 1, 2014. https://www.youtube.com/watch?v=8c6bng93j2U (accessed 6/20/16).

Wall, Joseph Frazier. *Iowa: A Bicentennial History*. New York: W.W. Norton & Company, 1978.

Walter, John. "People Pages: He Blends Computers and Cows." *Agriculture.com* (Fall 2000): 16.

Ward, Mindy. "How Do You Measure Perfect?" *Wallaces Farmer,* September 9, 2016, Show-Me Life. http://farmprogress.com/blogs-how-measure-perfect-11319#authorBio (accessed 9/21/16).

Wilcox, JoAnn. "Every High-Tech Thing I Know I Learned in Farmer Kindergarten." *Successful Farming* (December 2000): 24.

———. "If There's a Rattle and a Sputter, Aren't You Going to Fix It?" *Successful Farming* (May-June 2000): 33.

7

"THE DISTRICT OF HICKS"

Persistent Urban Views of Farmers as Backward

> That's my middle-west--not the wheat or the prairies or the lost Swede towns but
> the thrilling, returning trains of my youth and the street lamps and sleigh bells,
> in the frosty dark and the shadows of holly wreaths thrown by lighted windows
> on the snow.
>
> *F. Scott Fitzgerald*, The Great Gatsby, *1925*[1]

In my life as a professional saxophonist, I have played a wide range of performances, some of which have been much more glamorous than others. While sitting
down to edit this book, I received a call from a pianist for a gig I never could have
anticipated. The show was for a six-night production of the play *Copacabana* by
Barry Manilow at a large and respected local theater. The musical aired first on
television in 1985 in which Manilow starred and wrote many of the songs. James
Lipton (the New York actor and screenwriting also famous for the show *Inside the
Actors Studio*) wrote the story for the play.[2] Manilow himself was born in Brooklyn, New York and attended the prestigious Julliard School of Music in New York
City. He has spent his entire life working in show business and living in America's
largest metropolis.[3] It did not take me long to realize that the production prominently features the view urban New Yorkers have of the Midwest. For six nights,
I blared big band swing music on my saxophone while quietly observing rube stereotypes. The story of the entire play revolved around a naïve and wide-eyed young
woman named "Lola" (a name taken from Manilow's famous song "Copacabana")
who moves to New York City from Tulsa, Oklahoma in the late 1940s with dreams
of becoming an actress and showgirl. Much of the comic relief comes at the expense of the Midwest depicted as unworldly and unsophisticated. In one scene,
Lola auditions for a part by horribly singing a bawdy song leading to repeated

DOI: 10.4324/9781032637952-8

rejection. After her third attempt at impressing the cynical New York club manager and failing, Lola states, "Maybe I should try it again," at which the professional urban piano accompanist quips "Maybe you should try it in Cleveland sister!" The audience laughed at this joke on all six nights of the performance.

Lola finds herself lost in the big city until she is saved by a talented but struggling New York songwriter, Tony (also taken from Manilow's song "Copacabana") but not before he mocks her rural background. In the scene where Tony and Lola meet, she tells him that her tune was written by her music teacher back home, Mr. Schminckel, at which he jokes "Who's Mr. Schminckel, your wrestling coach!?" Tony, who is likely meant to represent a young Manilow himself, then proceeds to direct the piano accompanist to speed up the bawdy song at which point the big band (with me on sax) launches into an energetic swing tune. With the help of the worldly urban songwriter, the hayseed from Tulsa gets the job and, predictably, the two fall in love. In another scene, Lola's Midwest parents are portrayed as silly and bumbling. Her father, who comically wears a badly fitting toupee and frumpy old-fashioned clothing, is depicted as cheap and slovenly. He rebukes Lola's mother for tipping the cab driver in New York too much declaring, "Why do I work, why don't I just throw my money away!" The couple comically bickers in which Lola's mother accuses her father of scarfing down "three hot dogs and a knish." As a Midwest woman, Lola's mother appears bored and unhappy.

The music in the play also inserts urban stereotypes about the Midwest. In Lola's introductory song entitled "Just Arrived," she begins with the fermata "Just arrived track 17, all the way from Tulsa Okla" she then pauses, thinks a bit and instead of pronouncing the full word "Oklahoma" she sings "Okla-nowhere." Ashamed of coming from "nowhere," the hick vocalist expresses a desire to "take the town" and "leave my past behind." Exiting the train, Lola then proceeds to sing about all the ambitious, but still naïve, dreams she has asking, "Where's the chauffeured limousine? [chuckle] where's the marching band?" Throughout the song, the dreamer from Tulsa goes from fear of the big city to displaying a silly optimism beginning with, "Just in time to take the town, and though my knees a shaking, I'm making my stand" and concluding with "It all starts now, I can see my name in neon!" The song slowly builds in tempo until Lola declares, "Before too long, I'll be on some marque! New York or bust, I've just, arrived!" The chorus features Lola repeating the line "Just arrived track 17, all the way from Tulsa, Oklahoma" while other female vocalists sing "Just arrived track 24, all the way from St. Paul, Minnesota" while yet another singer answers "All the way from Fargo, North Dakota."[4] The entire song assumes that women from these Midwest towns dream of leaving their boring surroundings lacking high culture and coming to a more glamorous New York City to make it in show business. By the conclusion of the play, New York City itself had become the hero, the Midwest had been sufficiently humiliated, and I had gotten much better at sight-reading music.

The stereotypical urban depictions of the Midwest are not unique to Manilow's predictable play but exist as a common trope in the contemporary American culture

of both rural and city folk. The production very literally presents an urban performance of identity through mocking the rural and, for Manilow, relatively unknown
Midwesterner. Other urban-produced theatrical productions have presented this
stock character, originating in vaudeville and minstrel plays of the 19th century,
of a wide-eyed rural yokel seeking to make it in New York City or Hollywood but
woefully unprepared for his or her sophisticated surroundings. For example, in a
dance number of the 1952 film *Singin' in the Rain*, Gene Kelly plays a country
bumpkin who plays a fiddle and wears an outdated suit. He soon decides to try
his talents on Broadway. As a male version of Lola in Manilow's play, he dons
ill-fitting clothing and a comical hat with black-rimmed glasses and catches a train.
Upon arriving, Kelly dances through the urban station with a suitcase mesmerized
by the fast-paced city, only to be quickly rejected by several talent agents who regard him as a country yokel.[5] As with Manilow's play, New York City is depicted
as the most exciting place in America where everyone desires to live as opposed to
the provincial backwater of the country.

Thus, despite efforts by farmers in the Corn Belt to perform their identities as
modern through technological use from the early 20th century to the present, negative urban views about rural life have stubbornly persisted. In Chapter 2, I outlined
the rube stereotype in early 20th-century America and argued that this pejorative
view of Midwest farmers played a crucial social function during the rural-urban
conflict of the 1920s. To be sure, Midwest farmers have not faced the status of
"degenerate" reserved for residents of the rural South and Appalachia, who urban
Northerners have regarded as evolutionarily backward for over a century. Nevertheless, Corn Belt residents have not attained their desired level of sophistication in
the eyes of many urbanites.[6] In this chapter, I will explore these unfavorable urban
views of the rural Midwest after the rural-urban conflict of the 1920s and 1930s in
domains occupied by more serious thinkers than Oliver Wendell Douglas in *Green
Acres* (discussed in Chapter 5).

I contend that Jeffersonian agrarianism affects not only the *performers* in rural
America. Rather, I see the Jeffersonian yeoman myth as impacting *observers* of
rural life as well. Additionally, these urban discourses reflect resentment of farmers
as greedy and untrustworthy, reminiscent of how city-dwellers saw farmers during
the rural-urban conflict from 1910 to 1930. From 1940 to the present, urbanites still
tend to hurl rube stereotypes at the rural Midwest as a way of highlighting what
urban audiences see as unjust political or economic advantages given to farmers,
such as farm subsidies. The persistence of urban stereotypes in the face of efforts
by farmers to act out an ultramodern identity has intensified rural resentment. Rural Midwesterners since the 1930s have exhibited both a sense of inferiority and
an obsession with debunking urban stereotypes, a desire that continues to drive
performative technology use.

This persistent urban view of the Jeffersonian agrarian has expressed itself
in one of two ways since the rural-urban conflict of the 1920s. First, some urban actors have advanced the image of the Midwest "rube" re-packaged as a

rural dweller that is more moral than his Southern "hillbilly" cousin, but also less sophisticated – a "yokel," "hick," or "bumpkin" when compared to city dwellers. The second discourse, promoted by organic reformers and urban dwellers, sees the industrialized farmer as a dupe fooled by agribusiness and too unworldly to appreciate environmental concerns. Ironically, this second urban and organic stereotype sees pre-technological farmers much the same way a Jeffersonian would as the most moral type of farmer. Unlike rural modern identities, however, the second urban and organic view also regards the modern technological farmer as greedy, ignorant, and beholden to corporate interests. This image of technological farmers as naïve colors how rural people regard organic reformist identities and explains why organic discourses have alienated many farmers.

The Effect of the Yokel Stereotype on the Rural Psyche

Urban stereotypes of rural people in the decades after the rural-urban conflict of the 1920s and 1930s took a variety of rhetorical forms including rube, hick, yokel, bumpkin, and hayseed, all of which denoted an unsophisticated and unintelligent rural person whose lifestyle, from dress to work activities, suggested an unwillingness to embrace change and modern sensibilities. Interestingly, the word "bumpkin" arose from the 16th and 17th centuries word "bunkin" denoting a humorous short stumpy "Dutchman." The term "Dutchman" evolved in the U.S. by the later 18th century from originally meaning an inhabitant of the Netherlands to a racist slur for Germans. Thus, the term bumpkin may have originally functioned to make fun of German American farmers.[7]

The first printed use of the word "hick" to denote an ignorant or silly country person occurred in 1565 but it did not gain widespread use until the 1920s in the U.S. The word "hick" comes out of the first name "Richard," but it is less clear why this first name became associated with rural people.[8] In any event, the 1920s conception of rural people as inherently stupid, closed minded, and old-fashioned became the norm in the mind of city-dwellers. The depiction of rural people found in *Green Acres* serves as only a vivid visual expression of a common urban view after the early 20th century. Hollywood would rehash the basic *Green Acres* plot in which a displaced urban dweller would encounter yokel Midwesterners as late as the 2000s. Urban sitcom writers have continued to produce shows making fun of the Midwest as backward, such as *Ed* taking place in the fictitious Ohio town named "Stuckyville" to suggest that the Corn Belt is not modern but "stuck" in time.[9]

Indeed, as late as 1980, residents of Columbus, Ohio endured "endless wisecracks about cows grazing near the Ohio capital building." Residents of the city noted their town had a "bumpkin image" even though at that time Columbus topped both Atlanta, Georgia and Denver, Colorado in population by at least 50,000 people. The yokel stereotype bothered business leaders of Columbus so much that they formed the Central Ohio Economic Development Counsel to promote economic and cultural development. In spite of such efforts to make Columbus, Ohio more modern, *The Wall Street Journal* noted in their headline about the city, "Cultural Focus Still Football."[10] City dwellers, by contrast, must have occupied themselves with more sophisticated amusements.

Nor does the hick stereotype only occupy the minds of urbanites. One finds an obsession with combatting urban yokel assumptions among Midwesterners throughout the region. As one rural author writing for the *New York Times* in 1953 complained "the country bumpkin, who went to New York a generation ago and shortly acquired the Brooklyn Bridge for his very own was properly recognized as a stock character."[11]

The intensified image of the rube became so ubiquitous in urban discourses that by the middle of the 20th century, it found expression in a wide variety of cultural domains and even in daily interactions between rural and urban people. For city and country denizens, the persistent yokel stereotype resided in the most mundane events of daily life. For example, *The Los Angeles Times* often published minor stories that seemed to carry little significance except for perpetuating rube stereotypes about the rural Midwest. The paper reported in June of 1949 that the notorious criminal Sigmund Engle described the Minnesota women he was conning as "yokels" in an interview from his jail cell.[12] In another column in the late 1940s headlined "Bride Fleeing 'Hick' G.I. Gets Child Back," the paper told the story of a London woman who married an American soldier from rural Missouri during World War II but left him because she hated living in the "hick town" of Sikeston, Missouri which she described as having no running water. The urban paper reported with embarrassment that the London woman's "tale about her disillusionment with America were the big news for millions of newspaper readers in this metropolis [London] today," reflecting that urban people in Los Angeles may have been ashamed of the perception abroad of America as full of "hicks." The writer also related that the London woman had been reunited with her daughter after a long custody battle with her "hick" husband from Missouri, who is given little voice in the story.[13] While *The Los Angeles Times* likely published these columns in part because they presented interesting human-interest stories, they also focused on using such tales to confirm yokel stereotypes about the Midwest.

Other urban publications such as *The New Yorker* often employed the term rube and hayseed as late as the 1990s and 2000s to describe anyone from the rural Midwest or living in the Corn Belt regardless of their talents or occupations. These urban writers and commentators often speak of native Midwesterners as having some kind of inherent inferiority that cannot be cured even by moving to urban centers or succeeding in urban-based jobs. For urban observers, while the Midwest rube is not biologically degenerate like the Southern rural hillbilly, he or she certainly possesses an inborn stupidity, backwardness, and closed mindedness making him or her perpetually less sophisticated than native urbanites.

One of the most famous examples of a Midwest background deemed as conferring some type of inborn "rubeness" was the way urban sports reporters discussed the emergence of French Lick, Indiana native Larry Bird as a basketball star in the later 1970s and early 1980s. While Bird embraced the "hick from French Lick" moniker devised by urban reporters, city-based papers clearly used the Midwest rube stereotype to create more reader interest in Bird, particularly as a foil to his more urbane rival, Magic Johnson (who ironically was born in Michigan). One story about Bird from *The Los Angeles Times* noted urban notions that the basketball star was unintelligent and boring, and Bird often attempted to frame his rural Midwest background in positive terms in interviews to dispel yokel stereotypes held by urban reporters. Bird often tried to sell his Midwest background as a set of "small town values" such as hard work and thriftiness rather than as a cause for negative views about him or his hometown. But, for many of these urban observers, Bird's basketball skills and huge salary still did not dispel rube prejudices. The overriding theme of such stories about Bird in urban newspapers was, "how could such a hick from the Midwest be so good at a glamorous sport like basketball?"[14]

Similarly, New York writers have gone as far as assigning yokel stereotypes to admired actors starring in acclaimed Broadway plays as late as the 1990s, simply because they spent significant time in the Midwest as if some kind of inborn lack of sophistication has attached to them as a result of the tenure in the Corn Belt. For example, *The New Yorker* described actor Jarrod Emick as a rube and hayseed in 1994 when reviewing his performance as Joe Hardy in "Damned Yankees," a role for which he eventually won a Tony Award, simply because Emick was a product of the South Dakota State theater department.[15] Broadway plays themselves (as well as off-Broadway plays such as *Copacabana*) also continued to depict characters from the Midwest as unsophisticated and unintelligent hicks too backward to navigate city life well into the late 20th century notwithstanding the rapid modernization of American agriculture.[16] Urban discourses even employed yokel stereotypes about Midwest farmers in consumer reports that seem to have little to do with agriculture. For example, *Washington Post* columnist Brock Yates reported on the surprising popularity of pickup trucks among the "preppies" and "yuppies" of the nation's capital in the late 1980s, not just among "farmers" and "bumpkins" of the Midwest. Yates considered the pickup truck "a low-rent subject" and a "country Cadillac" unfit for "polite embassy-party conversation," and he associated Midwest farmers not with modern urban business owners but with what he regarded as lowly blue-collar workers such as plumbers. Yates writes "In the old days, your plumber drove a truck; today your boss might be heading to the mountains in a $20,000 pickup with six-speaker stereo, air conditioning and a plush interior," implying that only upper-class urban residents of Washington. D.C. could drive such luxurious and truly modern vehicles.[17] Surely, the bumpkin farmer and the plumber, Yates thought, would drive something junkier and more old-fashioned than this new breed or sophisticated urban buyers. Yates perpetuates this hick image even though by the late 1980s, most Midwest farmers operated combines with sophisticated electronics to maximize efficiency and monitor yields, hardware far more technologically complex than anything used by most Washingtonians.[18]

Even when not using terms such as "rube" and "yokel" by name, urban commentators from 1920 onward mocked the Midwest as backward and provincial. For example, *Wall Street Journal* staff reporter Jim Hyatt painted farmers in rural Ohio as bumpkins by choosing to highlight a tractor pull in 1971. One may glean the condescending tone of Hyatt's article by its title, "Summer in Sandusky is not the Dull Scene You City Boys Think: With Gaylord Zechman, Farmer, at the Wheel, Most Anything Can Happen-Up to 8 MPH." The remainder of Hyatt's piece depicts farmers as unsophisticated hicks who enjoy "racing" loud and dirty machines at low speeds in the mud and who are too unintelligent to appreciate the risks associated with dangerous equipment. After one tractor wheeled in the air and crushed one of the contestants, Hyatt mocked that the injured farmer "is philosophical about it: 'Somebody had to be a guinea pig, I guess.'" Hyatt opens his article imitating what, to him, is a less sophisticated dialect to frame life in the rural Midwest as painfully boring, "probably many's the night when you've sat around with a drink in your hand and idly wondered 'What do they do for entertainment in a place like Sandusky, Ohio?' Here's where you find out."

In addition, Hyatt inserted a subtle political slight at farmers in a section sarcastically entitled "Poverty Down on the Farm" in which he wrote, "Some of the poor men spend $15,000 on their machines."[19] In doing so, Hyatt sought to fan distrust of farmers among his urban audience who, since at least the beginning of the 20th century, had suspected agrarians of claiming poverty to achieve government support, including subsidies. Hyatt's imagery referenced the old rube stereotype of rural people as both unsophisticated and dishonest. The New York writer calls on the urban reader to be thankful that he or she does not live in such a backward place and to not trust farmers in the political sphere.

Urban stereotypes about rural America as old-fashioned, slow, and bucolic have even infiltrated American law in statutes and decisions ranging from nuisance, child custody, torts, criminal law, land use, and bankruptcy. For example, up until the late 1970s, many courts held rural doctors to a lower standard of care in medical malpractice suits than their urban counterparts because of a conception that the country offered fewer modern resources and learning opportunities for physicians. Courts have also given rural areas more nuisance and land use protections than urban areas on the assumption that rural people oppose development. Further, many criminal statutes and decisions apply less stringent legal standards to rural residents because of a Jeffersonian idyll positing that rural communities are close-knit and self-regulating. In addition, the Jeffersonian pastoral has operated to grant rural land greater protection, in eminent domain cases for example.

While not employing terms such as rube or yokel, many judges and legislators have nevertheless conceived of "rural" as a less modern lifestyle when they must define the term to reach a legal decision. For instance, when determining tax or zoning status, courts have found the selling of eggs, growing hay, hunting, fishing, chopping wood, or living far from paved roads as all persuasive of granting rural status.[20]

Further, the yokel stereotype from the rural perspective shows remarkable persistence notwithstanding recognition among urban people of the increasing technological sophistication of farmers. From the rural point of view, it is almost as if something is consistently blocking the central message of their performative use. Even when Midwest farmers held a convention in 1958 of the National Flying Farmers Association (NFFA) in New York City to discuss use of the latest airplanes on farms, for example, the *Christian Science Monitor* based there reported that "A couple of hundred 'hicks' from the sticks' who could tell New Yorkers a few things about fast streamlined living, have dropped into town." The farmers themselves enjoyed little success in promoting their modern image. "In an hour long press conference," the article states, "the board of directors and a handful of other members of the NFFA tried to communicate their enthusiasm and some of the facts of modern aviation to reporters of this sophisticated metropolis but only with partial success." Rather than giving the farmers a stage for performing their modern identities, the New York reporters asked condescending questions rooted in yokel stereotypes. "A bit incredulously," the reporter noted, "the newsmen kept coming back to the same questions: How could they pay for an airplane on a farm? How about landing fields, maintenance, and navigation? How did they find time for learning all that?"

The cartoon accompanying the article (Figure 7.1) mockingly featured a yokel farmer with a cowboy hat chewing on a piece of straw in an airplane accompanied by a cow headed to New York, hardly the image of a modern agrarian. The author of the article speculated that, "reporters snatched enough to see that life on the farm must have changed more than they thought in the last 10 years."[21] Yet, yokel stereotypes formed a significant aspect of urban discourses regarding the rural Midwest well after 1958. A decade later, Richard Orr writing for the *Chicago Tribune* still found it necessary to write an article entitled "Farmer Is No Country Bumpkin; He's an Affluent Business Man." Orr wrote, "The typical family farmer isn't a rube in bib overalls, a straw hat on his head, and a straw in his mouth, following a team of horses." The remainder of the article recounted evidence of the farmer's modernity including "a working knowledge of engineering, biology, chemistry, conservation, and assorted other specialties…" Orr also reasoned that his Chicago readers, even in 1968, had little idea that their farming neighbors "own automobiles, televisions, and other modern home appliances."[22]

The yokel stereotype asserts itself most directly when rural people address urban audiences or when recent urbanites discuss their childhood in the Midwest. Calvin Trillin, writer for *The New Yorker* who grew up in Kansas City, Missouri for example, identified in 1984 a "Midwestern ailment I have called rubophobia – not fear of rubes but fear of being thought a rube…"[23] The yokel stereotype of Corn Belt farmers had become so ingrained in American culture by the middle of the 20th century that rural residents of the Midwest had internalized it.

Many other contemporary commentators on life in the Corn Belt note what can only be described as an inferiority complex among its maligned denizens. Indeed,

FIGURE 7.1 The cartoon accompanying a 1958 article on farmers using airplanes on their farms. The drawing reflects persistent yokel stereotypes despite farmer's use of a technology not employed by most urban residents. Frederick W. Roevekamp, "Flying Farmers Save Cash," *Christian Science Monitor*, August 15, 1958, 3.

natives from the Corn Belt exhibit an exaggerated level of insecurity about their identities as Midwesterners. These writers expressed as a strange combination of a deep desire to combat yokel stereotypes and a tendency to perpetuate them at the same time. In recalling a childhood church trip from Iowa to Washington, D.C. in 1961, author Douglas Bauer, for example, exhibited an acute awareness of the rube type. Bauer, who's church group raised money to tour the nation's capital by selling homemade hamburgers at the Iowa State Fair, both adopted pejorative views of rural people as unsophisticated and sought to combat them when he perceived them in the minds of urbanites he encountered. For instance, Bauer recalled, "We reached the Capitol and followed our chaperones through hallways, on elevators, as the guides and operators condescendingly counted our number. It was as if the more there were of you, the higher the rube quotient of your group." While Bauer resented this condescension, he also held a deep desire to be accepted by the urban tour guide as more urbane, "I yearned to be, and to be able to be, alone in Washington. To be by yourself meant, by definition, that you knew your way, regularly walked there."

When Bauer's group reached the "Speaker's dining room" to have lunch with a congressman from Iowa, he recalled

Finally, this was the Washington I'd imagined as a I'd stirred the beefburger [*sic*] at the State Fair stand, the place of ornamental excess beyond the range of a plains imagination. I was sitting inside my fantasy timidly fingering its heavy utensils, when a large red-faced man entered the room, smiled and broadly waved." Bauer wondered, "Who was this?

and then wrote a description of the man that reveals a great deal about how yokel stereotypes had become integrated into the rural mind,

His grey hair was combed inelegantly straight back and shone from Brylcreem [*sic*]. And his forehead, in contrast with the rest of his pink complexion, was milk-white. The unmistakable markings of a Farmer's Tan. He was doubtlessly in the wrong room, though it was understandable why he'd be so obviously at ease with us. He was as identifiably a farmer as we were. But here he was, making his way along the table to its head. How fitting. I moaned: a farmer-tourist gone amok in the Speaker's Dining Room. We had to get him out of here before Congressman Simpson arrived.

Of course, the man turns out to be Congressman Simpson in Bauer's story.[24] On finding out the identity of the "farmer-tourist gone amok," Bauer declared

Impossible. Congressmen were of this place, of Washington. Congressman Simpson, therefore, would be a man of urbanity and grace. Of course, I would not have recognized urbanity or grace if Averell Harriman had walked into the room, but if I didn't know exactly what I was looking for in the personality of the congressman, I knew full well what I saw before me. I saw one of us...[25]

The whole experience of seeing "one of us" in a luxurious urban setting caused Bauer to confront his own rural identity based, in part, on an assumed dichotomy between the urban sophisticate and the rural rube,

Just as I thought I was beginning to learn the steps and recognize the fabrics, I'd be undermined by the congressman from the District of Hicks. His words came intermittently to the far end of the table on a flat South Iowa twang.

When Congressman Simpson left the room, Bauer noted with surprise "he strode out, smiling, taking big long awkward steps, shoulders dipping alternately, the barnyard gait I'd seen on the sidewalks of Prairie City every day of my life."[26]

Bauer's experience visiting the Capitol building demonstrates that farmers by the middle of the 20th century interpreted even the most mundane details of performance, such as the way someone walked, through the lens of urban yokel stereotypes about rural people. Paradoxically, Bauer's story also shows that Mid-westerners identified themselves as having the very stereotypically rural traits that they sought to debunk. As a result, the use of technology to show modernity is not just about performing for an external audience but also for oneself as a rural person. Such a continued working out of what Trillin called "rubophobia" explains why the performative use of new technology must be continuous and daily in the Corn Belt. Even though Bauer observed a congressman with a farmer's tan, a twang, and a "barnyard gait," he and his Iowa friends still concluded, "We were not, in our minds' experience, just a couple hours from Washington. We lived several cultures away..."[27]

As late as the 1990s and 2000s, writings produced by urban authors raised in the Corn Belt, such as Terry Teachout's *City Limits: Memoirs of a Small-Town Boy*, disparaged people they left back in the Midwest as yokels. Teachout openly wonders why anyone from New York City where he currently lived would ever want to go to the anti-modern rural Missouri where he grew up. In her review of the book, Jane Smiley interpreted Teachout's perpetuation of rube stereotypes as a means of constantly reaffirming the validity of his decision to move to the city, much like her grandfather justified his move to urban St. Louis by referring to people in his native rural Missouri as "hayseeds, farmers, hicks."[28]

Although rural people seem to internalize rube stereotypes, at other times they appear desperate to debunk them. Even Midwest authors admitted to "rural boos-terism," a term used in the middle of the 20th century for the act of bragging ex-cessively about one's town. Rural people in Corn Belt towns would even form community clubs to boost the town's image and economic status. As one Midwest author confessed, boosterism resulted from "inward insecurity" requiring "the sort of 'pep talks' a man gives himself when he is afraid or at least unsure of himself."[29] Trillin even admonished his own hometown's effort to dispel rube stereotypes through boosterism noting, "people are a lot more likely to consider you sophisti-cated once you quit going around saying how sophisticated you are."[30]

Many commentators noted the rural penchant for boosterism as further evidence of lack of intelligence and worldliness among farmers. An excerpt of a poem in a 1963 edition of the *New Yorker* by Jon Swan, a poet from originally rural Iowa, entitled "A Portable Gallery of Pastoral Animals" portrayed a boostering bumpkin as no more intelligent or refined than an obnoxious rooster,

> Or, often, in autumn, a pumpkin-
> Stands that typical bumpkin
> Turned booster,
> The ROOSTER.
> That alert bird
> Always has the first word,
> And says it excessively

By the dawn's sure light.
He's incurably bright,
Like a breakfast-club type or m.c.[31]

Other authors noted the rural obsession with debunking the Midwest rube image. One rural writer for the *Los Angeles Times* in 1952, for example, lamented "The common concept of the American farmer is the rube-the hick-the hayseed from the country who buys the Brooklyn Bridge" even though

> The average Farmer has a bigger capital investment than the average shoe retailer – and the big business of feeding 155 million Americans and their dogs calls for some pretty fancy cost accounting and more savvy than it takes to run a TV network.

The rural columnist went on to point out that "Most executive farmers, today, are college graduates." Despite evidence of modernity, the author complained,

> Many farm papers and magazines are edited for the vodvil [*sic*] stage concept of the farmer…and the farmer's wife is particularly misunderstood. The average farm journal still tells the farm wife how to make a clever little dress out of feed sacks, whereas, the true farm wife reads *Vogue* and knows her *Dior*.[32]

Another author, Kenneth S. Davis, from rural Kansas in a 1949 editorial for the *New York Times* lamented "…the East, with its dominance over mass communications, has managed to impose a fairly accurate picture of itself on the Midwestern subconscious whereas the Midwest remains, in Eastern stereotypes, a raw, pioneering land." As evidence of this Eastern urban stereotype, Davis cited the attitudes of soldiers and civilian employees from New York City he encountered when stationed at Fort Riley in Manhattan, Kansas during World War II:

> Some of these sojourners expressed surprise at the presence of indoor plumbing and the absence of Indians in Kansas. Others, better informed, were astonished to encounter, here and there, Midwesterners who read books and gave other signs of being as 'cultured' as themselves.

Davis spends the remainder of the article seeking to debunk the urban yokel stereotype of the Midwest. Corn Belt residents, Davis argues, retain not only more moral qualities (hailed by Jeffersonian agrarianism) such as friendliness and practicality, but also a greater modernity than Eastern urbanites. For example, since Midwesterners know more about urban lifestyles through mass communication emanating from the cities and urban people are relatively uniformed about the Corn Belt, "the Midwestern mind might be deemed less 'provincial' and more 'sophisticated' than the urban Eastern one." Further, he claims that, "one finds, too, a youthful vigor, an uncritical enthusiasm for Progress, which is less evident in the longer-settled East," which Davis attributes to "cultural residue of frontier days." Despite these greater moral attributes and capacity for "Progress," Davis notes

> People out here feel that the East has traditionally regarded the Midwest as an inferior province to be exploited: that railroad rate structures have discriminated against Midwest industry, and that the interests of the East have generally been better cared for by Washington than have those of this region.

Davis then tells a story of when Gloria Vanderbilt, the 18-year-old great-great daughter of the railroad tycoon Commodore Vanderbilt, lived in Manhattan, Kansas briefly during World War II and "snubbed" the welcoming community. Vanderbilt avoided "having cokes with the other girls her age" at the drug store on Main Street and even installed a phone unlisted on the town's party line, "the only one [offline phone] we've ever had in this Manhattan." Davis commented that among the Manhattan town folk "there were even some snide references to the piratical methods by which old Commodore Vanderbilt was alleged to have established the family fortune in the first place," a reference to the common Midwest belief that Eastern urban industrialists like Vanderbilt became rich taking advantage of western farmers. To Davis's, the "snubbing" of the community by Vanderbilt confirmed that arrogant urban people from places like New York still held unjust yokel stereotypes about the Midwest.

The reaction of residents of Manhattan, Kansas to the glamorous heiress revealed that rural resentments of urban elites lasted well after the 1920s. With writing that could have come from Thomas Jefferson, Davis ends his piece claiming that people who work with the earth and their hands were "wiser" than people like Vanderbilt, and the future of freedom and democracy depends on the Midwest farmer rather than the city dweller.[33]

Nor can one consider yokel stereotypes about the Midwest a relic of a less enlightened Cold War age. As late as 1998, students at Ohio State in Columbus, Ohio jeered Secretary of State Madeleine Albright during a speech in which she began "We are very pleased to be here in America's heartland." As one Midwest editorialist explained

> Heartland has become an overworked word, rarely used by Midwesterners but beloved by visiting reporters and politicians. They like the way it's warm and fuzzy sound masks what they really mean: hinterland. It is a euphemism favored by those who see the nation's midsection as a bucolic American Siberia – a nice place to pass through on the way to either coast, but not a place you'd really want to live.

The writer then cites a long list of sources that "assiduously cultivate a hayseed image of cornfields and cows, barns, and bib overalls" from advertisements on a can of Red Gold tomatoes, to insulting comments regarding the Midwest from a passenger from the East Coast that she overheard on an airplane flight. The Midwesterner even mocked a chain of Machine Shed restaurants scattered throughout the Midwest which the owner describes as "a New Yorker's vision of an Iowa restaurant." The author even found presidential candidates taking photo ops in Iowa with hogs and silos stereotyping. She then pleads with politicians to stop asking, "the patronizing old question, 'Will it play in Peoria?'" and to "declare a moratorium on the word heartland." She wanted people from the East and West Coasts to "recognize that the Midwest is as richly complex and varied as any other part of the country" and, the rural author declared, "so are the people who live there."[34]

Government leaders even recognized the widespread urban adoption of yokel stereotypes, which often seemed to intensify when rural and urban interests conflicted over farm policy. In October of 1955, for example, the Secretary of Agriculture Ezra Taft Benson initiated a "Farm-City Week" for the purposes of "improving relations between farmers and city residents" which had decayed because of "a severe cost-price squeeze" on farmers due to overproduction. Benson organized a committee to travel in a "country-wide tour to sound out sentiment among farmers in anticipation of the opening of Congress" at the start of 1956 to support policies "to bolster sagging farm incomes." The article noted that a recent farm survey showed "a widespread resentment among farmers against labor unions" because they perceived urban workers as receiving high wages.[35]

At a luncheon of farm and urban industry leaders, Benson chose to open his political campaign to bolster farm prices not with economic data but by dispelling yokel stereotypes held by his urban audience. "Mr. Benson told the farm-city luncheon this afternoon," the article states, "that 'The hick' of a generation ago probably has a college degree and you can't tell him from a 'city slicker' in a dressed-up crowd." Benson went on claiming "When you cross the city limits you may find a different tax structure but there are no social or recreational advantages enjoyed by one group which are not available to all."[36] Benson recognized that in order to convince urban business leaders to aid the farmer economically, he had to dispel views of Corn Belt agrarians as an anti-modern "other." Ironically, while Benson was born in rural Idaho, he went onto become a controversial Secretary of Agriculture that advocated reducing agricultural subsidies as protecting inefficient farmers from market forces.[37]

Benson's concern with yokel stereotypes on a political stage seems less strange when one appreciates its social context since the 1920s. Namely, urban discourses tend to invoke negative images about rural life to a greater degree when city dwellers perceive farmers as having disproportionate political power or have "gotten too much of the pie" by way of subsidies or price controls on commodities. After the 1920s and 1930s, urban newspapers continued the practice of bemoaning the political power of Midwest farmers who city dwellers believed could force politicians to pass subsidies, price supports, tariffs on agricultural imports, discounted loans, and low-cost crop insurance. Many of these policies designed to boost prices for agricultural commodities annoyed urbanites who saw farmers as greedy and only concerned with narrow economic self-interest. This urban resentment, having roots in the rural-urban conflict, fans yokel stereotypes. City dwellers often use prejudices to frame rural people as underserving of political or economic power. In this way, the rural-urban conflict of the 1920s tended to repeat itself in such debates over farm policy. As one rural op-ed writer explained in 1963, "My city friends seem to believe that the countryside is teeming with affluent bumpkins who drive Cadillacs to town to pick up government checks."

Urban newspaper editors in the 1960s fanned the perception of farmers as greedy and unjustly receiving support by perpetuating the myth that "programs are shoved down city folks' throats by a highly organized, all-powerful, monolithic Farm Bloc." In reality, the Farm Bloc had fractured after the 1920s, giving way to competing farm interests vying for often conflicting government support. Further, urban-based politicians often voted for farm programs in order to gain backing from rural colleagues on bills serving urban interests.[38] Still, urban resentment over subsidies for farmers and other programs did kill farm bills in Congress, such as a one-year bill to control the production and surplus distribution of corn defeated in the House in 1957 by urban representatives.[39] Thus, urban discourse still depicted farmers as shrewd but unsophisticated hicks wielding too much political influence. In 1961, for example, the *Washington Post* reported that a "hick senator from Chicago" presented five farmers from Illinois at a news conference to accuse one Senator from New York of producing a "fake" farmer who pretended to arrive in Washington, D.C. driving a Cadillac he purchased with farm subsidies. The *Washington Post* reporter antagonized the five farmers at the press conference to such an extent that one farmer asked him "How would you like to work seven days a week?" The story implies that farmers attempted to acquire high no-grow payments by shrewdly fooling city people.[40]

Similarly, *The New York Times* reported that many traditionally Republican farmers had turned against President Gerald Ford in 1976 in his race against challenger Jimmy Carter simply because Ford and his Secretary of Agriculture, Earl I. Butz, had refused to raise loan support rates and boost exports to raise grain prices. The urban reporter noted that grain prices had recently fallen as "a result of their [the farmer's] own planting decisions and cattle and herd expansions" but that Corn Belt agrarians still unjustly blamed Ford for their economic problems.[41] Four years earlier, the same reporter noted that farmers in the Midwest cared little about the Watergate scandal or the Vietnam War because prices were high, implying that farmers place their own narrow pecuniary interests above more noble concerns.[42]

From the urban point of view, rather than combatting rural interests by taking on the morality of the Jeffersonian agrarian, which is deeply entrenched in American culture, it is often more effective to portray rural people as ignorant bumpkins. For example, the Washington, D.C.-based sports journalist Tony Kornheiser delved into politics with a *Washington Post* article in 2000 entitled "Cuts to the Hicks" in which he disparaged the American presidential primary system because "a few

thousand hicks in New Hampshire and Iowa decide our nominee for president."
"How did New Hampshire and Iowa come to speak for America?" Kornheiser
asked, "This is like asking Beavis and Butt-head [two notoriously stupid cartoon
characters popular at the time] to pick the Kennedy Center honorees."

Kornheiser then espoused a long tirade disparaging Iowa in language that cap-
tured urban discourses about the rural Midwest. The article seems completely una-
ware of rural performances of modernity. He writes,

> It's winter 11 months out of the year... Who in their right mind would live
> there? I mean aside from someone in any state that ends in Dakota. The fact is,
> most residents of Iowa and New Hampshire-who aren't farm animals-have been
> forcibly relocated there by the witness protection program.

Kornheiser then singled out Iowa,

> Let's talk about Iowa (Which one is Iowa? I always get it confused with Indiana,
> Illinois and Indonesia. Oh I remember). To get to Iowa, just click your heels
> three times and think of a grain silo. Iowa is such a big producer of pork that the
> state bird is bacon. The largest city is Des Moines, taken from the French phrase
> meaning 'These Moines.' All you have to do to win over the voters in Iowa is
> not keep kosher.[43]

Kornheiser's sarcasm suggests that urbanites often acknowledge their own stereotyping about the Midwest. Nevertheless, the article demonstrates how such
pejorative views often emerge when city dwellers see rural people as holding disproportionate political sway.

Again, in an interview of Missouri farmer Matt Neustadt in 2008 regarding his
starring in a new reality television series "Farmer Wants a Wife," the rural bachelor
expressed a desire "to break down ugly stereotypes that have prevented many potential mates from believing that he is, in fact, just your average sod buster." The article, written for an urban audience in the *New York Times*, reported that Neustadt's
muscular physique causes many women to view his discussion about his farm as
"trying out a line instead of honestly and earnestly just trying to find a companion."
The article noted that the "ample, corn-fed bellies," of the bachelor's neighbors
were much more typical of Midwest farmers than Neustadt's "rippling, washboard
abs." Further, the article described the rural bachelor as "lonely," a typical urban
view of Corn Belt life as isolated, whereas the urban bachelorettes in the show had
simply "tired of the dating life in the city" because "it's hard to find someone who
is willing to commit." As if leveling stereotypes at Neustadt were not enough, the
article's urban author, Edward Wyatt, implied that the surprisingly fit farmer's family had unjustly received $639,000 in farm subsidies and federal insurance between
1995 and 2006.[44] Wyatt's discussion of agricultural policy seems strangely out of
place in an article about a comparatively frivolous reality television show, unless

one appreciates the historical connection between yokel stereotypes and urban resentment of perceived rural economic or political advantages.

More recently, editorialist Earl Ofari Hutchinson noted an urban tendency in the 2016 election to construct a stereotype of the typical Donald Trump supporter as a "poorly educated" farmer in the Midwest living in the "backwoods" who "fears and loathes blacks, Latinos, gays, liberals of all stripes, Obama and big government." Trump's defeated opponent, Hillary Clinton, advanced this urban view of the Trump supporter in a speech in Mumbai, India on March 10, 2018, in which she stated

> All that red in the middle, where Trump won, what the map doesn't show you is that I won the places that represent two-thirds of America's gross domestic product. So I won the places that are optimistic, diverse, dynamic, moving forward.

The article went on to report, "Mr. Trump's 'Make America Great Again' campaign was looking 'backwards,' playing on what she said were feelings in the non-urban United States of voters who 'didn't like black people getting rights,' or women getting jobs." Many opponents of Clinton's immediately attacked her as "dismissing America's Heartland."[45]

Hutchinson theorized that by associating Trump with the marginalized Midwest yokel, urban opponents of Trump could convince themselves that the candidate was also not part of contemporary mainstream America. Hutchinson disputed the idea that only yokel Midwest farmers with these backward ideas supported Trump in that he claimed that others must back the candidate given his electoral success. On the other hand, the Boston editor did not reject the hick stereotype itself or the idea that such rubes in the Corn Belt indeed vote in favor of racist or xenophobic ideas.[46] Again, the yokel type asserts itself as an expression of urban political resentment of a perceived unjust (and in this case immoral) rural political power.[47]

Since the 2016 election, many scholars have highlighted the importance of intense resentment of urban elites for producing the cult-like support for Trump in many rural areas, including in parts of the Midwest. Much like the 1920s, the growing cultural and economic power of urban America combined with economic decline in rural areas fuels rural resentment and a feeling, as journalist Al Cross describes, of "Stop Overlooking Us!" Cross and these other authors, however, overlook the fact that this rural-urban resentment constitutes a repeated cycle that has driven interactions between people in the city and country since at least the early 20th century and even as far back as the 1700s when Benjamin Rush observed German – American farmers.[48] Such resentment reflects the politicized version of the pattern of audience. While the current period may present a newly intensified rural-urban divide, Trump is simply one recent figure through which this familiar pattern and rehashed resentment operate.

Views of the Modern Farmer as Duped by Agribusiness and Unable to Grasp Environmental Concerns

Some current policy advocates and environmental activists have also adopted a more subtle view of farmers as anti-modern as a way of critiquing agribusiness.[49] By creating a romanticized nostalgia about early-20th-century farming, activists, such as novelist Wendell Berry and botanist and geneticist Wes Jackson, construct a false image of the family farm as a place where people practiced environmental stewardship with the goal of keeping their farms small and building strong communities.[50] Food advocates have adopted pure Jeffersonian agrarianism while combining it with a more contemporary environmentalist ethos creating an updated notion of morality. More importantly, these critics of modern agriculture have assumed that farmers themselves also strive, or should strive, for this same blend of pure Jeffersonian and environmentalist virtues against a menacing, all-powerful, and immoral agribusiness. Farmers, as powerless victims, find themselves blocked by industrial interests from re-capturing a pre-technological Jeffersonian golden age.

A less positive version of this urban view of farmers is that the naïve agrarian has been duped into modernization by business interests. This notion sees farmers as too infected by greed or ignorance to appreciate the same environmental concerns so important to more enlightened urbanites. To be sure, not all environmental or organic advocates have adopted this "dupe" image of the farmer, but the pattern of audience causes many Corn Belt agrarians to see the discourse of reformers as presenting such a patronizing stereotype. Berry, for example, describes the typical farm prior to World War II in utopian terms as:

> The farms were generally small. They were farmed by families who lived not only upon them, but within and *from* them. These families grew gardens. They produced their own meat, milk, and eggs. The farms were highly diversified... In those days the farm family could easily market its surplus cream, eggs, old hens, and frying chickens. The power for field work was still furnished mainly by horses and mules. There was still prevalent pride in workmanship, and thrift was still a forceful social ideal. The pride of most people was still in their homes, and their homes looked like it.[51]

Berry characterizes this image of pre-World War II agriculture as the dream of the Jefferson because it promoted communities of small farmers who eschewed what Jefferson referred to as "artificers," but what would later be called manufacturers of technology.[52]

Berry sees the history of 20th-century agricultural change as a *moral devolution* following a common pattern within white European society in which the strong exploits the weak. Just as European settlers took advantage of Native Americans or slave owners used their strength to dominate slaves, Berry argues, "institutions of agriculture" such as "the university experts, the bureaucrats, and the 'agri-businessmen," ruined these moral rural communities by forcing technology and "bigness" upon them in the name of efficiency and progress.[53] The farmer in Berry's narrative clearly lives as a victim of technological change. In one piece, Berry succinctly states, "The predicament of the industrial farmer is precisely that of helpless economic dependence."

The farmer, according to Berry, lacks any agency in deciding which technologies to adopt or how to use them, nor does he have the savvy to pierce the veil of official statements from the United States Department of Agriculture (USDA) and other institutions extolling progress and faith in technology. A hybrid rural identity devised by farmers combining Jeffersonian ideals with modern sensibilities does not exist or was at least dictated to farmers by corporate interests. Berry describes the technologies that agribusiness firms sell farmers as "ready-made thoughts," by which he means farmers mindlessly adopt the reductionist agricultural practices such as monoculture and synthetic fertilizers demanded by contemporary technologies. Such discourse implies that modern farmers have immorally abandoned less technological and more localized farming practices simply because they are too weak or obtuse to question the motivations of agri-business. Rural people are either forced to adopt bigger and newer artifacts or are fooled into doing so at their own expense ultimately leading to the ruin of their communities, environment, and Jeffersonian way of life.[54]

Berry laments the farmer's closed mindedness and inability to imagine any way of producing other than what large companies sell her. Thus, he sees the modern farmer as foolishly sowing his own demise. As Berry explains, "This industrial version of agriculture is paid for--and this is my point--not by money only, but also by the intelligence of the buyers."[55] Here, while Berry does not use the terms rube or yokel, he still views farmers as unintelligent compared to urban advertisers and industries. The backward hick has become a mechanized dupe. For Berry, pure Jeffersonian agriculture formed the only possible moral way of farming and technology, purely the tool of urban industrialists and their institutions, always threatening this morality.

Further, Berry contends that "food is a cultural product; it cannot be produced by technology alone." Berry regards culture as familiarity that "can grow only among a people soundly established upon the land; and it nourishes and safeguards a human intelligence of the earth that no amount of technology can satisfactorily replace."[56] Berry and many other critics of contemporary mainstream agriculture associate less technological forms of production with a greater connection to place and a more intimate knowledge of the land. These authors assume that some immoral act has occurred on the part of an outside governmental or economic power, which has broken these sacred rural connections between farmers and the land by introducing technology and modern farming methods.

Of course, for some of these advocates, modern farming technology also destroys the natural environment. As expressed by one small Ohio farmer,

Where else can one be so much a part of nature and the mysteries of God, the unfolding of the seasons, the coming and going of the birds, the pleasures of planting and the joys of the harvest, the cycle of life and death? ... Here on our 120 acres I must be a steward of the mysteries of God.

The small farmer, David Kline, credits his Amish and Mennonite friends with this spiritual insight about land.[57] Juxtaposed with essays by Berry and others lamenting the blind faith in technology proffered by institutions of modern agriculture justifying "the destruction of every aspect of rural life," Kline's essay clearly impresses the reader with the message that less technological Amish-style farming preserves this sacred bond with the land whereas modern mechanized agriculture affronts not only communities but God.[58] Again, technology and morality stand in an inverse relationship.

In addition, from the perspective of Berry and some organic advocates, technology and culture form completely separate and even opposing domains. Rather than viewing technology as socially constructed, policy advocates critiquing modern agriculture sometimes frame advanced technologies always as undermining culture. "The best farming requires a farmer-a husband-man, a nurturer-not a technician or businessman," Berry argues. The know-how required to farm comes only as accumulated knowledge through established communities and families whereas technicians and businessmen can come out of training in any university. Thus, farming culture requires the preservation of place where knowledge can accumulate, which also calls for resisting the encroachment of advanced technologies and "community-killing agriculture."[59]

Additionally, technology often carries with it capitalistic competition detrimental to communities and the environment. Rather than seeking to preserve lasting farming communities and resources, institutions of agriculture have fooled farmers into seeking relief from dignified forms of manual labor by purchasing the newest technologies.[60] For example, in discussing the rural town of Matfield Green, Kansas, which had fallen into decay by the mid-1990s, Wes Jackson highlights the New Century Club, a social organization of women in the town from 1923 to 1964. For Jackson, modern capitalistic farming methods and technologies destroyed this virtuous community. He writes, "Despite the daily decency of the women in the Matfield Greens, decency could not stand up against the economic imperialism that swiftly and ruthlessly plowed them and their communities under."[61] Therefore, Jackson advocates the rejection of contemporary mainstream farming as big business and a return to the simpler, more environmentally conscious times exiting prior to World War II. The moral course points towards promoting the Jeffersonian agrarian, his "pastoral garden," and his rural community oppressed by the interests and technologies of "industrial agriculture." The organic advocate and environmentalist can thus "save" the oppressed Jeffersonian farmer.

I will return to a discussion of how this narrative of Jeffersonian agrarians sub-mitting to technological change imposed by immoral agribusiness has helped to construct an *organic reformist identity* in Chapters 9 and 10. While this organic identity does not in actuality represent all organic advocates, it does present a bundle of views that seems threatening when seen from the perspective of many farmers. Modern capitalistic farmers in the Corn Belt often regard organic or sus-tainable discourse as advocating a view of technology as an immoral male domain and the farmer as not modern enough to understand environmental concerns. These perceived aspects of organic identities tend to rekindle old rural resentments over rube stereotypes and constructions of rural femininity. When attempting to view the world from the farmer's perspective, one can also better understand why men and women in the Midwest react to new technologies in certain positive and nega-tive ways as they appear in the rural landscape.

Notes

1 F. Scott Fitzgerald, *The Great Gatsby*, 1920 (New York: Scribner Books, 1995), 184.
2 Barry Manilow et al., "Barry' Manilow's Copacabana," *CBS* (New York: Dick Clark Productions/Stiletto Entertainment, 1985), http://www.imdb.com/title/tt0088951/ (ac-cessed 5/13/18).
3 Biography.com website, "Barry Manilow Biography," *A&E Television Networks*, April 5, 2017, https://www.biography.com/people/barry-manilow-9542490 (accessed 3/13/18).
4 Barry Manilow, "Just Arrived," in play *Copacabana*, recorded in London, 1994 on You-tube: https://www.youtube.com/watch?v=IL-pm1dqXO4 (accessed 5/13/18).
5 "I Gotta Dance," *Singing in the Rain*, directed by Stanley Donen and Gene Kelly (Metro-Goldwyn-Mayer, MGM, 1952), on YouTube: https://www.youtube.com/watch?v=Np4drWBhelE (accessed 5/13/18).
6 For comprehensive histories of negative cultural views of rural dwellers in the American South, see Harkins, *Hillbilly*.
7 *Oxford English Dictionary*, s.v. "bumpkin;" *Oxford English Dictionary*, s.v. "Dutchman."
8 *Oxford English Dictionary*, s.v. "hick."
9 Nancy Franklin, "Dopey and Grumpy," *New Yorker* (December 11, 2000), 110.
10 Mark N. Dodosh, "Hello Columbus: Thriving Ohio Capital Seeks to Shed Image as a Country Bumpkin," *Wall Street Journal* (December 8, 1980), 1–2.
11 John Gould, "When a City Yokel Buys a Farm," *New York Times* (April 19, 1953), SM 25, 34, 37.
12 For example, see "Engel Found Minnesota Women Yokels," *Los Angeles Times* (June 26, 1949), 3 in which the notorious criminal Sigmund Engle described the Minnesota women he was conning as "yokels" in a 1949 interview from his jail cell.
13 "Bride Fleeing 'Hick' G.I. Gets Child Back," *Los Angeles Times* (February 9, 1949), 7.
14 Mike Littwin, "Larry Bird: The World Champions' MVP May Be a Little Short on Cha-risma, but He Sure Is One Hick of a Player," *Los Angeles Times* (February 14, 1982), C1, C12–14.
15 "Hayseed Catalogue," *New Yorker* (April 11, 1994), Goings on About Town: The Thea-tre, 14; Awardcrazy, "Jarrod Emick wins 1994 Tony Award for Best Featured Actor in a Musical," (November 23, 2003), https://www.youtube.com/watch?v=EY7B4IZhPWU (accessed 9/1/17).
16 "Newsmakers-Star is Born: Lindsay Is Cheered as a Hick," *Los Angeles Times* (March 25, 1973), B2.

17 Brock Yates, "Dream Machines: Why the Hottest-Selling Car's a Truck," *Washington Post* (April 12, 1987), 47.

18 Terri Queck-Matzie, "The Combine-King of the Harvest," *Successful Farming*, February 7, 2019, https://www.agriculture.com/machinery/harvesting/the-combine-king-of-the-harvest (accessed 5/23/19).

19 Jim Hyatt, "Summer in Sandusky Is Not the Dull Scene You City Bays Think: With Gaylord Zechman, Farmer, at the Wheel, Most Anything Can Happen-Up to 8MPH," *Wall Street Journal* (April 5, 1971), 1.

20 Lisa R. Pruitt, "Rural Rhetoric," *Connecticut Law Review* 39, no. 1 (2006): 159–240.

21 Frederick W. Roevekamp, "Flying Farmers Save Cash," *Christian Science Monitor* (August 15, 1958), 3.

22 Richard Orr, "Farmer Is No Country Bumpkin; He's an Affluent Business Man," *Chicago Tribune* (June 24, 1968).

23 Calvin Trillin, "Department of Amplification," *The New Yorker* (March 19, 1984), 118–119.

24 Simpson is likely a fictitious character in Bauer's story or an alias for another congressman, as no U.S. Representative from Iowa was named Simpson in the 1960s.

25 W. Averell Harriman was the governor of New York from 1954 to 1958 as well as a diplomat and statesman from the 1940s through the 1960s. Amy Tikkanen, "W. Averell Harriman American Diplomat," *Britannica*, https://www.britannica.com/biography/W-Averell-Harriman (accessed 10/27/23).

26 Douglas Bauer, "Rites of Spring," *The Washington Post* (April 6, 1980), SM14.

27 Ibid.

28 Jane Smiley, "Wide-Eyed in the Big City: City Limits of a Small-Town Boy," *New York Times* (November 10, 1991), BR14.

29 Kenneth S. Davis, "East Is East and Midwest Is Midwest: A Midwesterner Is Grateful that Standardization has Failed to Kill Healthy Regional Differences," *New York Times* (November 20, 1949), SM17.

30 Trillin, "Department of Amplification," 118.

31 Jon Swan, "A Portable Gallery of Pastoral Animals," *New Yorker* (May 4, 1963), 44.

32 Fred Beck, "Farmers Market Today…With Mrs. Fred Beck," *Los Angeles Times* (December 3, 1952), 4.

33 Davis, "East Is East and Midwest Is Midwest," SM17–19, 57–58.

34 Marilyn Gardner, "Beltway, Beware-The 'Heartland' Shows Its Edge," *The Christian Science Monitor* (February 26, 1998), 13.

35 See also William M. Blair, "Senators Hear Farmer Plaints," *New York Times* (October 25, 1955), 22.

36 William M. Blair, "Farm-City Accord Backed by Benson," *New York Times* (October 23, 1955), 52.

37 John P. Rafferty, "Ezra Taft Benson: American Religious Leader," *Britannica*, https://www.britannica.com/biography/Ezra-Taft-Benson (accessed 10/27/23).

38 Jack Heinz, "Those Annoying Farmers: Impossible But Not Really Serious," *Bill Harper's Magazine* (July 1, 1963), 61–68.

39 "Aiken Says Large Farm Subsidies Create a Stir," *Los Angeles Times* (March 25, 1957), 15.

40 George Dixon, "Washington Scene: Agrarian Revolt," *Washington Post* (July 19, 1961), A15.

41 Seth S. King, "Midwest Voters See Farm Prices, Not Butz, as the Campaign Issue," *New York Times* (October 10, 1976), 42.

42 Seth S. King, "Midwest Farmers' Minds on Big Crops and Not Politics," *New York Times* (October 15, 1972), 44.

43 Tony Kornheiser, "Cuts to the Hicks," *Washington Post* (January 30, 2000), F1–2.

44 Edward Wyatt, "A Farmer, Lonely, Holds Auditions," *New York Times* (April 29, 2008), E1.

45 "Clinton Says Trump Won on vows to Take Country 'Backwards,'" *CBS News*, March 13, 2018, https://www.cbsnews.com/news/hillary-clinton-middle-america-looking-backwards-lost-election-donald-trump/ (accessed 5/13/18).

46 Earl Ofari Hutchinson, "Trump's Supporters Aren't Just Old, Racist White Guys," *Boston Banner* (May 19, 2016), Opinion, 1–5.

47 For an analysis of how urban yokel stereotypes and rural resentment of them underlies the current red-blue political divide in the U.S., see Cliff Cobb, "Editor's Introduction-Fighting for Rural America: Overcoming the Contempt of Small Places," *The American Journal of Economics and Sociology* 75, no. 3 (May 2016): 569–588.

48 Cross, "Stop Overlooking Us!"; Cramer, *The Politics of Resentment*; Jose A. Del Real and Scott Clement, "Rural Divide," *Washington Post*, June 17, 2017, https://www.washingtonpost.com/graphics/2017/national/rural-america/ (accessed 8/1/23).

49 Danbom, *Born in the Country*, 256–257.

50 Wendell Berry, *The Art of Common Place: The Agrarian Essays of Wendell Berry* (Berkley, CA: Counterpoint Press, 2002); Wendell Berry, *The Unsettling of America: Culture & Agriculture* (San Francisco: Sierra Club Books, 1977); Wes Jackson, *Consulting the Genius of the Place: An Ecological Approach to a New Agriculture* (Berkley, CA: Counterpoint Press, 2010); Wes Jackson, *New Roots for Agriculture* (San Francisco, CA: Friends of the Earth, 1980).

51 Berry, *The Unsettling of America: Culture & Agriculture*, 39–40.

52 Ibid., 13, 143–144.

53 Ibid, 4–6.

54 Ibid., 32–33.

55 Wendell Berry, "Sowing Disaster," *Washington Post* (April 1, 1984), 16–17.

56 Berry, *The Unsettling of America: Culture & Agriculture*, 43–44.

57 David Kline, "An Amish Perspective," in *Rooted in the Land: Essays on Community and Place*, ed. William Vitek and Wes Jackson (New Haven, CT: Yale University Press, 1996), 35–39.

58 Berry, "Conserving Communities," in *Rooted in the Land*, 78.

59 Berry, *The Unsettling of America: Culture & Agriculture*, 41.

60 Ibid., 12, 45–47, 59.

61 Jackson, "Matfield Green," in *Rooted in the Land*, 98–103, 95–103.

Bibliography

"Aiken Says Large Farm Subsidies Create a Stir." *Los Angeles Times,* March 25, 1957, 15.

Awardcrazy. "Jarrod Emick wins 1994 Tony Award for Best Featured Actor in a Musical," November 23, 2003. https://www.youtube.com/watch?v=EY7B4IZhPWU (accessed 9/1/17).

Bauer, Douglas. "Rites of Spring." *The Washington Post*, April 6, 1980, SM14.

Beck, Fred. "Farmers Market Today…With Mrs. Fred Beck." *Los Angeles Times*, December 3, 1952, 4.

Berry, Wendell. *The Unsettling of America: Culture & Agriculture*. San Francisco: Sierra Club Books, 1977.

———. "Sowing Disaster." *Washington Post*, April 1, 1984, 16–17.

———. *The Art of Common Place: The Agrarian Essays of Wendell Berry*. Berkley, CA: Counterpoint Press, 2002.

Biography.com website. "Barry Manilow Biography." *A&E Television Networks*, April 5, 2017. https://www.biography.com/people/barry-manilow-9542490 (accessed 3/13/18).

Blair, William M. "Senators Hear Farmer Plaints." *New York Times*, October 25, 1955, 22.

"Bride Fleeing 'Hick' G.I. Gets Child Back." *Los Angeles Times,* February 9, 1949, 7.

"Clinton Says Trump Won on Vows to Take Country 'Backwards.'" *CBS News*, March 13, 2018. https://www.cbsnews.com/news/hillary-clinton-middle-america-looking-backwards-lost-election-donald-trump/ (accessed 5/13/18).

Cobb, Cliff. "Editor's Introduction-Fighting for Rural America: Overcoming the Contempt of Small Places." *The American Journal of Economics and Sociology* 75, no. 3 (May 2016): 569–588.

Cramer, Katherine J. *The Politics of Resentment: Rural Consciousness in Wisconsin and the Rise of Scott Walker*. Chicago, IL: University of Chicago Press, 2016.

Cross, Al. "Stop Overlooking Us!:' Missed Intersections of Trump, Media, and Rural America." In *The Trump Presidency, Journalism, and Democracy*, 231–256. New York: Routledge, 2018.

Danbom, David B. *Born in the Country: A History of Rural America*. Baltimore, MD: The Johns Hopkins University Press, 1995.

Davis, Kenneth S. "East Is East and Midwest Is Midwest: A Midwesterner Is Grateful that Standardization Has Failed to Kill Healthy Regional Differences." *New York Times*, November 20, 1949, SM17–19, 57–58.

Dixon, George. "Washington Scene: Agrarian Revolt." *Washington Post*, July 19, 1961, A15.

Dodosh, Mark N. "Hello Columbus: Thriving Ohio Capital Seeks to Shed Image as a Country Bumpkin." *Wall Street Journal*, December 8, 1980, 1–2.

"Engel Found Minnesota Women Yokels." *Los Angeles Times*, June 26, 1949, 3.

Fitzgerald, F. Scott. *The Great Gatsby*. 1925. New York: Scribner Books, 1995.

Franklin, Nancy. "Dopey and Grumpy." *New Yorker*, December 11, 2000, 110.

Gardner, Marilyn. "Beltway, Beware-The 'Heartland' Shows Its Edge." *The Christian Science Monitor*, February 26, 1998, 13.

Gould, John. "When a City Yokel Buys a Farm." *New York Times*, April 19, 1953, SM 25, 34, 37.

Harkins, Anthony. *Hillbilly: A Cultural History of an American Icon*. Oxford: Oxford University Press, 2004.

"Hayseed Catalogue." *New Yorker*, April 11, 1994, Goings on About Town: The Theatre, 14.

Heinz, Jack. "Those Annoying Farmers: Impossible But Not Really Serious." *Bill Harper's Magazine*, July 1, 1963, 61–68.

Hutchinson, Earl Ofari. "Trump's Supporters Aren't Just Old, Racist White Guys." *Boston Banner*, May 19, 2016, Opinion, 1–5.

Hyatt, Jim. "Summer in Sandusky Is Not the Dull Scene You City Bays Think: With Gaylord Zechman, Farmer, at the Wheel, Most Anything Can Happen-Up to 8MPH." *Wall Street Journal*, April 5, 1971, 1.

"I Gotta Dance." *Singing in the Rain*. Directed by Stanley Donen and Gene Kelly. Metro-Goldwyn-Mayer, MGM, 1952. YouTube: https://www.youtube.com/watch?v=Np4drW BhelE (accessed 5/13/18).

Jackson, Wes. *New Roots for Agriculture*. San Francisco, CA: Friends of the Earth, 1980.

———. "Matfield Green." In *Rooted in the Land: Essays on Community and Place*, edited by William Vitek and Wes Jackson, 95–103. New Haven, CT: Yale University Press, 1996.

———. *Consulting the Genius of the Place: An Ecological Approach to a New Agriculture*. Berkley, CA: Counterpoint Press, 2010.

King, Seth S. "Midwest Farmers' Minds on Big Crops and Not Politics." *New York Times*, October 15, 1972, 44.

————. "Midwest Voters See Farm Prices, Not Butz, as the Campaign Issue." *New York Times*, October 10, 1976, 42.

Kline, David. "An Amish Perspective." In *Rooted in the Land: Essays on Community and Place*, edited by William Vitek and Wes Jackson, 35–39. New Haven, CT: Yale University Press, 1996.

Kornheiser, Tony. "Cuts to the Hicks." *Washington Post*, January 30, 2000, F1–2.

Littwin, Mike. "Larry Bird: The World Champions' MVP May Be a Little Short on Charisma, but He Sure Is One Hick of a Player." *Los Angeles Times*, February 14, 1982, C1, C12–14.

Manilow, Barry. "Just Arrived." In play *Copacabana*. Recorded in London, 1994. Youtube: https://www.youtube.com/watch?v=IL-pm1dqXO4 (accessed 5/13/18).

"Newsmakers-Star Is Born: Lindsay Is Cheered as a Hick." *Los Angeles Times,* March 25, 1973, B2.

Orr, Richard. "Farmer Is No Country Bumpkin; He's an Affluent Business Man." *Chicago Tribune*, June 24, 1968.

Oxford English Dictionary. Oxford: Oxford University Press, 2017.

Pruitt, Lisa R. "Rural Rhetoric." *Connecticut Law Review* 39, no. 1 (2006): 159–240.

Queck-Matzie, Terri. "The Combine-King of the Harvest." *Successful Farming*, February 7, 2019. https://www.agriculture.com/machinery/harvesting/the-combine-king-of-the-harvest (accessed 5/23/19).

Rafferty, John P. "Ezra Taft Benson: American Religious Leader." *Britannica*. https://www. britannica.com/biography/Ezra-Taft-Benson (accessed 10/27/23).

Roevekamp, Frederick W. "Flying Farmers Save Cash." *Christian Science Monitor*, August 15, 1958, 3.

Smiley, Jane. "Wide-Eyed in the Big City: City Limits of a Small-Town Boy." *New York Times*, November 10, 1991, BR14.

Swan, Jon. "A Portable Gallery of Pastoral Animals." *New Yorker*, May 4, 1963, 44.

Tikkanen, Amy. "W. Averell Harriman American Diplomat." *Britannica*. https://www. britannica.com/biography/W-Averell-Harriman (accessed 10/27/23).

Trillin, Calvin. "Department of Amplification." *The New Yorker*, March 19, 1984, 118– 119.

Wyatt, Edward. "A Farmer, Lonely, Holds Auditions." *New York Times*, April 29, 2008, E1.

Yates, Brock. "Dream Machines: Why the Hottest-Selling Car's a Truck." *Washington Post*, April 12, 1987, 47.

8

"INBORN INNOVATORS" OR "HOG HOUSE JANITORS?"

The Acceptance or Rejection of Technologies and Rural Globalized Ultramodernity

> See those lights? They don't stop.... They're going all the time. Daytime, white. Nighttime, red. And then the sound. Anytime the windmills are turning, if the wind is blowing, you got a hum or a drone or a whooshing sound, depending on the speed of the wind.
>
> Joyce Manley, retired schoolteacher, Palm Springs, CA, 2007[1]

I generally avoid attending family reunions, but in 2004, a year after my wife and I married, it seemed like the right time to go to the Brinkman reunion in Greene, Iowa. Driving down I-35W from our home in Minneapolis, MN in the middle of June, I had remembered fondly my father looking out the window and saying "Son, there's Brinkman land" many years before. The corn and the horizon looked the same except for several large groupings of wind turbines after crossing the Minnesota-Iowa border. Arriving at my uncle's farm, we pulled into the long drive-way buttressed by a field of high green corn on the left and his prosperous home on the right with an American flag waiving in the wind.

Everything about my uncle's property looked clean, neat, and immaculately maintained. One could spend the entire afternoon trying to find a weed in his yard or field. His pickup truck and tractors appeared so clean as to appear almost un-used. Somehow, everything seemed to gleam in the sunlight. A clean line of stone landscaping surrounded his home. The driveway and garage floor looked similarly pristine and unstained. After meeting a legion of relatives who I thought I had never met (although they somehow remembered me), we all piled back in the car with my uncle's pickup truck in the lead. I followed the shining red Chevrolet through several turns wondering how my uncle knew how to navigate miles of grid-patterned roads through identical flat cornfields.

DOI: 10.4324/9781032637952-9

Then, in the distance, it appeared. A huge white wind turbine jutted up into the blue flat horizon. It had all of the magnificence of the Washington Monument where I grew up but was more sleek and futuristic. We turned down a dirt road that cut through another cornfield and moved toward the huge rotating object. We stopped next to the base of the "monument" and my uncle declared, "This is my new turbine!" The family members got out of their cars and marveled at the sheer massiveness of the device: it seemed to go up in the sky forever as if to say "I'm the future and no one can stop me!" As with my uncle's farm, everything appeared straight, clean, and new. The turbine itself had sleek lines and rotated almost effortlessly. The ominous "whooshing" sound struck me as quiet relative to the hugeness of the blades. The road to the turbine similarly featured a nice clean border between the dirt of the field and the gravel, which almost looked like small grey geometric pieces. The turbine sat on a clean concrete base with a gravel border that created a perfect line with the bordering field. The field, like all the fields around us, featured rows of corn in almost a straight line. Each corn stalk grew in the same size and shape. The dirt between the rows even looked like one uniform brown-black color. Literally everything appeared neat, clean, ordered…prosperous.

My grandfather, then in his late 70s, looked up at the huge blades passing in front of the sun and smiled with an expression as if to say "I can't believe we've come this far." I thought, "I got through college, but I've done nothing like all this. I mean, wow!" My wife, an engineer, was even impressed. The whole family felt visibly proud to be Brinkmans.[2]

The massiveness of wind turbines, immediately noted by me and my family members, and the noticeable profiles of the devices across the Corn Belt's horizon makes them artifacts offering rich insight into how farmers use technology to perform an ultramodern rural identity. Interestingly, the same characteristics of wind turbines that make people dislike them in other parts of the country – their huge size, their metallic appearance, and their connection to vast technical networks, render them particularly suited to form and reinforce farmers' self-images as ultramodern producers.

In this chapter, I analyze how unarticulated discourses and identities shape the relationship farmers have with *new* technologies. I argue that Midwest farmers still accept or reject a technology according to how that object fits into rural modern identities. First, my narrative shows how rural globalized ultramodernity explains the welcoming of wind turbines that have grown rapidly across the landscapes of the Midwest in recent years. One can rarely travel in middle America without noticing the miles of enormous rotating machines that break up the horizon during the day and emit an ominous duo of pulsating red lights and low rumbles at night. Iowa, for example, obtained almost 60% of its electricity in 2020 from wind turbines.[3] Second, I contend that several other contemporary technologies illicit opposition or rejection among Corn Belt farmers notwithstanding their economic benefits because they threaten ultramodern rural identities. New tractors containing software locks, precision agriculture using big data, and large hog production systems all garner opposition in rural America for non-economic reasons. I conclude with a discussion of the "Cow Wars" in Iowa in 1931 to demonstrate that the reasons for farmer's resistance to tuberculosis eradication efforts by the Federal government mirror the motives of contemporary agrarians in rejecting some recent technologies.

"Please in My Backyard" (PIMBY): The Welcoming Acceptance of Wind Turbines in the Corn Belt[4]

The worldwide growth of the wind energy industry since the 1970s has prompted scholars to explore public perceptions of electricity-producing turbines that dot a diverse range of landscapes. While people generally recognize the value of the machines as sustainable forms of power, surveys reveal that many residents oppose local wind turbine installations because of their noise and impact on wildlife or because they constitute symbols of an undesirable industrialization of natural environments.[5] Several studies document a "Not in My Backyard" (NIMBY) phenomenon in various locales, and fervent opposition to construction of turbines remains common. In Massachusetts, for example, citizens have sought to dismantle monumental machines in the Town of Falmouth and prevent construction of gigantic turbines in the ocean off the coast of Cape Cod.[6] Such wind energy projects have only recently moved forward after wind companies have invested substantial funds in community outreach programs and donated millions of dollars to adjacent towns to counter local opposition.[7] As the quotation at the beginning of this chapter demonstrates, residents have also voiced opposition to wind farms in the San Gorgonio Pass in California for several reasons including the light and sound pollution emitted by the giant turbines. In discussing the proposed turbines in the San Gorgonio Pass, another neighboring homeowner citied the impact of the wind farms on "natural" landscapes stating, "They want to take this national monument and turn it into an industrial park," while the local Sierra Club chapter noted "The windmills are a known source of avian mortality; they kill thousands of birds every year. The last count I had was 6,800 per year with the windmills we already have."[8]

Less widely publicized, denizens of certain rural communities have welcomed erection of huge wind turbines. In fact, the positive reception has created a new term that mocks the moniker used by opponents. Instead of a NIMBY reaction, these cheerleaders of the mammoth generators exhibit a "Please in My Backyard" (PIMBY) response, as observed by policy analyst Lester Brown, among ranchers in Colorado and dairy farmers in upstate New York.[9] Likewise, geographer Jacob Sowers documented a PIMBY dynamic among residents of northwestern Iowa in 2002.[10] By the time Sowers conducted his series of interviews, utility-scale wind turbines had already become prevalent on private land rented from farmers. Sowers found strong support for the technology across communities – from city officials to farmers – even among landowners without turbines on their properties. Similarly, the Wind Energy Foundation reported a 2014 poll in which "87% of Midwesterners support increasing the use of wind energy."[11] In several Midwest states such as Nebraska and Iowa, the growth of wind energy is often celebrated and supported across the political spectrum.[12]

American scholars of the PIMBY phenomenon in rural communities have generally adopted a model viewing wind turbines as isolated technologies conferring distinct articulable benefits, highlighting the economic interests of stakeholders.[13] Geographers Martin Pasqualetti and Cleveland Cutler noted, for example that "farmers have learned that wind power can make them money and help them to keep their land."[14] Sowers acknowledged the symbolic value of the turbines as icons of community pride in an otherwise monotonous landscape, but he attributed PIMBY attitudes primarily to the belief that the payments to farmers helped the community as a whole and allowed residents to maintain their rural lifestyles.[15] Such an analysis that frames wind turbines as technologies conferring *recognized* benefits with an emphasis on economic gain, however, overlooks the *unexpressed* social and cultural views about technology that are deeply and historically embedded in rural farming communities discussed in this book. In short, studies relying on only expressed rational explanations for PIMBY responses to wind turbines in the Corn Belt ignore performative use of technology.

I argue that PIMBY attitudes among farmers in the American Midwest are driven not only by articulated benefits of turbines, but also by a strong cultural tendency to implant values such as prosperity and modernity within all machinery used for productive purposes. Wind turbines constitute simply one artifact farmers use to construct and maintain their ultramodern identities. While recognizing that older identities of traditional Jeffersonian and German agrarianism and rural capitalistic modernity still exist in the social milieu of rural America, I also demonstrated that farmers have rehearsed a cultural practice since the early 20th century of forming rural modern identities through technological use. Wind turbines serve as only the latest props in this familiar performance. One should also remember an important caveat before employing performative use in the context of wind energy. Midwest farmers, after all, have never existed as one monolithic group and identities and discourses of modernity were often contested and used strategically by rural Americans. For example, rural residents in Minnesota resisted the erection of power lines as late as the 1970s based on a discourse citing "perceived traditions" among farmers. Thus, discourses and identities of farmers regarding modernity were contested and possibly co-existing with discourses of traditional Jeffersonian agrarianism.[16] This multiplicity of rural identities, however, is an old phenomenon. Harold Briemyer, in writing about growing up in rural Ohio in the 1920s, for example, notes "early discord" by "traditionalists" over efforts by 4-H leaders to teach "scientific agriculture" at his local high school.[17] Nevertheless, as I have shown in the proceeding chapters, something resembling a "culture of the machine" also developed and still exists in the Corn Belt carrying strong associations between technology, progress, and distinctly rural versions of modernity. The courses in "scientific agriculture" in Briemyer's town, after all, continued and many farmers also embraced its methods.[18] The PIMBY attitude toward wind turbines exists at the end of this long history of forming rural identities as ultramodern users of technology. Wind turbines, in short, have become well accepted in some (but not all) circles because they symbolize the ultramodern manner by which farmers think of themselves.[19]

I aim in this book to add to the historiography of American agricultural modernization by focusing on contested discourses and the use of technologies as a social practice to form rural identities.[20] The PIMBY attitude toward wind turbines results as a manifestation of the ultramodern culture that rural residents in the Corn Belt helped create. Farmers employ an ultramodern discourse to hail the benefits of wind turbines within a rural ethos while discounting drawbacks highlighted by residents opposing the erection of wind turbines in other parts of the U.S. While the landscapes in Iowa and the San Gorgonio Pass in California *do* look different, this actual difference in terrain cannot fully explain the divergence in attitudes exhibited by residents in these two places toward wind turbines. The machines, after all, kill birds and emit the same noise and light pollution in both places. Attributing NIMBY or PIMBY attitudes to the relative beauty of scenery is simply to reify or "black box" socially constructed aesthetics. There is nothing "inherently" more attractive about one landscape over the other as they exist outside of a social and cultural lens.[21] Alternatively, I argue that unarticulated historically determined differences in the way people in these two places view themselves and nature offer richer explanations for why people embrace or reject wind turbines.

Of course, the fact that wind turbines remind farmers of the windmill once ubiquitous across the Midwest aids in their acceptance. Jeffersonian ideas about the morality of rural life have shaped such aesthetics becoming part of a distinctly Midwestern modernity. For example, in a 2006 article, farmers Elaine Robertson, Tom Watne, and Deloris and Everett Smith of Blairsburg, Iowa declared that they overlooked the constantly blinking red lights emitted by the 135 wind turbines surrounding their land as well as the disruption caused by the installation process, which damaged several drainage tiles.[22] The farmers saw the turbines as simply the latest in a long line of progressive energy technologies beginning in 1812 when their ancestors installed windmills. As a result, the farmers in the article ignored inconveniences to them and emphasized benefits that they did not personally gain, such as the tax dollars generated by the wind turbines for the local governments, the jobs created, and the ecological benefits of renewable energy. The inclusion of wind turbines within a suite of technologies that reinforce a discourse of rural capitalistic modernity underlies a statement made by Watne in the article: "I thought the towers would be more irritable to your sight, but now they seem stately, quite pretty even."[23] Similarly, Iowa corn and soybean farmer David Ausberger welcomed the erection of wind turbines on his farm even though a "bulldozer came through the most beautiful field of beans I'd ever had."[24]

The wind turbine serves as a perfect symbol of rural ultramodernity. Due to the cultural memory of windmills, turbines reference rural legacies and Jeffersonian morality but also fetishize the new and the high-tech. The huge size of the object also lends itself to performance of identity and, of course, the turbine produces electricity meaning it is a moral display of wealth. Many farmers expressed the association of wind turbines with modernity and progress directly. For example, Wisconsin crop farmers Charles Hammer and Nancy Kavazanjian went as far as obtaining a USDA grant to install their own wind turbine, which would sell electricity back to "the grid." In a 2009 article, Hammer explained his motivations for writing the grant proposal: "We feel like we've always been innovators." Kavazanjian adds "Charlie was the first farmer to no-till soybeans into corn-stalks back in the 1980s... Now it's a common practice." The farmers continued that "[w]hen you're an innovator, some things you try don't always work out, but you learn and go on... Now we're the first farmers in the area to be zone tilling and using GPS to apply fertilizer and weed control."[25] Wisconsin dairy farmer Cory Holig, who also holds a university soil science degree, included among his reasons for installing a wind turbine (and solar photovoltaic panels) the desire to preserve his recently deceased father's legacy as a technological pioneer. He observed "[m]y dad was an innovator, and my grandfather was, too. We are always looking for new opportunities. When they come your way, you can't wait. You have to take a chance."[26] Thus, these farmers embrace these machines because they reinforce their view of themselves as part of families of "inborn innovators," not just because of future financial gain.

In a 2007 article, Illinois farmer Kurt Williams overlooked compaction of soil caused by the wind turbine construction company because the impressively massive machine represented a "legacy" for his five- and seven-year-old sons much as the 1851 windmill still standing on his farm represented his ancestors. In the article's accompanying front page photo, Williams stands in front of his wind turbine with a piece of complex mechanical hardware displaying wires (possibly a portion of the connected power station). The caption of the picture reflects the ultramodern tendency to frame innovation as an inborn trait residing within farm families, "ON BOARD: Ellsworth farmer Kurt Williams is part of a new legacy of Prairie State farmers who are cashing in on a commercial wind farm, Twin Groves." Inside the pages of the *Prairie Farmer* issue, Williams poses with his wife and two boys with a vast expanse of Illinois farmland in the background and a barn far in the distance.[27] Again, the article demonstrates a link commonly expressed in discourses of rural capitalist modernity between technology, ancestry, and legacy that reinforces rural identities as ultramodern users of technology.

Residents and farmers of King City, Missouri expressed a similar view of turbines as symbols of progress and modernization by describing the coming utility-scale wind farm as something "cutting edge."[28] One resident even postulated that, in the future, the town would build a "museum honoring wind power."[29] As Minnesota farmer Theodore Scharden stated simply, "I think the windmills are neat."[30] Similarly, the Iowa Farm Bureau web site in 2015 proffered the motto "People.

Progress. Pride," boasting that "Iowa has also capitalized on its geography and high prevailing winds to harvest the newest and most bio-friendly energy crop yet – wind power." While the article extols the benefits to farmers in the form of lease income, more importantly, it observes that turbines lead to progress by "adding to the energy independence of Iowa and America."[31]

The association of technology with rural modernity pervades statements from farmers praising wind turbines in the 1990s and 2000s.[32] In addition, farm magazine articles reflect a strong production ethos that views the wind as another "crop" and the wind turbines as serving the same purpose as the tractor or combine. As stated by Missouri farmer Mike Waltemath after receiving a wind turbine, "[n]ow we are farming the wind."[33] Iowa farmer Roger Kadolph also viewed the wind turbines through the lens of a modern capitalist producer: "I didn't really expect them [the power company] to come all the way out here in northern Iowa to start a wind farm," he says. "But this is really great. Now we grow corn on the ground and generate power in the air – all on the same piece of property."[34] Such discourse suggests that even if windmills had never existed across the Midwest, farmers would still use turbines to perform their modern identities. Photos of farmers and their families with wind turbines often include crops or the harvest to associate the technology with productive capacity (Figure 8.1).

FIGURE 8.1   "Doug Brinkman with truck full of corn and Wind Turbines in the background, taken by Dennis Brinkman" Author's personal records. Charles City, Iowa, 2009.

Further, while farmers with turbines on their land obtain lucrative economic benefits, their support for the devices (and the support gained from townsfolk who do not have turbines on their property) stems, in large part, from a conception derived from an amalgamation of the familiar rural-urban conflict and a new globalized sensibility. These farmers view the turbines as symbols of their technological savvy, represented not just by these machines, but also by their GPS- and laser-guided tractors, their yield monitoring systems, and the computerized networks they develop themselves and employ in global commodity markets (Figure 8.2).[35]

Many farmers view wind turbines not simply as profit-making devices, but as symbols of an in-born ultra-modernization that they inherited from their ancestors and will pass onto their children. Hence, the front cover of *Progressive Farmer* in Figure 8.2, typical of the way rural publications display wind farms, shows six large turbines jutting gloriously into a bright blue sky with a verdant expanse of crops below. The top of the machine in the foreground grazes the bottom of the title *Progressive Farmer*. Thus, similar to the way electric utility ads in the 1920s directed the eyes upward to associate the artifact with the future and progress (see Figure 3.3), the wind turbines visually lead the viewer upwards toward becoming

FIGURE 8.2   Front page of *Progressive Farmer*, 2009, showing the link between wind turbines and ultra-modern identity. *Progressive Farmer* (October 2009): Front Cover.

a more "progressive farmer." As I have argued in this chapter, this type of positive depiction of wind turbines only makes sense if one understands it within the context of an ultramodern rural identity. Such an agrarian self-image gives a particularized meaning to all artifacts used for productive purposes, not just wind turbines. According to farmers' view of themselves, the sophisticated technological knowledge they have acquired supersedes, at least in their minds, the know-how of city folk who delight in the latest smartphones but who rarely design or operate equally complex hardware and software.

In sum, this cultural analysis may explain why many Midwestern farmers embrace wind turbines and PIMBY views, even though they did not build or design the machinery themselves. The turbines, along with other modern and sophisticated technologies, symbolize the most recent step in a long-term transition of farmers who have contributed to creating an ultramodern identity, one that still goes largely unappreciated by their relatively backward urban cousins.

Rural Ultramodern Identity Leading to *Rejection* of Technologies

While the ultramodern way farmers have come to view themselves creates an acceptance of wind turbines and the most advanced machines throughout the Corn Belt, it also has led to rejection or resistance of other contemporary technologies, systems, or work processes. As with the factory farms in the 1920s such as Hawthorn Farm, many of these contemporary technologies offer economic or technological advantages but garner resistance among farmers because they violate certain features of rural identity. Once again, the way an object fits into the user's sense of self offers a more satisfactory explanation for its adoption or rejection than its potential financial or design benefits. Additionally, the newness of the technology does not alone determine its acceptance among users. Midwestern agrarians only embrace an object with enthusiasm when they perceive it both as being the "latest" *and* as advancing their control over work and property, preserving family-based production, and fitting into a legacy of innovativeness. Recently, a minority of Midwest farmers have even rejected wind turbines and pressured counties to pass construction bans even though farmers receive significant rent income and installation payments. Corn Belt farmers most often express NIMBY attitudes when wind turbine or solar companies own the equipment and export the electricity outside the community and when the devices become so numerous that the landscape resembles an urban setting. Thus, the same technology can garner resistance among rural Midwesterners when it takes control away from farmers and their families for the benefit of a far off urban other and fails to reinforce the agrarian's self-image as an innovator.[36] When technologies hinder these basic tenants of rural ultramodernity they sometimes meet resistance notwithstanding their newness or economic benefits.

Corn Belt farmers recently have also resisted the most advanced tractors and combines when the machines contain locking technology intended to prevent the reprogramming of hardware and software. From the farmer's perspective, such devices, while new, challenge ultramodern rural identities because they prevent the agrarian from fully controlling work and property. In addition, the software locks impede the practice of tinkering with the latest hardware, a performative practice which the farmer has used since at least the 1920s to reinforce his or her notion of modernity. According to the Iowa-Nebraska Equipment Dealers Association, the locks prevent the damage of the latest combines and tractors costing hundreds of thousands of dollars and ensure the machines comply with safety and environmental regulations.[37] Rather than viewing the technology as necessary to ensure the safety and efficiency of an enormously complex network of machinery and big data, ultramodern farmers regard them as only the latest threat to their identities from a distant and urban "other."

One hog farmer in Nebraska, Danny Kluthe, modified his tractors to run on pig manure in 2017. Kluthe explained

> I take the hog waste and run it through an anaerobic digester and I've learned to compress the methane. I run an 80 percent methane in my Chevy Diesel Pickup and I run 90 percent methane in my tractor. And they both purr. I take a lot of pride in working on my equipment.

Thus, Kluthe found the locks so objectionable not primarily because they reduced his profits but because they prevented him from using his machinery in a way that reinforced his identity as an ultramodern re-designer of complex technology.[38]

As a result, many farmers go as far as using pirated software from Eastern Europe to hack into the engines of tractors and combines as a means of working with the complex machinery. These ultramodern agrarians sought out this hacking solution on their own on the Internet from black markets in Poland and Ukraine without assistance from computer or mechanical engineers. Given the importance of the newest tractors in reinforcing the farmer's view of himself as an independent capitalist combating Soviet collectivization during the Cold War, it is perhaps ironic that the contemporary agrarians seek out pirating software from the former Eastern Block. On the other hand, such willingness to embrace post-Soviet technology highlights my contention that ultramodernity has somewhat abandoned the nationalistic concerns of rural capitalistic modernity prevalent in the 1950s–1960s in favor of a more global outlook.

Farmers see themselves as more modern than urban dwellers, in part, because they operate at the cutting edge of complex technological systems in a globalized network. As such, the fact that the farmer can locate and use software developed as far away as Eastern Europe reinforces, rather than weakens, her ultramodern identity. In fact, heads of consumer and repair advocacy groups expressed ignorance that such pirated software even existed. Nevertheless, one such advocate, Gay Gordon-Byrne, head of the Repair Association, almost expected such ingenuity on the part of farmers in finding ways to fix their machinery given their ultramodern identities. Of the practice of hacking, Gordon-Byrne declared, "It's not extreme. I mean, if you have something and you can't fix it and you get it on the Internet and find a way to fix it, I don't blame anyone for trying it." Farmers also petitioned some Midwest state legislators, including Illinois and Nebraska, to introduce bills requiring the legal sale of repair software.

Another Nebraska farmer, Kyle Schwarting, described the software lock on his new combine in 2017, "You're paying for the metal but the electronic parts technically you don't own it. They do." Schwarting resented the fact that "they" installed a device that required him to call his John Deere dealer every time his engine broke down rather than fixing it himself. While citing the "hundreds" of dollars it costs him in such situations to have the dealer with the "software key" fix the engine, such cost seems out of proportion to the vitriol with which Schwarting regarded the lock. The combine itself, after all, costs the Nebraska farmer $200,000 to $300,000 used.[39] Given such an expense, surely the few hundred dollars in repairs, which likely represents a tax write-off, seems minor. In addition, the pirated European software that farmers seek to hack the lock software often costs €499 (roughly $529–590) or more to download.[40] More than the money, Schwarting expressed annoyance at the fact that "they" prevented him from using the advanced mechanical knowledge he gained running an auto shop prior to farming. The article exasperatedly noted Schwarting "thought he'd be a natural to do the mechanical work himself" given his experience as an ultramodern user of machines.[41] Thus, the restraint the lock posed to his identity, not just the economic cost of repair, explains why Schwarting objected so vehemently to the device.

Schwarting did not find himself alone in regarding the software locks on the newest farm machinery as an affront to his ultramodern identity. After speaking with several more Corn Belt farmers, another rural columnist reported that Midwest agrarians saw the software locks "as an attack on their sovereignty." "The nightmare scenario, and a fear I heard expressed over and over again in talking with farmers," the reporter related, "is that John Deere could remotely shut down a tractor and there wouldn't be anything a farmer could do about it." One farmer using pirated Ukrainian software expressed similar distrust of the manufacturer when he stated "What happens in 20 years when there's a new tractor out and John Deere doesn't want to fix these anymore? Are we supposed to throw the tractor in the garbage, or what?" Upon joining a forum for the sale of hacking software, the columnist "found dozens of threads from farmers desperate to fix and modify their own tractors." Corn Belt agrarians especially resented license agreements whereby the user typically agrees not to repair the machine and waives the right to sue the manufacturer for lost profits in the event of disrepair.

The advanced programs purchased by farmers to hack the latest tractor engines also reflect an ultramodern identity. Not only do these farmers understand how to fix the advanced mechanical hardware after hacking into the engine, but also how to operate the software. For example, farmers download the "John Deere Service Advisor" pirated from highly trained John Deere technicians to recalibrate tractors and diagnose mechanical failures. One rural resident described the Service Advisor program casually as "It can program payloads into different controllers. It can calibrate injectors, turbo, engine hours and all kinds of fun stuff." Farmers also download payload files "that can customize and fine-tune the performance of the chassis, engine, and cab" as well as data link drivers allowing the farmer's "Service Advisor laptop to actually communicate with the tractor controllers." The article describing the "sketchy-looking" Eastern European hacking forums further reported "Also for sale (or free download) on the forums are license key generators, speed-limit modifiers, and reverse-engineered cables that allow you to connect a tractor to a computer." Many farmers watch demos of the complex software on YouTube and implement them on their farms themselves.[42]

More generally, farmers have debated whether the use of precision farming technologies enhances their ultramodern identity or threatens it. On the one hand, some farmers like Jeff Heepke and his brothers of Edwardsville, Illinois embrace precision farming because "We're always looking for that next thing." Heepke participated in a trial of the Monsanto Fieldscripts program in which new software directed a GPS planter to plant different hybrid corn based on historic yield, soil type, and elevation data. By varying the type and amount of seeds planted every 30 feet, Heepke could push his yield higher in better parts of his field. Other farms adopted the Fieldscripts software after it was released in Illinois, Iowa, Indiana, and Minnesota in 2014.

Another farmer in Illinois, Shelly Finfrock, participated in a trial of a service called Climate Basic where she received a text message every morning telling her how much rain each field received in the last 24 hours down to hundredths of an inch. In addition, she constantly reviewed Google satellite data with rainfall estimates for each field. Based on the information, Finfrock knew when and where not to plant her fields, which spread over five counties. She also accessed rainfall and temperature data on her land going back several decades. Finfrock gladly entered in her soil test results and yield results into the program to attain the most useful data possible. From its launch throughout the Midwest in 2014 through 2017, farmers used Climate Basic on more than 1 million acres.

On the other hand, some farmers regard the collection and mining of big data about their farms by corporations like Monsanto, which has acquired Climate Basic, as "big brother." However, such concerns that big data could serve as a vehicle for an outside "other" to impede farmer's independence rarely leads to a wholesale rejection of precision farming, only resistance to services such as Climate Basic or Fieldscripts. For example, Bellville, Illinois farmer Greg Guenther elected to only share his combine's yield data to a collection of experts and specialists he employed to help him make decisions about his crops, but avoided using a program like Climate Basic, which would have moved that information onto the cloud. Guenther perceived of any cloud-based program as diminishing his control over his land by giving an outside "other" access to yield data. As Guenther explained, "Once it's off my computer who knows who's going to get that data?" and "This is my data, but I'm very nervous about letting it out of my hands." Even Finfrock expressed trepidation with using Monsanto-owned Climate Basic as "Having big brother watching."[43]

Farmers have also resisted the adoption of large corporate hog farms despite their economic benefits because they threaten to undermine their rural ultramodern identities. One may view these hog farms as a contemporary technological system involving infrastructure, medicines, and animals to increase efficiency and production of pork. Aside from the bad odor and physical demands of the work, farmers especially object to the contract arrangements of these corporate hog operations in which larger companies own the hogs and dictate the conditions of care.[44]

USDA agricultural economist Nigel Key has provided empirical evidence that "growers have a strong preference for autonomy" even though "contract operators earned significantly more on average than independent operators." Key expressed his conclusions in the language of an agricultural economist as, "We find over a wide range of assumptions that the autonomy premia is positive, large enough to be economically important, and of the same order of magnitude as the risk premia."[45] In plainer language, Key surmised that even when accounting for many other variables, farmers across the U.S. reject more profitable and less financially risky technological systems because they impede their independence and control. The lack of control on the farmer's part and the removal of production from the purview of the family farm render these new hog operations immoral when viewed through the lens of rural ultramodernity.

Discourse expressed by Midwest farmers regarding these new hog production systems confirms Key's findings. Jeff Seabaugh, a hog farmer in Montgomery County, Illinois described himself as "an indentured servant." While framing his objections in economic terms, Seabaugh based the hatred of his job on his lack of control over work,

> The top dollar comes to the guy that owns the pigs, not us that raise the pigs. We are at the mercy of them. You have to go along with whatever they say. You have no voice in it. It is getting worse.

Harold Steele, an Illinois hog farmer who helped to develop new hog confinement systems that increased productivity saw his work, regrettably, as destroying rural globalized ultramodernity. "Simply put," Steele explained, "corporations took over the hog business. Where does this put 'we the people'? Into a garbage can." Many of these farmers, including Steele, also tend to view the environmental air and water pollution emitted by the large hog lots as caused by a lack of control by the farmers over work processes and ownership by corporations often located in far off China or Brazil. From these farmers' point of view, the environmental damage is not an inherent feature of large hog farming, but a result of violating the morality inherent in rural identities. Contamination of the environment stands in for the pollution of moral family-based production processes and farmer autonomy. For example, when agrarians controlled work processes prior to the takeover by foreign corporations, the farmers described the hog farming as follows,

> As their facilities expanded to hold thousands of pigs, they built earthen lagoons to stockpile the manure, then applied that natural fertilizer to nearby fields. The corn and grain fed their livestock, creating a 'virtuous cycle' of reused waste — the very definition of sustainable agriculture.

After the "seismic shift" in which Chinese and Brazilian firms purchased U.S. corporations and took control away from farmers, Corn Belt residents saw large hog farming as immoral, polluting, and contrary to rural ultramodernity. Steele expressed this sentiment of farmers that such contract hog farms violated the independence so central to rural modern identities when he stated, "To be a farmer, you're responsible for labor, capital and management. Those three steps are essential. All you've got left is labor."

The hog farm contracts bring sizeable pecuniary benefits, often allowing farmers to insulate themselves from risk and obtain bank loans to expand. Nevertheless, many farmers like Seabaugh reject such economic advantages because the contracts damage their identities as independent ultramodern producers farming in a family-based system. Steele's son quit large hog raising because under the contracts "we were a hog house janitor." Thus, farmers have rejected this technological system that they helped to develop and once thought economically advantageous because they reduce their self-image to mere labor, the same reasons their great-grandfathers rejected the efficient factory farms of the 1920s. Of course, many contract hog farmers can cite financial disadvantages such as the inability of the contracts to pay enough per hog to keep pace with inflation and depreciation of capital. However, unlike wind turbines, the negative effects on rural identity led farmers to emphasize the disadvantages of these technological systems.[46] Farmers can raise the same objections, after all, to crop farming which often exposes farmers to low corn or soybean prices, and few farmers respond by giving up their ultramodern combines.

Some Midwest farmers even formed cooperative associations, such as the Fresh Air Pork Circle in Alta Vista and Elma, Iowa, to avoid contract hog farming by selling to consumers preferring free-range pork. These farmers often noted a desire to preserve their rural ultramodern identities by resisting urban industrialization rather than an economic motivation. As one member of the Fresh Air Pork Circle in the late 1990s explained, "We don't want to get rich off this, we just want to keep raising hogs!" Mark A. Grey, a scholar who worked closely with Fresh Air Pork as part of an ethnographic study observed,

> Farmers are a diverse lot, who value their independence. In the face of structural change that clouds the future of their way of life, these farmers all kept farming

in large part to maintain their independence, to 'be my own boss.' They had a 'they-can-go-to-hell-if-they-don't-like-it' attitude.

Indeed, when asked about whether it would be more economically advantageous to adopt contract farming rather than a free-range co-op, the founder of Fresh Air Pork simply responded, "Those bastards can go to Hell!"

To resist the threat of industrialization from an outside urban "other," these farmers went through considerable efforts to form a cooperative and market their products themselves to new buyers even though their business plan projected only the sale of 100 hogs for a few thousand dollars in its first year, a small amount for most Iowa pig farmers. As Grey said after spending about a year with the cooperative "Raising hogs was more than an economic activity, it was a lifestyle, and a lifestyle they loved." Grey noted widespread criticism of corporate concentration of hog farming among farmers in northeast Iowa, even among agrarians that did not belong to the cooperative. While observing conversations among farmers at the Howard Country Equity in Iowa where farmers sold grain, bought supplies, and visited over coffee, Grey heard widespread criticism of other farmers who had signed production contracts and gossip often centered on immoral or illegal activities carried out by distant owners.

Significantly, Grey noted all these farmers

> Also embraced aspects of industrialized agriculture. Indeed, most were grain farmers who sought the largest yields possible, regardless of the environmental impacts from application of fertilizer and pesticides or the availability of hybrid seeds from only a handful of corporations. Some were also dealers for seed companies.

This support for large grain farming while rejecting large hog production only raises confusion if one views agriculture through the lens of rational economics or urban, organic, or environmentalist worldviews. My model of rural identities and notion of performative use reconcile this conundrum. Namely, farmers viewed the planting of hybrid seeds as reinforcing their identities because they could do so while controlling work and machines within a family-based production process. Large contract hog farming on the other hand threatens this rural sense of self. Environmental or economic reasoning only stand in as proxy arguments in support of ultramodernity. Such vehement resistance of some contemporary technologies or systems, like integrated hog operations, while at the same time embracing others such as wind energy or large combines only makes sense if one understands how farmers use objects to perform ultramodern rural identities and discourses. As this book has shown, it is not industrialization per se but a whole suite of ideas, views, and ideologies formed over many years and comprising a distinct Midwest rural version of modernity that farmers form and reinforce through technology. A deeper understanding of ultramodernity in the Corn Belt resolves this apparent paradox, which Grey found so vexing while eavesdropping at the Howard County Equity.[47]

Contemporary hog-raising systems do not present the first instance in which rural modern identities clashed with notions of modernity and industrialism from outside the farm. Farmers' resistance in defense of rural modernity has existed in the Midwest at least since the 1910s. For example, federal programs to eliminate bovine tuberculosis from 1917 to 1941 met considerable resistance in the Corn Belt. This opposition culminated in the "Cow Wars" in Iowa in 1931 in which violent protests among farmers led the governor of Iowa to declare martial law. The crisis began in 1917, when Federal health authorities devised a utopian plan of banishing bovine tuberculosis to make milk supplies safer. Health officials planned to test all cattle by injecting each cow with a small amount of the disease and then culling the ones that reacted. These efforts required what one historian has called "the unprecedented peacetime use of police power" as officials tested every cattle farm in the nation and destroyed 3.8 million cows.

Farmers with the help of the American Farmer's Union and the American Medical Liberty League mounted failed political and legal challenges to the anti-tuberculosis campaign throughout the 1920s. In 1929, the Iowa legislature made all testing in compliance with the federal program mandatory, further antagonizing many farmers. On March 8, 1931, in Tipton, Iowa, about 1,000 farmers confronted government inspectors accompanied by a police force to test the herds on the W.C. Butterbrodt and E.C. Mitchell farms and a violent clash erupted. On March 19, 1931, Milo Reno of the Iowa Farmer's Union and Jacob W. Lenker of the Iowa Farmers Protective Association organized a march of 1,500 farmers on the Iowa state capitol and many farmers spoke on the House floor to express their arguments against the anti-tuberculosis program.

At first, one may view the Iowa Cow Wars as simply motivated by economic self-interests of dairymen. However, an examination of the farmer's testimony and the realities of the program reveal a non-economic motivation. According to the speeches delivered to the House, the farmers regarded public health justifications for the program as simply rhetoric cloaking a system of graft among health officials and packing plants. After killing a cow without the consent of the farmer, the health inspector would send the carcass to a USDA-approved packer who would sell the meat after trimming the diseased parts. Farmers thus saw the program as a threat from an outside "other" because they suspected the health inspectors to "be in cahoots" with the USDA packers. This collusion seemed even more likely to farmers since many also questioned the scientific integrity of research showing that the test revealed a health problem in otherwise healthy-looking cows. Many Corn Belt agrarians also doubted scientists' warnings that bovine tuberculosis could transmit to humans through milk.

Furthermore, the federal indemnity process worked in a way such that the change in prices farmers received from the government to compensate them for their loss of cattle lagged the change in market prices for beef. By 1931 when the

Cow Wars ignited, the price of beef fell such that farmers actually received more money from the government in indemnity for a culled animal than they received selling that cow on the market. Many farmers were even accused of attempting to doctor the tuberculosis test in the early 1930s to receive compensation rates above market price. In addition, from 1927 to 1932, only 1.7% of the tested animals in Iowa reacted and the culled animals in 1930 represented only about 0.012% of Iowa's total farm property according to the 1930s census.

Thus, the farmers leveled such fierce opposition to the anti-tuberculosis program not for rational economic reasons but because the health officials and government testers represented a threatening urban actor attempting to impose an outside version of modernity which took control and decision-making away from the family. As two historians have noted,

Many farmers were alarmed that outsiders claiming expert status were meddling in their day-to-day operations; opponents also bitterly complained that they had few avenues for appeal if their animals were condemned. These concerns often carried an individualistic or libertarian theme deriding the heavy-handed actions of government bureaucrats who threatened individual liberty, property, and (if the science was wrong) even lives.[48]

Indeed, by 1941, the federal anti-tuberculosis program generated returns to the livestock sector of more than ten times the program's costs and saved tens of thousands of human lives. By all measures, the health officials attained a resounding medical and economic success that made all farmers better off. One may argue that a single farmer who lost several cows because of culling may not appreciate the overall economic or health benefits of the program. But this argument fails to support a rational economic motivation for the Cow Wars because, again, by 1931, such a farmer received higher compensation rates for killed cows than she would have received on the beef market. The sight of a huge Federal police force traveling the countryside testing and culling cattle elicited the reaction among many farmers of "Not on my farm!" regardless of economic considerations.[49]

Hence, as with rural opposition to contemporary hog operations or tractor locks, the Cow Wars of 1931 ultimately occurred not because of farmers' concerns with money, but because farmers viewed the federal culling program as a threat to their identities as modern agrarians with control over work and property on a family farm.

More recently, the North Dakota Farmers Union (NDFU) even protested a proposed state bill that exempted corporate pork and dairy farms from a 1932 law banning corporate ownership of farms and farmland. Farmers in the union gathered enough signatures to force a vote on the bill and in 2016 voters voted "no" to the referendum by 76%. The farmers overwhelmingly won the referendum debate by appealing not only to economic or environmental arguments, but also directly to rural identities. Mark Watne, the president of the NDFU expressed his opposition to the corporate farming bill as, "We always believed that the people of North Dakota would agree that the family farm structure is best for our state's economy and our communities."[50] Once again, the way farmers react negatively to integrated hog farming, a technological system, which at first glance would seem to attract an industrialized grower seeking to maximize profit and reduce risk, demonstrates this culturally specific form of performative use. Since the 1970s, farmers have also used technology to resist organic reformist discourses perceived as threating their rural ultramodern identities.

Notes

1 Bill Redeker, "Blow Back from Neighbors Over Wind Farms," *ABC News*, May 6, 2007, http://abcnews.go.com/WNT/story?id=3065474&page=1 (accessed 5/15/16).

2 A variation of this story is published in Brinkman and Hirsh, "The Effect of Unarticulated Identities and Values on Energy Policy."

3 Brinkman and Hirsh, "The Effect of Unarticulated Identities and Values on Energy Policy;" Dar Danielson, "Iowa Moves Up to Second in Electricity Created by Wind Power," *Radio Iowa*, January 28, 2016, http://www.radioiowa.com/2016/01/28/iowa-moves-up-to-second-in-wind-power-energy/ (accessed 5/13/16).

4 Part of this section is published in the article, Brinkman and Hirsh, "Welcoming Wind Turbines and the PIMBY ('Please in my backyard') Phenomenon."

5 Maarten Wolsink, "Invalid Theory Impedes Our Understanding: A Critique on the Persistence of the Language of NIMBY," *Transactions of the Institute of British Geographers* 31 (2006): 85–91; Patrick Devine-Wright, "Beyond NIMBYism: Towards an Integrated Framework for Understanding Public Perceptions of Wind Energy," *Wind Energy* 8 (2005): 125–139; Richard F. Hirsh and Benjamin K. Sovacool, "Wind Turbines and Invisible Technology: Unarticulated Reasons for Local Opposition to Wind Energy," *Technology and Culture* 54 (October 2013): 705–734.

6 Martin J. Pasqualetti, "Opposing Wind Energy Landscapes: A Search for Common Cause," *Annals of the Association of American Geographers* 101, no. 4 (2011): 907–917; Patrick Devine-Wright, "Place Attachment and Public Acceptance of Renewable Energy: A Tidal Energy Case Study," *Journal of Environmental Psychology* 31 (2011): 336–343; Michael Dear, "Understanding and Overcoming the NIMBY Syndrome," *Journal of the American Planning Association* 58 (1992): 288. For information on the problems encountered in Falmouth, see WGBH, "The Falmouth Experience," at http://www.wgbh.org/wcai/turbine.cfm, obtained January 23, 2015). The offshore wind debate in Massachusetts has been explored in Wendy Williams and Robert Whitcomb, *Cape Wind: Money, Celebrity, Class, Politics, and the Battle for Our Energy Future on Nantucket Sound* (New York: Public Affairs, 2007).

7 Carolyn Fortuna, "Cape Cod Offshore Wind Moves Ahead — Despite Controversy," *CleanTechnica*, August 28, 2022, https://cleantechnica.com/2022/08/28/although-controversy-continues-cape-cod-offshore-wind-moves-ahead/ (accessed 10/27/23).

8 Redeker, "Blow Back from Neighbors Over Wind Farms."

9 Lester R. Brown, *Plan B 4.0: Mobilizing to Save Civilization* (New York: The Earth Policy Institute, W.W. Norton & Company, 2009), 116–117; Ron Pernick and Clint Wilder, *The Clean Tech Revolution: Winning and Profiting from Clean Energy* (New York: Harper Collins, 2009): 62–64; Michael C. Slattery, Becky L. Johnson, Jeffrey A. Swofford and Pasqualetti, "The Predominance of Economic Development in the Support for Large-Scale Wind Farms in the U.S. Great Plains," *Renewable and Sustainable Energy Reviews* 16, no. 6 (2012): 3690–3701; Jeffrey Swofford and Michael Slattery, "Public Attitudes of Wind Energy in Texas: Local Communities in Close Proximity to

Wind Farms and their Effect on Decision-Making," *Energy Policy* 38, no. 5 (2010): 2508–2519; Pasqualetti, "Wind Power: Obstacles and Opportunities," *Environment* 46, no. 7 (2004): 22–38.

10 Jacob Sowers, "Fields of Opportunity: Wind Machines Return to the Plains," *Great Plains Quarterly* 26, no. 2 (2006): 99–112. See also Pasqualetti, Robert Righter, and Paul Gipe, ed. Cleveland J. Cutler, *History of Wind Energy*, "Rejuvenated North America," vol. 6, *The Encyclopedia of Energy*, ed. Cutler J. Cleveland and Robert U. Ayres (Amsterdam: Elsevier, 2004), 430.

11 Wind Energy Foundation, "Polls," at http://www.windenergyfoundation.org/ wind-at-work/wind-consumers/polls (accessed 6/7/15).

12 Brinkman and Hirsh, "The Effect of Unarticulated Identities and Values on Energy Policy;" "Wind Energy," *Iowa Environmental Counsel*, https://www.iaenvironment.org/ our-work/clean-energy/wind-energy (accessed 10/19/23).

13 Sowers, "Fields of Opportunity," 99–112; Pasqualetti, Righter, and Gipe, "Rejuvenated North America," 430.

14 Pasqualetti, Righter, and Gipe, "Rejuvenated North America," 430.

15 Sowers, "Fields of Opportunity," 109; Slattery, Johnson, Swofford, and Pasqualetti, "The Predominance of Economic Development in the Support for Large-Scale Wind Farms in the U.S. Great Plains," 3698.

16 John Byczynski, "My Father's Past, My Children's Future: Agrarian Identity and a Powerline in Minnesota, 1974–1980," *Agricultural History* 88, no. 3 (2014): 313–332.

17 Breimyer, Over-Fulfilled Expectations, 63.

18 Ibid.

19 See also Brinkman and Hirsh, "The Effect of Unarticulated Identities and Values on Energy Policy."

20 Kline, *Consumers in the Country*, 6; see also Hal S. Barron, *Mixed Harvest: The Second Great Transformation in the Rural North, 1870–1930* (Chapel Hill: University of North Carolina Press, 1997), 7–16, 243–245; Curtis S. Beus and Riley E. Dunlap, "Endorsement of Agrarian Ideology and Adherence to Agricultural Paradigms," *Rural Sociology* 59, no. 3 (1994): 462–484; Anderson, *Industrializing the Corn Belt*; Kendra Smith-Howard, *Pure and Modern Milk: An Environmental History since 1900* (New York: Oxford University Press, 2014), 3–11; Peter D. McClelland, *Sowing Modernity: America's First Agricultural Revolution* (Ithaca, NY: Cornell University Press, 1997).

21 For a discussion of social constructions of nature, see Peter Coates, *Nature: Western Attitudes since Ancient Times* (Berkeley: University of California Press, 1998), 3–17.

22 Agricultural drainage "tiles" normally come in the form of perforated tubes buried in the ground to remove excess water from the soil profile to enhance crop production. "Ag 101-Drainage," "Ag 101-Drainage," U.S. Environmental Protection Agency, 94. http:// www.epa.gov/agriculture/ag101/cropdrainage.html (accessed 8 May 2015).

23 Melissa Hemken, "Wind Aids Local Economy," *Wallaces Farmer* (February 2006): 42.

24 Susan Thompson, "Iowa's Turn," *Wallaces Farmer* (February 2008): front page.

25 Fran O'Leary, "Taking 'a Chance' on Wind," *Wisconsin Agriculturalist* (August 2009): 8.

26 Ethan Giebel, "Teamwork Fuels Success at Dairy," *Wisconsin Agriculturalist* (November 2013): 50.

27 Anna Barnes, "Harnessing the Wind," *Prairie Farmer* (April 2007): Front Page, 8–9.

28 Matthew Lablanc, "Change in the Wind," *Columbia Daily Tribune* (November 12, 2006), 1–4.

29 Jerilyn Johnson, "Wind Power Takes Off," *Missouri Ruralist* (March 2008): 6; Johnson, "Wind Energy Is Now a Reality in Rock Port," *Missouri Ruralist* (March 2008): 7.

30 Scharden meant "wind turbines." Douglas Jehl, "Curse of the Wind Turns to Farmers' Blessing," *New York Times* (November 26, 2000), 1–2.

31 "Energy: Harnessing the Power in and Above Iowa Fields," *Iowa Farm Bureau*, http:// www.iowafarmbureau.com/public/114/ag-in-your-life/energy (accessed 6/1/15). See

also Rod Swoboda, "A Wind Energy Lesson at the Fair," *Wallaces Farmer* (September 2007): 30; Tom J. Bechman, "Rural Revival in Wind," *Indiana Prairie Farmer* (February 2008): Front Page.

32 See for example Mauricio Espinoza, "Wind Blows Dollars into Northwest Ohio," *Ohio Farmer* (May 2012): 26; Johnson, "Wind Power Takes Off, 6; Johnson, "Wind Energy Is Now a Reality in Rock Port," *Missouri Ruralist* (March 2008): 7.

33 Johnson, "Wind Power Takes Off," 6.

34 Abraham McLaughlin, "Reaping the Wind," *Christian Science Monitor* (March 9, 1999), 2; see also "Putting Wind in the Rotation," *Wallaces Farmer,* April 23, 2012, http://farmprogress.com/story-putting-wind-rotation-9-58874.

35 See for example, *Progressive Farmer* (October 2009): Front Cover.

36 Jennifer Hiller, "'Over Our Dead Bodies': Backlash Builds Against $3 Trillion Clean-Energy Push," *Wall Street Journal,* May 8, 2023, https://www.wsj.com/articles/inflation-reduction-act-backlash-clean-energy-wind-solar-f3d4d900 (accessed 8/1/23). Hiller indicates that NIMBY attitudes towards wind turbines is still a minority view in the Midwest and highly localized, existing most heavily in Kansas while other Midwest states like Iowa still show significant PIMBY views; see also Taylor Fisher, "The Power of the People: Resisting Big Wind in Rural Iowa," *The Georgetown Environmental Law Review,* January 24, 2023, https://www.law.georgetown.edu/environmental-law-review/blog/the-power-of-the-people-resisting-big-wind-in-rural-iowa/ (accessed 10/19/23).

37 Grant Gerlock, "Farmers Look for Ways to Circumvent Tractor Software Locks," *All Tech Considered: NPR,* April 9, 2017, 1–11, http://www.npr.org/sections/alltech-considered/2017/04/09/523024776/farmers-look-for-ways-to-circumvent-tractor-software-locks (accessed April 10, 2017).

38 Jason Koebler, "Why American Farmers are Hacking Their Tractors with Ukrainian Firmware," *Motherboard,* March 21, 2017, https://motherboard.vice.com/en_us/article/xykkkd/why-american-farmers-are-hacking-their-tractors-with-ukrainian-firmware (accessed 3/22/17).

39 Gerlock, "Farmers Look for Ways to Circumvent Tractor Software Locks."

40 Koebler, "Why American Farmers are Hacking Their Tractors with Ukrainian Firmware."

41 Gerlock, "Farmers Look for Ways to Circumvent Tractor Software Locks."

42 Koebler, "Why American Farmers are Hacking Their Tractors with Ukrainian Firmware."; Gerlock, "Farmers Look for Ways to Circumvent Tractor Software Locks."

43 Maria Altman, "More Illinois Farmers Are Embracing High Tech Ag-But Is 'Big Data' Too Much Like 'Big Brother?,'" *St. Louis Public Radio,* May 6, 2014, http://news.stlpublicradio.org/post/more-illinois-farmers-are-embracing-high-tech-ag-big-data-too-much-big-brother-1#stream/0 (accessed 2/2/17).

44 For a discussion of citizens coalitions forming in the Midwest to oppose corporate hog farming based on environmental impacts see Heather Williams, "Fighting Corporate Swine," *Politics and Society* 34, no. 3 (2006): 369–397; for a study of attitudes among non-farm residents of rural and urban Ohio regarding large corporate hog farming see Jeff Sharp and Mark Tucker, "Awareness and Concern about Large-Scale Livestock and Poultry: Results from a Statewide Survey of Ohioans," *Rural Sociology* 70, no. 2 (2005): 208–228.

45 Nigel Key, "How much do Farmers Value their Independence?" *Agricultural Economics* 33 (2005): 117–126.

46 David Jackson and Gary Marx, "Illinois Contract Pig Farmer: Work Is Low-Paying, Physically Punishing," *Chicago Tribune* (August 8, 2016).

47 Mark A. Grey, "'Those Bastards Can Go to Hell!' Small-Farmer Resistance to Vertical Integration and Concentration in the Pork Industry," *Human Organization* 59, no. 2 (2000): 169–176.

48 Alan L. Olmstead and Paul W. Rhode, "Not on My Farm! Resistance to Bovine Tuberculosis Eradication in the United States," *The Journal of Economic History* 67, no. 3 (2007): 768–809.

49 Ibid.
50 Robin Shreeves, "North Dakota Says 'No' to Corporate Farms," *Mother Nature Network*, June 16, 2016, https://www.mnn.com/your-home/organic-farming-gardening/blogs/north-dakota-says-no-corporate-farms (accessed 1/31/17).

Bibliography

"Ag 101-Drainage." *U.S. Environmental Protection Agency*, 94. http://www.epa.gov/agriculture/ag101/cropdrainage.html (accessed 8 May 2015).

Altman, Maria. "More Illinois Farmers are Embracing High Tech Ag-But Is 'Big Data' Too Much Like 'Big Brother?'" *St. Louis Public Radio*, May 6, 2014. http://news.stlpublicradio.org/post/more-illinois-farmers-are-embracing-high-tech-ag-big-data-too-much-big-brother-1#stream/0 (accessed 2/2/17).

Anderson, J.L. *Industrializing the Corn Belt*. DeKalb: Northern Illinois University Press, 2009.

Barnes, Anna. "Harnessing the Wind." *Prairie Farmer* (April 2007): Front Page, 8–9.

Barron, Hal S. *Mixed Harvest: The Second Great Transformation in the Rural North, 1870–1930*. Chapel Hill: University of North Carolina Press, 1997.

Bechman, Tom J. "Rural Revival in Wind." *Indiana Prairie Farmer* (February 2008): Front Page.

Beus, Curtis S. and Riley E. Dunlap. "Endorsement of Agrarian Ideology and Adherence to Agricultural Paradigms." *Rural Sociology* 59, no. 3 (1994): 462–484.

Breimyer, Harold F. *Over-Fulfilled Expectations: A Life and an Era in Rural America*. Ames: Iowa State University Press, 1991.

Brinkman, Joshua T. and Richard F. Hirsh. "Welcoming Wind Turbines and the PIMBY ('Please in My Backyard') Phenomenon. The Culture of the Machine in the Rural American Midwest." *Technology and Culture* 58, no. 2 (2017): 335–367.

———. "The Effect of Unarticulated Identities and Values on Energy Policy." In *The Handbook of Energy Transitions*, edited by Katherine Araújo, 71–85. London: Routledge, 2022.

Brown, Lester R. *Plan B 4.0: Mobilizing to Save Civilization*. New York: The Earth Policy Institute, W.W. Norton & Company, 2009.

Byczynski, John. "My Father's Past, My Children's Future: Agrarian Identity and a Powerline in Minnesota, 1974–1980." *Agricultural History* 88, no. 3 (2014): 313–335.

Coates, Peter. *Nature: Western Attitudes since Ancient Times*. Berkeley: University of California Press, 1998.

Danielson, Dar. "Iowa Moves Up to Second in Electricity Created by Wind Power." *Radio Iowa*, January 28, 2016. http://www.radioiowa.com/2016/01/28/iowa-moves-up-to-second-in-wind-power-energy/ (accessed 5/13/16).

Dear, Michael. "Understanding and Overcoming the NIMBY Syndrome." *Journal of the American Planning Association* 58 (1992): 288–300.

Devine-Wright, Patrick. "Beyond NIMBYism: Towards an Integrated Framework for Understanding Public Perceptions of Wind Energy." *Wind Energy* 8, no. 2 (2005): 125–139.

———. "Place Attachment and Public Acceptance of Renewable Energy: A Tidal Energy Case Study." *Journal of Environmental Psychology* 31 (2011): 336–343.

"Doug Brinkman with Truck Full of Corn and Wind Turbines in the Background, Taken by Dennis Brinkman" Author's personal records. Charles City, Iowa, 2009.

"Energy: Harnessing the Power in and Above Iowa Fields." *Iowa Farm Bureau*. http://www.iowafarmbureau.com/public/114/ag-in-your-life/energy (accessed 6/1/15).

Espinoza, Mauricio. "Wind Blows Dollars Into Northwest Ohio." *Ohio Farmer* (May 2012): 26.

Fisher, Taylor. "The Power of the People: Resisting Big Wind in Rural Iowa." *The George-town Environmental Law Review,* January 24, 2023. https://www.law.georgetown.edu/environmental-law-review/blog/the-power-of-the-people-resisting-big-wind-in-rural-iowa/ (accessed 10/19/23).

Fortuna, Carolyn. "Cape Cod Offshore Wind Moves Ahead — Despite Controversy." *CleanTechnica,* August 28, 2022. https://cleantechnica.com/2022/08/28/although-controversy-continues-cape-cod-offshore-wind-moves-ahead/ (accessed 10/27/23).

Gerlock, Grant. "Farmers Look for Ways to Circumvent Tractor Software Locks." *All Tech Considered: NPR,* April 9, 2017, 1–11. http://www.npr.org/sections/alltechconsidered/2017/04/09/523024776/farmers-look-for-ways-to-circumvent-tractor-software-locks (accessed April 10, 2017).

Giebel, Ethan. "Teamwork Fuels Success at Dairy." *Wisconsin Agriculturalist* (November 2013): 50.

Grey, Mark A. "'Those Bastards Can Go to Hell!' Small-Farmer Resistance to Vertical Integration and Concentration in the Pork Industry." *Human Organization* 59, no. 2 (2000): 169–176.

Hemken, Melissa. "Wind Aids Local Economy." *Wallaces Farmer* (February 2006): 42.

Hiller, Jennifer. "'Over Our Dead Bodies': Backlash Builds Against $3 Trillion Clean-Energy Push." *Wall Street Journal,* May 8, 2023. https://www.wsj.com/articles/inflation-reduction-act-backlash-clean-energy-wind-solar-f3d4d900 (accessed 8/1/23).

Hirsh, Richard F. and Benjamin K. Sovacool. "Wind Turbines and Invisible Technology: Unarticulated Reasons for Local Opposition to Wind Energy." *Technology and Culture* 54, no. 4 (2013): 705–734.

Jackson, David and Gary Marx. "Illinois Contract Pig Farmer: Work Is Low-Paying, Physically Punishing." *Chicago Tribune,* August 8, 2016.

Jehl, Douglas. "Curse of the Wind Turns to Farmers' Blessing." *New York Times,* November 26, 2000, 1–2.

Johnson, Jerilyn. "Wind Energy Is Now a Reality in Rock Port." *Missouri Ruralist* (March 2008a): 7.

———. "Wind Power Takes Off." *Missouri Ruralist* (March 2008b): 6.

Key, Nigel. "How Much do Farmers Value their Independence?" *Agricultural Economics* 33 (2005): 117–126.

Kline, Ronald R. *Consumers in the Country: Technology and Social Change in Rural America.* Baltimore, MD: Johns Hopkins University Press, 2000.

Koebler, Jason. "Why American Farmers Are Hacking Their Tractors with Ukrainian Firmware." *Motherboard,* March 21, 2017. https://motherboard.vice.com/en_us/article/xykkkd/why-american-farmers-are-hacking-their-tractors-with-ukrainian-firmware (accessed 3/22/17).

Lablanc, Matthew. "Change in the Wind." *Columbia Daily Tribune,* November 12, 2006, 1–4.

McClelland, Peter D. *Sowing Modernity: America's First Agricultural Revolution.* Ithaca, NY: Cornell University Press, 1997.

McLaughlin, Abraham. "Reaping the Wind." *Christian Science Monitor* (March 9, 1999): 2.

O'Leary, Fran. "Taking 'a Chance' on Wind." *Wisconsin Agriculturalist* (August 2009): 8.

Olmstead, Alan L. and Paul W. Rhode. "Not on My Farm! Resistance to Bovine Tuberculosis Eradication in the United States." *The Journal of Economic History* 67, no. 3 (2007): 768–809.

Pasqualetti, Martin J. "Wind Power: Obstacles and Opportunities." *Environment: Science and Policy for Sustainable Development* 46, no. 7 (2004): 22–38.

———. "Opposing Wind Energy Landscapes: A Search for Common Cause." *Annals of the Association of American Geographers* 101, no. 4 (2011): 907–917.

Pasqualetti, Martin J., Robert Righter, and Paul Gipe, ed. Cleveland J. Cutler. *History of Wind Energy*, "Rejuvenated North America." Vol. 6, *The Encyclopedia of Energy*. Amsterdam: Elsevier, 2004.

Pernick, Ron and Clint Wilder. *The Clean Tech Revolution: Winning and Profiting from Clean Energy*. New York: Harper Collins, 2009.

Progressive Farmer (October 2009): Front Cover.

"Putting Wind in the Rotation." *Wallaces Farmer*, April 13, 2012. http://farmprogress.com/story-putting-wind-rotation-9-58874 (accessed 6/1/15).

Redeker, Bill. "Blow Back from Neighbors Over Wind Farms." *ABC News*, May 6, 2007. http://abcnews.go.com/WNT/story?id=3065474&page=1 (accessed 5/15/16).

Sharp, Jeff and Mark Tucker. "Awareness and Concern about Large-Scale Livestock and Poultry: Results from a Statewide Survey of Ohioans." *Rural Sociology* 70, no. 2 (2005): 208–228.

Shreeves, Robin. "North Dakota Says 'No' to Corporate Farms." *Mother Nature Network*, June 16, 2016. https://www.mnn.com/your-home/organic-farming-gardening/blogs/north-dakota-says-no-corporate-farms (accessed 1/31/17).

Slattery, Michael C., Becky L. Johnson, Jeffrey A. Swofford and Martin J. Pasqualetti. "The Predominance of Economic Development in the Support for Large-Scale Wind Farms in the U.S. Great Plains." *Renewable and Sustainable Energy Reviews* 16, no. 6 (2012): 3690–3701.

Smith-Howard, Kendra. *Pure and Modern Milk: An Environmental History Since 1900*. New York: Oxford University Press, 2014.

Sowers, Jacob. "Fields of Opportunity: Wind Machines Return to the Plains." *Great Plains Quarterly* 26, no. 2 (2006): 99–112.

Swoboda, Rod. "A Wind Energy Lesson at the Fair." *Wallaces Farmer* (September 2007): 30.

Swofford, Jeffrey and Michael Slattery. "Public Attitudes of Wind Energy in Texas: Local Communities in Close Proximity to Wind Farms and their Effect on Decision-Making." *Energy Policy* 38, no. 5 (2010): 2508–2519.

Thompson, Susan. "Iowa's Turn." *Wallaces Farmer* (February 2008): Front Page.

Williams, Heather. "Fighting Corporate Swine." *Politics and Society* 34, no. 3 (2006): 369–397.

"Wind Energy." *Iowa Environmental Counsel*. https://www.iaenvironment.org/our-work/clean-energy/wind-energy (accessed 10/19/23).

Wind Energy Foundation. "Polls." http://www.windenergyfoundation.org/wind-at-work/wind-consumers/polls (accessed 6/7/15).

WGBH. "The Falmouth Experience." http://www.wgbh.org/wcai/turbine.cfm (accessed 1/23/15).

Williams, Wendy, and Robert Whitcomb. *Cape Wind: Money, Celebrity, Class, Politics, and the Battle for Our Energy Future on Nantucket Sound*. New York: Public Affairs, 2007.

Wolsink, Maarten. "Invalid Theory Impedes Our Understanding: A Critique on the Persistence of the Language of NIMBY." *Transactions of the Institute of British Geographers* 31, no. 1 (2006): 85–91.

9

"COMPANY IN THE COMBINE"

Gender, Farming, and Comparing Organic Reformist and Rural Ultramodern Identities

> The corn picker had put an end to the tall corn. Uniformity was now the goal and Ioway was the place where the tall corn grew-once-but now conformed to the demands of machines. Uniformity-that was the criterion of excellence. In Iowa. Ioway was gone.
>
> Winifred M. Van Etten, "Three Worlds," 1978[1]

In the fall of 2015, the Chatterdon family began their corn harvest as they had for four years on their Illinois farm. The Chatterdons had farmed for many generations. The family, consisting of Erin (Chatterton) Featherlin, her mother Charlotte, her uncle Brett, her cousin Jason, and her brother Josiah had all inherited their Illinois land from Erin's father, Greg, after he died in a car accident. Brett Chatterdon had previously worked in a partnership with his brother Greg on the farm. In a recorded interview by *Prairie Farmer* editor and *Wallaces Farmer* blogger Holly Spangler, the family took a break from the corn harvest to discuss the production process. Spangler opened the interview with a short introduction about Greg Chatterdon, a friend of her and her husband's, stating

> Greg was the kind of person who would drop everything to help somebody. And his family was his life. And I've watched his family over the last four years and I have marveled at how they have held each other up and how they've reorganized their farm. And how in many ways they've made it look easy and, yet, we know it wasn't.

The interview took place in one of the Chatterton's corn fields toward the end of the harvest. The family stood in work clothes next to their combine in a tan field littered by recently harvested corn stalks with winter fast approaching:

DOI: 10.4324/9781032637952-10

Spangler: What would you say has changed since Greg's death?

Brett: Opportunities for Erin and Jason, who are very motivated to be involved in the family business. There have been doors that have been
opened because of Greg's absence, which is awkward and uncomfortable in a lot of ways, but yet we're just continuing the family legacy.
And I think legacy is probably one of the biggest things my wife and
I talk about a lot, is just leaving a family legacy. And it's not about
the land, and it's not about…farming even. It's about family and faith
and teaching your kids about what's important, and your family about
what's important.

Erin: So when you talk about legacy, it's not necessarily even about farming
per se, something that you planned on, it just happened in your family
that way?

Brett: Right, it's absolutely not about farming. It's about family… My grandfather was killed by a train, in '45? Right after my father was in World
War II he came back and was in Bradley University majoring in engineering. My dad always said he wanted to build bridges. And, uh,
when my grandfather was killed my father came back to the family
farm. He took over for his father to help out his mother. And then my
father raised his family running a farm.

   After explaining that part of the family "legacy," included the wise
execution of several "buy out agreements" and "trust plans" by his
grandfather, father, and between himself and his brother Greg, to distribute land among family members in a fair way that preserved "love"
and avoid family disputes, Brett continued to explain:

Spangler: So you have ownership in it, that's your thing…

Brett: And people, people, function better that way. Erin wants to have goals
and have successes on her own. If I'm tellin' her what to do all the
time, it takes away human desire to achieve things and to have success
and ownership and stuff.…

Erin: Have some independence.

Brett: Yah, Erin, share about why you like farming. And that's part of why it
works for your legacy that you can work…

Erin: Right, that I love it. The schedule of farming allows for me to still help
be a part of the continued legacy but be able to help provide for my
family too. I still get to be a mom, and I have my children with me in
the field, but I still get to be available for my sick children, and to have
the laundry done, and supper on the table, and supper in the field.

Spangler: And plenty of company in the combine?!

Erin: Exactly, exactly I'm never alone! [laughter by several people and in-
audible comments]

   Erin continues to discuss her involvement in the business of the
farm after the passing of her father:

Erin: And I was really excited about all this stuff I was getting to do. I've always loved to 'play' business, I've always loved to drive tractors.... But the thing that ran over in my mind that I thought if I had the opportunity to say it was gonna' be to talk about the relationships that Brett, and Jason, and I share within the business because Brett, and Jason, and I are family and we're business partners but for my end of things though I feel like, ya know, we're best friends too and we, we, at the end of the day when it's all said and done we can all hold hands with each other and pray. That relationship I have with them is huge, ya know? I don't go more than twenty-four hours without talking to them, that's a very odd thing. I appreciate that relationship that that we have. Everybody is very honest and loving with each other and I think that's what makes us work.[2]

During the short interview, Brett and Erin's voices tremble as if to cry with emotion at times. This wavering of the voice occurs when Brett discusses his father and grandfather and at the end when Erin states

> At the end of the day when it's all said and done we can all hold hands with each other and pray. That relationship I have with them is huge, ya know? I don't go more than twenty-four hours without talking to them, that's a very odd thing.

At the conclusion of the interview Erin, Brett, and Jason return to driving trucks and combines to finish the corn harvest.

My theory of performative use not only offers greater insight into why many rural residents exhibit PIMBY attitudes toward wind turbines or reject large hog farming. It also reveals how farmers' experience with urban industrialism in the 1920s and the pattern of audience leads to unarticulated disagreements between some organic and sustainable food advocates and mainstream Corn Belt farmers in the debate over the reform of agriculture. I have argued that the unexpressed use of technology reinforces identity. This chapter and the next will argue that this identity influences how people react to policy debates about agriculture.

More specifically, one seeking to account for rural perspectives should view the debate over "industrial" agriculture posed by the organic and sustainable food movement as a controversy between two identities: organic reformism and rural modernity/ultramodernity. I contend that both organic reformism (at least the version that alienates Corn Belt farmers) and rural modernity often use technology performatively to establish and reinforce a set of conflicting ideas about work, gender, technology, history, nature, and morality.[3] Many of the actors on the two sides offer rational proxy arguments masking these clashing collections of ideas because both identities have become deeply ingrained through historically contingent cultural factors to such an extent that neither side gives them much thought. In other words, performative use of technology for both identities is embodied.

But the inability of both sides to confront these unexpressed fundamental differences (which I outline below) leads to the familiar "pattern of audience" (discussed in Chapter 1) whereby rural residents perceive an outside "other" as threatening and the outsider regards the farmer as backward. This repeated social practice of performance and "othering," I argue here and in my concluding chapter, creates a roadblock to agricultural reform by simply solidifying rural resentment.

One important factor in understanding the unarticulated notions driving the resentment of organic discourse among ultramodern farmers is the way female farmers see the organic view of the relationship between gender and technology. A careful reading of the interview of the Chatterton family above reveals that Erin, the wife, has always enjoyed driving tractors and combines and the manager, Brett, gives her, not the male family members, control over the introduction of the newest precision technologies on the farm. Nor do the Chatterdons see a dichotomy between Erin "being a mom" and using the most advanced technology. In fact, she relates that she often has her small children with her while operating some of the world's most up-to-date farm machinery in a field containing mono-cultured corn.

In this chapter, I argue that many scholars and organic food advocates have incorrectly interpreted the historical relationships between gender, technology, work, and family on farms in ways that prevent both producers and reformers from participating in useful dialog about agricultural reform. Many advocates maintaining an organic reformist identity and rural denizens with an ultramodern sense of self engage in performative use with different material objects to reinforce opposing ideas about technology, work, and gender. Through an analysis of farm journals and farmers' memoirs, this chapter demonstrates that Corn Belt farmers hold a strong cultural view of only *some* technologies as masculine, particularly those mechanized objects related to crop production. Moreover, I contend that this slight gendering seems to arise from the division of work processes on the family farm beginning with the increasing use of technology in American agriculture early in the 20th century. While not essentializing organic or environmental thought as one monolithic set of ideas, I do intend to identify and explain aspects of discourse used by those critiquing mainstream agriculture that alienate farmers in the Corn Belt by providing a broader historical and social context for rural resentment. Nor do I intend to "take sides" in the debate, only to provide a richer understanding of rural perspectives and to make an interesting point about how the controversy relates to gender constructs in the Midwest.

Most importantly, Midwest farmers have never viewed technology as an *exclusively* male domain and have framed production outside the home as the realm of both sexes. In contrast to this rural experience of gendering the latest productive technologies as both male and female, some historians of technology have argued that the two main movements to reform American farming – the organic foods movement and sustainable agriculture – constitute feminized sentiments because of their appeal to emotion and an emphasis on the "family farm." As a corollary, academics view industrial agriculture as hyper-masculinized due to its over-mechanization and corporate control.[4]

Scholars in other fields have reinforced the idea that a feminine nurturing of nature formed the crucial aspect of the morality of pre-technological agricultural production. For example, historian Carolyn Merchant views the rise of science and technology in Western Europe in the 17th through the 19th centuries in terms of replacing a society more conducive to a feminine nurturing of nature by market-oriented structures promoting the male domination of nature and of women. In Merchant's view, modern science and technology enhanced the use of both women and nature as a resource as opposed to a "premodern organic world." Beginning in the 16th century, Europeans used science and technology to rationalize, objectify, and commodify the environment transforming nature from an organic cosmos with a living female at its center to a dead and passive machine controlled by men.[5] Some feminist scholars like Donna Haraway have rejected Merchant's tendency to essentialize nature based on gender. Other historians, such as Londa Schiebinger and Ruth Oldenziel, while recognizing the socially constructed nature of gender, offer theories supporting Merchant's claim of a Western tendency to view nature as female and technology as male.[6]

Throughout this book, my perspective has differed from this view of gender and work by demonstrating that *both* men and women on Midwest family farms participated in important "production" processes and formed a modern rural identity based on the use and adoption of technology. Further, I seek to challenge the dichotomy between a feminized "family farm" and a masculinized "industrial farm" by showing that *both* technology and glorification of "the family" form important aspects of farmers' identities and discourses as modern producers for men and women.[7] Consistent with my user perspective of technology, I aim not to take a reductionist view by claiming that all organic advocates promote the same identity. Rather, I seek to understand how organic discourse sounds from the perspective of Midwest farmers as well as historicize today's rural resentment toward agricultural reformers.

The debate over agricultural reform most often takes the form of proxy arguments over science or economics that stand in for the real underlying clash between rural organic and ultramodern identities. Both sides often engage in the performative use of technology to reinforce notions of self and perform identities for others. These two competing forms of performative use have deeply embedded historical and cultural trajectories.

Organic/Sustainable Reformist Identity and Discourse

Scholars and policy advocates urging the reform of American agriculture toward a more sustainable model or based on organic food production do not constitute one monolithic group. Nevertheless, many do organize themselves around a dominant discourse that assumes clear dichotomies between characteristics of "family farms" and "industrialized farms." Many organic advocacy organizations have promoted this dichotomy between a feminine farm using less technology and a masculine factory farm. Food and Water Watch's Food Systems issues page, for example, features a graphic showing a bright line between "Farm vs. Factory" in which the moral feminine nurturing sustainable farm features environmental stewardship, better living conditions for animals, small scale, and less technology and science use whereas immoral industrial agriculture brings only environmental degradation, inhumane treatment of animals, large scale production, and overuse of technology and science in an immoral male domination of nature. The page features a drawing of the moral farm in which a brightly lit farm features several fields with many people working the land growing vegetables with hand tools while animals freely graze in separate fields. The farm, nestled in between two mountains and next to a lake, also features an old-fashioned red barn and a farmer driving a 1930s-era tractor next to beekeeping hives and a contemporary plastic greenhouse. Such a farm has never actually existed in the rural Midwest. In contrast, the immoral factory is a dimly lit and drab landscape of long animal sheds spewing animal waste and a large field being irrigated by a large system with no people shown and the farm home only far in the background to reinforce the idea that this cannot be a "family farm." Except for farmers entering into large hog lot contracts, few farmers would say that this image resembles their modern farm either.[8]

According to the modern farmer's point of view, such a binary conception between "industrialized agriculture" and the "family farm" is a false urban construct and a way for outsiders to undermine the morality of the rural way of life. From a rural perspective, organic discourse tends to frame farmers' technology use as immoral and attacks female farmers' identities as producers by viewing modern farming practices as a form of male and corporate domination. While the organic food movement has origins in the early 20th century, the discourse I present here has existed at least since the 1960s and early 1970s.[9]

In addition, this dominant organic reformist discourse tends to view the history of American agriculture as devolving from a family-oriented production process closely resembling the ideals of the organic or sustainable foods movement to corporate-controlled industrial farming damaging both families and the environment. As such, one may think of these critics of contemporary agriculture as a subset of environmentalism. Organic discourse often views the history of agriculture in terms of a clear "break" between the first and second halves of the 20th century in which a previous moral bucolic ideal became swallowed by the rise of immoral factory farms antithetical to nature. This clear break often corresponds with the start of Norman Borlaug's Green Revolution.[10]

As with rural discourses of rural capitalistic modernity and ultramodernity, rural organic discourse often incorporates older notions of morality from traditional Jeffersonian agrarianism particularly the notion of the pastoral. Sustainable food advocate and writer John Seymour, for example has quoted Jefferson as stating

I have often thought that if heaven had given me choice of my position and calling, it should have been on a rich spot of earth well watered, and near a good market for the production of the garden. No occupation is so delightful to me as the culture of the earth.[11]

Seymour tends to ignore that on Jefferson's land, the "culture of the earth" was done mostly by the over 600 enslaved people Jefferson owned during his lifetime.[12]

Technological use forms the foundation for this dichotomy between the family farm and the industrialized farm, according to organic discourse. The family farm stands as a lost ideal for some organic and sustainable advocates because these farms used less technology and were, therefore, closer to nature than the later highly mechanized farm. Further, reformist discourses see American family farms in the early 20th century as the cousins of the idealistic organic or sustainable farm since the 1970s. "Back to nature" advocates tend to regard these "pre-industrialized" family farms as more feminine and moral than the current masculinized and immoral technological farm.

According to some organic reformers, the American family farm prior to mechanized agriculture was more "feminine" because, like current organic farms, it nurtured the environment by using less technology. In contrast, the industrialized farms of today are "masculine" because they dominate and control nature through mechanization. In short, the images of the nurturing female, the family, and nature all form a unified ideal underlying some organic/sustainable discourses, at least those that tend to garner rural resentment. This alienating organic discourse often reinforces a worldview that regards an essentialized feminine "nature" and a masculine "technology" as mutually exclusive (I have visually represented this binary organic discourse in Figure 9.1).

One can see evidence of this type of discourse used by advocates of organic and sustainable farming in a wide variety of domains. Critics of mainstream American agriculture writing for popular audiences such as Michel Pollan, Barbara Kingsolver, Wendell Berry, and E.O. Wilson articulate perhaps the most explicit expression of this identity. Pollan argues that "the process of industrialization [of organic agriculture] will cost organic its soul" and he advocates a return to the organic ideal which describes "a landscape of reconciliation that proposed to replace industrialism's attitude of conquest toward nature with a softer, more harmonious approach." He sees a clear distinction between a farm using technologies such as petroleum-driven machinery and synthetic fertilizers and chemicals and one not using such technology as "a way to feed ourselves more in keeping with the logic of nature, to build our food system that looked more like an ecosystem that would draw its fertility from the sun."[13]

Further, Pollan portrays mainstream farmers not as the moral family farmers of the past "but as 'agribusinessmen.'"[14] He sees the history of agriculture as "In the years immediately after the war, industrial agriculture (which benefitted from the peacetime conversion of munitions to chemical fertilizer and nerve gas research to pesticides) also consolidated its position: there would be no other kind." While Pollan remains ambiguous about his view of agriculture prior to World War II, he implies that it resembled a kind of "nurturing" of nature much more than it resembled the "domination" currently practiced by "industrial" agriculture.[15]

Barbara Kingsolver also expresses this concept of a clear break between a more "natural" form of agriculture and an overly mechanized farming. She writes, "Most

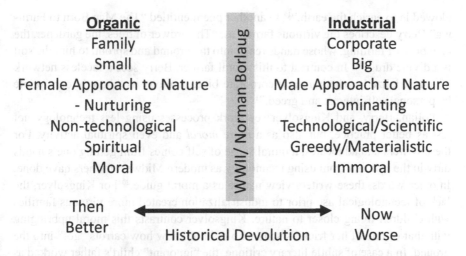

Organic	WWII/ Norman Borlaug	Industrial
Family		Corporate
Small		Big
Female Approach to Nature		Male Approach to Nature
- Nurturing		- Dominating
Non-technological		Technological/Scientific
Spiritual		Greedy/Materialistic
Moral		Immoral

Then ←——————————————→ Now

Better   Historical Devolution   Worse

FIGURE 9.1 The version of organic identity that alienates many mainstream Corn Belt farmers is often based on clear dichotomies between a moral "organic" and an immoral "industrial" way of farming. The horizontal line on the bottom represents the binary conception under such an organic discourse that assumes a more moral rural lifestyle the further backwards one goes back into history (a narrative of historical devolution). Organic discourses often identify World War II, which roughly corresponds to when microbiologist Norman Borlaug began his "Green Revolution" research, as the turning point where agriculture started to become immoral.

people of my grandparents' generation had an intuitive sense of agricultural basics" consisting of a long list of skills including

> When various fruits and vegetables come into season… What an asparagus patch looks like in August… Most importantly: what animals and vegetables thrive in one's immediate region and how to live well on those, with little else thrown into the mix beyond a bag of flour, a punch of salt, and a handful of coffee.[16]

Kingsolver's list of "agricultural basics" lacks knowledge pertaining to material use of "higher tech" artifacts such as how to repair a tractor engine or how to maximize the efficiency of combines or install GPS units. In fact, the most moral way to reinforce an organic identity, for Kingsolver, involves avoiding the use of these technologies completely. Few photos of Kingsolver on her small hobby farm feature advanced or contemporary technologies.

Popular novelist and environmental advocate Wendell Berry views agriculture prior to "industrialization" in similarly idealized terms. Berry has also farmed in Kentucky much of his life. In one poem entitled "Enriching the Earth," Berry states, "To enrich the earth I have sowed clover and grass/ to grow and die. I have plowed in the seeds/ of winter grains and of various legumes/ their growth to be plowed in to enrich the earth."[17] In another poem entitled "The Man Born to Farming" Berry describes the virtuous farmer as, "The grower of trees, the gardener, the man born of farming, whose hands reach into the ground and sprout, to him the soil is a divine drug."[18] In contrast to this moral farmer, Berry sees a faceless network of industrialized agriculture and corporate biotechnology as "directly corruptible by personal self-interest and greed."[19]

Pollan, Berry, and Kingsolver see work processes using less technology not just as better practice, but also as a more *moral* and even spiritual activity. For these writers and advocates, a moral sense of self comes from getting one's hands dirty in the soil, not from using technology as modern Midwest farmers have done. In other words, these writers view nature as a moral guide.[20] For Kingsolver, the lack of technological use prior to industrialization created more virtuous families with children living closer to nature. Kingsolver contrasts this moral upbringing with that of one of her friend's sons who did not know how carrots "got" into the ground. In a case of subtle literary critique, the "ignorant" child's father worked as a biology professor and grew up on an Iowa farm. Immoral science and industrialized farming stand in contrast to Kingsolver's own virtuous natural family farm.[21] Green environmental advocates Andrew Dobson and Peter Bunyard express the spiritual elements of Kingsolver and Pollan's discourse even more explicitly. "The search for self-sufficiency is, I believe, as much spiritual and ideological as it is one trying to reap the basic necessities of life out of the bare minimum of our surroundings," Dobson and Bunyard explain.[22]

This common discourse of popular authors advocating agricultural reform also appears in practical or self-help literature aimed at those wanting to "live the organic dream." This literature aimed at practical farming contains narratives, similar to Kingsolver's, where urban people experience a sort of spiritual awakening and choose a more moral lifestyle consistent with the organic ideal. The first sentence of *The Organic Farming Manual* clearly creates a dichotomy between moral non-technological, small family farms, and immoral mechanized corporate agriculture when it states:

> Organic farming is family-friendly, economically viable for small and midsized farms, and a boon to our stressed environment-and its products taste great. For the farmers themselves, there's a real joy that comes from working with the natural cycles of the soils, plants, and animals, instead of trying to beat them into submission for profit and convenience.[23]

The forward to perhaps one of most popular self-help books, John Seymour's *The Self Sufficient Life and How to Live It,* alleges "In the modern world, during the last hundred or so years, there has been an enormous and historically unique shift: away from self-reliance and toward organization" leading to a dangerous dependence of people on "complex organizations, on fantastic machinery, on large money incomes." In language resembling Berry's, Seymour suggests that allowing, "ourselves to be dependent on some vast 'Thing created by the Merchants of Greed' is madness." For Seymour, this "vast thing" is the globalized food system. Seymour uses not only nature as his guide to morality, but ancient work processes prior to this "historically unique shift." Seymour hails Namibian nomadic hunters, isolated olive farmers in Crete, and 19th-century peasants in England's Golfen Valley as living sustainable and moral lives closer to nature. In Seymour's view,

> To say that an invention is labor-saving is the highest praise, but it never seems to occur to anyone that the work might be enjoyable. I have plowed all day behind a good set of horses and been sad when the day came to an end![24]

All of these groups of pre-industrial producers, according to Seymour, "have found good, honest, and useful ways of making a living. Some are fairly well off with regard to money; others are poor in that regard but they are all rich in things that really matter."[25] E.O. Wilson similarly celebrates native agriculture preceding industrialization as better utilizing biodiversity. Wilson does not object to modern biotechnology per se. Rather, he argues that the Green Revolution and agrotechnology has "whittled down" diversity by not using enough genetic material in modifying organisms, rendering the entire food system vulnerable. Thus, Wilson advocates utilizing native and local knowledges rooted in a pre-technological world.[26]

One can observe the image of what a moral nature and relationship with technology looks like for Seymour and others forming a rural organic reformist identity in a drawing in his book *The New Complete Book of Self-Sufficiency*. In the picture, a family walks in between a patch of cabbages and onions. The family looks more urban than rural as they wear shorts, sandals, and polo shirts. The family dog runs in front of them and the cat sits beside the onion patch. Several glass boxes sit in the garden but no other artifacts appear in the drawing. While the vegetables grow in rows and have an ordered appearance, a less structured nature surrounds and encroaches on the garden. Several birds, two rabbits, a squirrel, a mouse, and a vole live among the vegetables. Overall, one notices the distinct lack of technology, the literal closeness of non-agricultural nature in the form of trees and wildlife, and the prominence of the family including family pets.[27] The less technical the process of food procurement, the more natural and virtuous it becomes in keeping with the discourse proffered by Pollan, Kingsolver, Dobson, and Berry.

STS scholars have drawn on or reinforced elements of organic discourse viewing technology as masculine in order moralize less technological (and, therefore, feminine) work processes of the past. In rejecting technology use, these older means of production imitated nature, promoted the family, and constituted a feminine nurturing sensibility, as opposed to technological, masculinized, and corporatized contemporary mainstream agriculture. For example, philosopher of technology Albert Borgmann expresses the same discourse as Seymour in a more philosophical rhetoric. For Borgmann, labor-saving technologies in agriculture remove people from a "focal practice," in this case, the arduous laboring over land. Technology, from Borgmann's perspective, leads to the loss of human engagement with nature and family. In language mirroring Kingsolver's, Borgmann also laments the decline of "pretechnological" knowledge and skills and their replacement by technologies and networks that do the work for humans without any "practice" or sense of accomplishment. As with Pollan, Kingsolver, Berry, and Seymour, less technological work processes become imbued with morality.[28] Once again, this organic discourse assumes a clear demarcation between a pre-technological agriculture in which prior farming methods aimed for a moral nurturing of nature, a female quality, as opposed to current immoral domination of nature, a male quality. Indeed, organic discourse often adopts contentions by some scholars that men immorally use science and technology for "the domination of both nature and women" under a corporate system of exploitive industrialized agriculture.[29]

In addition, these STS voices tend to view a moral pre-technological farming as more conducive to preserving families. Historian of technology Carroll Pursell has observed that "A study of farmers in Wisconsin who were dedicated to the practice of what they called 'sustainable agriculture' revealed that an overwhelming number of them did so in the name of family farms, or domestic rather than market values." Pursell then assumed that all farmers held a dichotomous view of technology and the family.

> They [the farmers] believe the principles of sustainable agriculture that could help preserve family farming (the reliance on small-scale, labor-intensive production using nonsynthetic chemicals, for example) are inseparably related to values that sustain farm families. These values include the integration of work life and family life, and environmental conservation.'[30]

Pursell, in arguing that the backlash in the 1980s against the sustainable foods movement resulted from the re-masculinization of society, adopts the same worldview exhibited by Pollan, Kingsolver, and Berry in which a set of ideas about morality, nature, work processes, and gender all form one consistent dominant discourse used to conceptualize and critique modern agriculture.[31]

Nor should one regard this type of organic identity as existing only among those participating in small hobby farming outside of the rural Midwest. Keeping in mind that rural Americans have always had contested identities, this organic sense of self constitutes one possible identity, albeit a less pervasive and dominant one, available to farmers in the Corn Belt. Distinguished Fellow for the Leopold Center for Sustainable Agriculture and Religion and Philosophy Professor Frederick L. Kirschenmann, for example, manages a 3,500-acre organic farm near Des Moines, Iowa in the heart of the Corn Belt. In discourse echoing that of Kingsolver and Pollan, Kirschenmann argues that

> Our modern industrial culture tends to view not only food but almost all of reality as a collection of fragments (things) rather than a web of relationships.

Modern philosophers trace this tendency to the 17th-century scientific revolution. Rene Descartes wanted science to become a "universal mathematics," which, of course, tended to reduce all of reality to measurable things and ignored dynamic relationships.

While Kirschenmann does not address technology use directly, he advocates Michael Pollan's call "to start thinking about food as less of a thing and more of a relationship" including reducing fossil fuel, irrigation, pesticide, and artificial fertilizer use. In rhetoric mirroring Berry's and Seymour's, Kirschenmann advocates for smaller, less mono-cultured, farms using more labor implying less use of technology. "The notion," Kirschenmann laments, "that farming is drudgery is still deeply engrained in our culture."[32] Thus, the dominant organic discourse that views a dichotomy between large, masculinized, reductionist, mechanized, and industrialized agriculture and small, feminized, relational, less technological, and family-oriented farming helps to form identities among some residents of the Corn Belt as much as it does among East Coast-based environmentalists. The crucial conceptual frame, therefore, is not to focus on rural versus urban or even Midwestern versus Eastern discourses, but to acknowledge that once rural Americans strategically choose an identity, they commit themselves to many unarticulated relationships, assumptions, and practices. These commitments are historically embedded and determine how people interact with technology.

Interestingly, both discourses adhere to *the same* original notions of independence and the morality of a family-based production process, but they have used technology differently and have constructed opposing notions of nature in incorporating these ideologies into their respective identities. Both discourses borrow ideas from Jeffersonian and German agrarianism, but they make use of them in opposing ways. Modern rural discourses and identities moralize the family and productive work, not nature untouched by technology. For instance, Mildred Armstrong Kalish recalls the following about growing up on an Iowa farm in the late 1920s and early 1930s,

> Besides making the world go around, those folks who did their assigned chores were identified as 'goodworkers.' They were respected, held up as an example to others, greeted with a smile and a hearty handshake, and privileged to enjoy a feeling of goodwill throughout the community.[33]

Kalish's account reflects the German agrarian practice of moralizing productive work for all members of the farm family. Thus, some organic advocates drawing from feminist perspectives assuming a gendering of technology and production as exclusively male domains, while assigning the family and nature to female spheres, have oversimplified the historical relationship between work, technology, and gender in the rural Midwest. In the process, this underlying assumption of duality has led critics of modern agriculture to overlook more important rural discourses and identities driving the relationship between technology and people.

My narrative has also demonstrated that although farm men and women both sought to construct their identities as modern producers through use of technology, gendering of work processes did occur on American family farms with operation of large field machinery which was designated as a masculine domain. Evidence suggests that what made fieldwork "men's work" was the large machinery itself, not the act of producing or even of farming crops. In addition to the fact that both men and women saw themselves as producers, women also wrote in farm journals to discuss how to improve crop production techniques. For example, a woman identified only as Mrs. E. J. Kirk of Ohio wrote to *Wallaces' Farmer* in 1925 on techniques for growing and harvesting barley to supplement the corn crop used to feed hogs. Kirk's editorial used the term "we" suggesting that she participated in the planting and harvesting process as well as the care for large animals.[34] Therefore, work with large field machinery appeared only *weakly* gendered.

To the extent that tractors and combines comprised a masculine domain, the machinery itself led to its gendering, not the activity of producing commercial agricultural products. While there is nothing inherently masculine about technology, there is something about large field machinery that seemed essentially masculine to Americans on farms in the early 20th century that had nothing to do with efforts to subjugate women or with different female or male approaches to nature. Rural Americans, in other words, saw tractors and combines as "male" even without a clear demarcation between a male sphere of productivity, technological use, and work and a female sphere of consumption, nature, family, and home. Organic discourse and some scholars often ignore this nuanced relationship between gender, technology, and productivity on Midwest farms.[35]

Kenneth Hassebrock's memoirs about growing up on an Iowa farm in the 1920s and 1930s sheds light on how and why rural Americans viewed field machinery as weakly masculine that goes beyond the anachronistic dualisms of technology and nature contained in some organic discourses. While Hassebrock does not state the age in which he joined his father in the fields, he does comment that, "Doing such work was definitely a milestone for those of us growing up in Iowa" and that in spite of the fatigue caused by plowing soft ground by walking behind a spike-tooth harrow and four horses, "there was no shot of slowing down or stopping-after all, I was now a man."[36]

Several aspects of Hassebrock's detailed narrative of farm practices reveal important cultural and technological influences on the gendering of large field machines on American farms in the early 20th century. First, Hassebrock discusses large farm machinery as a continuation of the draft horse, traditionally handled by men for many centuries in Western agriculture. While horses and tractors obviously do not have the same "hardware" in them, Hassebrock suggests a kind of *gender incumbency* in which similar features of use mean that prior gendering becomes embedded within the new technology.[37]

In this sense, Midwesterners constructed tractors as masculine not because of some essentialized or universal notion of gender and technology or nature, but because culturally-specific work processes and practices of use write cultural scripts influencing later users.[38] For Hassebrock, what made performance of field-work a male activity came from "not only the ability to control horses or drive a tractor, but also sufficient maturity to act in a rational manner when the unexpected happened, which was not infrequent."[39] The horse not only carried with it an incumbency passed on to tractors in terms of how farmers used it, it also carried similar risks that the user had to avoid through a kind of emotionless problem-solving ability, a trait regarded by Hassebrock, and other early 20th-century Americans, as male. For Hassebrock, just as a horse could become stuck or difficult to control, field machinery could break down at any moment when the time demands of harvest or planting required quick and creative solutions.

Even when machinery did not malfunction, early farm equipment often did not perform the intended task and farmers had to modify machinery. For example, Hassebrock recounts how in the early 1920s, his father purchased a corn picker that he planned to power with his old Hart Parr tractor, only to discover that the wheel spacing of the tractor prevented the picker forks from aligning with the corn rows. Hassebrock describes his father's creative solution to the problem, reflecting an impressive technical acumen:

> The engine in a 1918 Chevrolet car Dad had junked was still operable. Using the engine to power the mechanism of the picker while it was being pulled across the field with three horses appeared feasible. Dad and Wilber proceeded to mount this engine on the picker and make all the necessary modifications. A chain sprocket wheel was attached to the drive shaft that extended from the transmission of the engine, and another one replaced the small flywheel of the picker. Clutch and throttle controls were placed near the operator: the small front carriage was put back into place and the boom arm was removed. Picking would now proceed with Dad operating the picker and Wilbur the wagon.[40]

Finally, large farm machinery by its design, which placed the driver high on a seat, created a very different lived experience for the operator than walking behind a team of horses pulling a plow in a way that tended to romanticize male fieldwork. The horse or ox required the farmer to either attach himself to the plow and direct it on the ground while pulled by horses or sit on a small plow that did not elevate the operator above the back of the horse. Harold Briemyer as a boy hated this work with horses, writing that he "remembers to this day how tiresome it was to walk behind a spike-toothed harrow that his team dragged over a 20-acre field. The operation seemed interminable." Horse farming, down on the ground trudging endlessly with the horses, was "devoid of any artistic or spiritual uplift."[41] In contrast, in the extremely flat landscape of the Corn Belt, the new tractors that farmers encountered early in the 20th century placed them literally on top of the world in control of a symbol of strength and power in a way that accentuated American ideas about masculinity.

Boys growing up in the Midwest recalling tractor farming spoke of their fathers in heroic overly masculinized terms. David Hamilton described how his uncle appeared to him as a young boy helping with harvest on their Missouri farm:

> On the tractor, Unc stood straight, his worn fedora shading his eyes, one hand on the wheel, the other resting on his hip, his feet spaced for balance, and he rocked with the tractor as it picked its way across old furrows that had settled back into each other. The big wheels on either side held him high. I imagined him standing higher on them, straddling the two, stepping as they turned, and by a kind of moonwalk striding hugely across earth that ran back beneath the tractor to meet the plow.

Hamilton described work in the fields with a tractor as if writing about play or recreational activity commenting "And when have you ever known a boy who did not want to take the wheel of a tractor?"[42] Later in his memoir, Hamilton describes a recurrent dream he had of cultivating with a John Deere G where ease in operating machinery and an idealized nature created a heavenly experience. "I was mesmerized by the widening strip of black earth against the lime-green of new wheat in spring," Hamilton writes and adds,

> We could have named our daughter 'Chartreuse,' but we didn't think of that. I think instead of how visible work, with land, machines, and tools, has nudged me towards mysteries, often feminine, and how the boyishness of trying to master such work knocks on the door.

After mastering the tractor in a way that the operator and Hamilton became one efficient mechanism, he finished his work as "then with a smile that I'm glad no one was around to record, I could loaf and invite my soul."[43] Certainly, simple functionalist economic explanations for rural technological use cannot account for such emotive and poetic prose about machines.

Importantly, while fieldwork became a male-dominated domain and large machinery rendered masculine (although each remained only weekly gendered), both sexes regarded the use of tractors or combines as serving the family and constructed identities in which technology, nature, modernity, and family legacy became intimately intertwined. As the poet James Hearst recalled regarding the change he witnessed on his family's Iowa farm in the early 20th century, his family lost its capacity of self-sufficiency, but both men and women "welcomed the machine age with open arms. Engine power instead of muscle power, large farms, private telephones, balanced account books, asphalt roads, bathrooms, promise of a better, at least easier, life."[44] Similarly, the USDA report by agricultural economist Emily Hoag Sawtelle in 1924 quoted a Michigan farmwoman who described herself as the daughter of a pioneer farmer. She saw "progress" as a steady campaign of using technology to subdue untouched nature

I have lived to see towns and cities, where once I saw bear and deer and lynx. In the Wild Wood I married a young man who worked in the lumber yards and we hewed a house out of the wild woods. What was a wilderness when I was a child is now beautiful farms. The old log cabins have been replaced with fine dwellings; wild animals with fine stock. We farm women do work hard but we are not lonesome or discouraged. Fifty years ago my mother never heard of a gasoline engine. Most farmers use them now to saw wood, pump water, churn, run the cream separator, run the washing machine.[45]

Importantly, a non-technological agrarian existence under a rural modern identity has none of the morality of John Seymour's organic reformist discourse, but rather is what Hassebrock calls an "*Agony of Survival*" in a constant struggle against a harsh and unpredictable nature. Rural modern Corn Belt agrarians associate Seymour's organic ideal with the constant threat of financial ruin, not just for the male and female producers, but also for the entire family as a productive unit.[46] Hence, Hassebrock selects as the photo on the front cover of his memoir (written in the 1980s) an image of his family with an early 20th-century automobile, an old technology, as an appropriate representation of this "agony." Hassebrock aims to contrast this less modern agony with the much better rural existence that ultramodern technologies have allowed, a prosperous lifestyle that his struggling farm family had always hoped for.

Such views of technology as combatting the agony of rural life, or as Frederick L. Kirschenmann terms "drudgery," is a common feature of discourses of rural capitalistic modernity/ultramodernity. Characterizing pre-technological work as mundane enhances the idea that the family unit has evolved toward a more progressive state. Rather than extoling the spiritual aspects of working with the land without modern technology as many organic advocates do, for example, Harold Breimier recalled cultivating before mechanization at the dawn of 1920 as "boring" and a task his parents forced him to do.[47] This description differs greatly from the heroic description David Hamilton gave about he and his father mounted on tractors and his desire as a young boy to sit at the wheel.[48]

Hassebrock's memoir cover also reflects the rural tendency to view older technology as evidence of a progressive inborn modernity. Repeating this pattern, the farmers embracing wind turbines in Chapter 8 saw the progression from their grandfather's windmills to their contemporary large wind farms as proof that "we are born innovators." Importantly, this discourse of evolution strikingly opposes an organic discourse of devolving rural relationships between technology and nature.

In addition, while rural Americans regarded rural machinery as inherently male, women still used field machinery and regarded it as a sign of their family's progress and modernization. Winifred M. Van Etten recalled plowing fields in the 1920s as an important part of her childhood and bringing her closer with her father. "I regarded myself as an expert" at operating the plow, she recalled.[49] Another female author, Julie McDonald recalled how her family "was pleased to exchange them [the mules] for a John Deere tractor, and we all cheered when the mules went down the lane for the last time. The new tractor reminded me of a giant grasshopper."[50]

In addition, if rural Americans in the Corn Belt of both genders see nature as moral, they do so primarily because the landscape represents family legacy and productivity. While unaltered nature may have some inherent value, it carries less moral cache than it does under most organic identities. Under rural modernity/ ultramodernity, nature altered by technology signals family legacy and success through inborn innovativeness. One can even make the claim that both men and women on farms *see* a cornfield differently than urban denizens. For example, in Emily Hoag Sawtelle's 1924 report interviewing 8,000 farm women seeking to perform their rural modern identities, she viewed the family, the farm home, and the land in almost equivalent terms, all of which technology rendered *more* moral,

> The farm homes of America keep alive the sacred traditions of our land.... Every field has its story; this splendid old field, for instance, was drained with laborious toil by the grandfather and planted with high hopes by the father and is now tended with pride by the grandson who reaps the harvest of fruit and victory. In the older states, home traditions have sometimes accumulated until, the farm as it has been handed down from one generation to the next, has become a venerable spot... In newer sections of our nation the land becomes endeared to us because here have striven and conquered pioneer parents.[51]

Thus, rural capitalistic modernity imbues the land with meaning that differs from some organic discourse regarding unaltered nature as having an inherent value. The less humans have impacted untouched nature with technology, the more virtuous land and production processes become for many of those advocating agricultural reform. In contrast, farmers with rural modern identities view nature as moral because it documents family conquest or inborn capacity to initiate change in the landscape, often framed as "progress." Jane Adams writing about her own farm family in Illinois eloquently explains this rural view of nature as "The landscape encodes memories."[52]

I have represented identities of rural capitalistic modernity and globalized ultramodernity in visual form in Figure 9.2. One may particularly notice how these rural modern discourses and identities commit the farmer to a bundle of ideas about history, moral performances, gender, productivity, and technology that differ significantly from those proffered by some organic discourses and identities shown in Figure 9.1. By stating that "technology in general is non-gendered," I simply mean that not all technologies under a rural modern conception are viewed as an exclusively male sphere notwithstanding the fact that certain devices, such as large field machinery as I have discussed, have become gendered masculine (Figure 9.2). In contrast to the organic reformist discourse, rural capitalistic modernity and ultramodernity view history as an evolution or progress narrative where the family becomes more technological and, hence, better off. This view of history moralizes

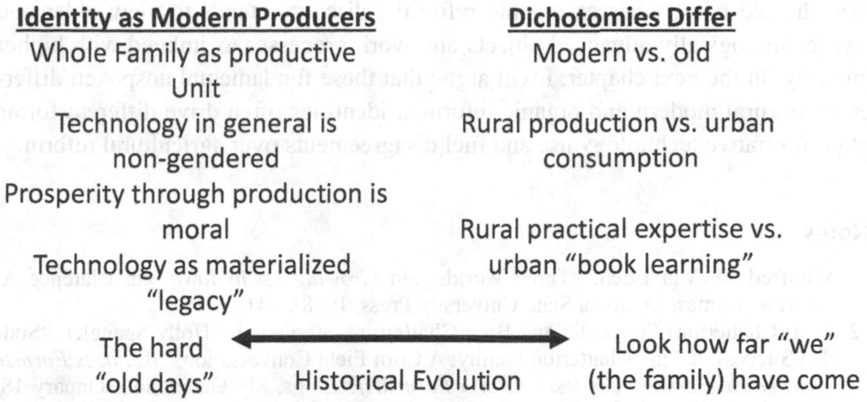

Identity as Modern Producers	Dichotomies Differ
Whole Family as productive Unit	Modern vs. old
Technology in general is non-gendered	Rural production vs. urban consumption
Prosperity through production is moral	
Technology as materialized "legacy"	Rural practical expertise vs. urban "book learning"

The hard ←——————————→ Look how far "we"
"old days" Historical Evolution (the family) have come

FIGURE 9.2   Rural modern identity as a general framework for technology use vs. organic reformist discourse.

FIGURE 9.3   "Glen Brinkman with Case International 3388." Author's personal records. Charles City, Iowa, 1980–1981.

the use of newer artifacts. As such, rural modern farmers often choose the largest and most contemporary technology on their farm to perform their identities (Figure 9.3). In Figure 9.3, for example, Iowa farmer Glen Brinkman in the early 1980s poses with his new Case International 3588 attached to a large disk implement. The enormity of the machine and its placement in front of prosperous farm buildings and grain silos continues the German agrarian traditions of performing identity through material objects used for production and highlights the farmer's modernity.

In addition, rural modern discourses emphasize dichotomies that differ from the organic vs. industrial binary of organic reformist discourse. Rather than evaluating morality according to a framework that seeks to categorize lifestyles and objects as either organic and moral or industrial and immoral, rural modern discourses view rural ways of living and artifacts as more virtuous than those that are urban. In addition, rural modern discourses view new objects or work processes as more moral than the old ones, whereas organic reformist discourse tends to view older and less technologically advanced objects and work processes as imbued with higher morality. In the next chapter, I will argue that these fundamental unspoken differences in rural modern and organic reformist identities often drive different forms of performative technology use and fuel disagreements over agricultural reform.

Notes

1 Winifred M. Van Etten, "Three Worlds," in *Growing up in Iowa*, ed. Clarence A. Andrews (Ames, IA: Iowa State University Press, 1978), 141.
2 Erin (Chatterton) Featherlin and Brett Chatterton, interview by Holly Spangler, "Sudden Succession" in "Chatterton Family: A Corn Field Conversation," *Wallaces Farmer and Prairie Farmer -Confessions of a Farm Wife* Blogs, My Generation, January 18, 2016, http://farmprogress.com/blogs-chatterton-family-corn-field-conversation-10567 (accessed 9/12/16); see also Erin (Chatterton) Featherlin and Brett Chatterton, interview by Holly Spangler, "Sudden Succession" in "Chatterton Family: A Corn Field Conversation," *American Agriculturalist*, January 17, 2016, http://www.americanagriculturist. com/blogs-chatterton-family-corn-field-conversation-10567 (accessed 1/18/17); I have edited this interview transcript for clarity.
3 See for example *Food and Water Watch*, "Food System," https://www.foodandwaterwatch.org/issues/food-system/ (accessed 9/28/23); *Toxic Free NC*, "Just and Sustainable Agriculture," https://toxicfreenc.org/programs/just-sustainable-agriculture/ (accessed 9/28/23).
4 Carroll Pursell, "The Rise and Fall of the Appropriate Technology Movement in the United States, 1965–1985," *Technology and Culture* 34, no. 3 (1993): 629–637.
5 Carolyn Merchant, *The Death of Nature* (New York: HarperCollins Publishers, 1980), 42–78.
6 Londa Schiebinger, "Why Mammals Are Called Mammals: Gender Politics in Eighteenth Century Natural History," *The American Historical Review* 98, no. 2 (1993): 382–411; Oldenziel, *Making Technology Masculine*, Introduction; Haraway, "A Cyborg Manifesto."
7 Challenging the dichotomy between "the family farm" and industrialized agriculture aligns with recent work by Smith-Howard, *Pure and Modern Milk*, 1–20.
8 Food and Water Watch, "Food System;" see also *Toxic Free NC*, "Women of Color Farmer's Network," https://toxicfreenc.org/fundraiser-2023-women-of-color-farmers-network/ (accessed 10/19/23).
9 See Timothy Vos, "Visions of the Middle Landscape: Organic Farming and the Politics of Nature," *Agriculture and Human Values* 17, no. 3 (2000): 246–247; William Lockeretz, ed., *Organic Farming: An International History* (Cambridge, MA: CABI Publishing, 2007), Chapter 3.
10 Prabhu L. Pingali, "Green Revolution: Impacts, Limits, and the Path Ahead," *Proceedings of the National Academy of Sciences of the United States of America* 109, no. 31 (July 2012): 12302–12308.
11 John Seymour, *The Complete Book of Self-Sufficiency* (London: Corgi Books, 1978), between 133 and 134.
12 Linda Mann, "The Enslaved Household of President Thomas Jefferson," *The White House Historical Association*, November 20, 2019, https://www.whitehousehistory. org/slavery-in-the-thomas-jefferson-white-house#:~:text=Despite%20working% 20tirelessly%20to%20establish,most%20of%20any%20U.S.%20president (accessed 10/28/23).
13 Michael Pollan, *The Omnivore's Dilemma* (New York: Penguin Press, 2006), Chapter 9.

14 Ibid., 52.
15 Michael Pollan, *In Defense of Food* (New York, Penguin Press, 2008), 100–101.
16 Barbara Kingsolver, *Animal, Vegetable, Miracle* (New York: HarperCollins Publishers, 2007), 8–9.
17 Wendell Berry, "To Enrich the Earth," in *New Collected Poems*, ed. Wendell Berry (Berkeley, CA: Counterpoint Press, 2012), 125.
18 Wendell Berry, "The Man Born to Farming," in *New Collected Poems*, ed. Wendell Berry (Berkeley, CA: Counterpoint Press, 2012), 115.
19 Wendell Berry, "Twelve Paragraphs on Biotechnology," in *Citizenship Papers*, ed. Wendell Berry (Berkeley, CA: Counterpoint Press, 2003), 53.
20 For a discussion of social constructions of nature, including nature as a guide, see Coates, *Nature*, 3–17.
21 Kingsolver, *Animal, Vegetable, Miracle*, 11–12.
22 Andrew Dobson, *Green Political Thought* (New York: Routledge, 1990.), 93–94.
23 Ann Larkin Hansen, *The Organic Farming Manual* (North Adams, MA: Storey Publishing, 2010), 2.
24 John Seymour and Will Sutherland, *The Self-Sufficient Life and How to Live It* (New York: DK Publishing, Inc., 2009), 6–13.
25 Ibid., 13.
26 E.O. Wilson, *The Diversity of Life* (New York: W.W. Norton & Co., 1999), 289–301.
27 John Seymour, *The New Complete Book of Self-Sufficiency*, Illustration (London: Dorling Kindersley Publishers, Ltd, 2003), 52–56. See also https://weedsuptomeknees. wordpress.com/2012/04/ (accessed 1/27/16).
28 Albert Borgmann, *Technology and the Character of Contemporary Life* (Chicago, IL: University of Chicago Press, 1984), 195–205.
29 Merchant, *The Death of Nature*, xxi. See also Haraway, "A Cyborg Manifesto," 149–181; Oldenziel, *Making Technology Masculine*, 10.
30 Pursell, "The Rise and Fall of the Appropriate Technology Movement in the United States, 1965–1985," 635.
31 Carroll Pursell, "The Construction of Masculinity and Technology," *Polhem* 11 (1993): 206–207.
32 Frederick L. Kirschenmann, "Food as Relationship," *Journal of Hunger & Environmental Nutrition* 3 no. 2 (2008): 106–121; *Leopold Center for Sustainable Agriculture*, Iowa State University, http://www.leopold.iastate.edu/about/staff (accessed 1/27/16).
33 Kalish, *Little Heathens*, 114–115.
34 Kirk, "Barley for Hogs Next Summer."
35 Ruth Oldenziel and others have recognized the socially constructed "Western tendency to view technology as an exclusively masculine affair" but have ignored the rural experience. Haraway, "A Cyborg Manifesto," 149–181; Oldenziel, *Making Technology Masculine*, 10.
36 Kenneth Hassebrock, *Rural Reminiscences: The Agony of Survival* (Ames: Iowa State University Press, 1990), 88–89.
37 Historian Robert Friedel discusses the concept of "path dependency" to describe how technological design features become embedded in later technical devices. Robert Friedel, "Why You Need to Understand Y2K," *Invention and Technology* 15, no. 3 (Winter 2000): 24–31.
38 The historian Finn Arne Jørgensen makes use of the slightly more abstract theory of cultural scripting in technological systems. Finn Arne Jørgensen, "The Backbone of Everyday Environmentalism: Cultural Scripting and Technological Systems," in *New Natures: Joining Environmental History with Science and Technology Studies*, ed. Dolly Jørgensen et al. (Pittsburg: University of Pittsburg Press, 2013), 72–73.
39 Hassebrock, *Rural Reminiscences*, 88.
40 Ibid., 31–33.

41 Breimyer, *Over-Fulfilled Expectations*, 49.
42 Hamilton, *Deep River*, 87.
43 Ibid., 144–146.
44 James Hearst, "Young Poet on the Land," in *Growing up in Iowa*, ed. Clarence A. Andrews (Ames: Iowa State University Press, 1978), 58–59.
45 Sawtelle, 9.
46 Hassebrock, *Rural Reminiscences*, Front Cover; see also Scot, *Prairie Reunion*, 57–58; Breimyer, 44–45.
47 Breimyer, 49.
48 Hamilton, 87.
49 Hassebrock, Front Cover; see also Scot, *Prairie Reunion*, 141.
50 Julie McDonald, "Growing Up in Western Iowa," in *Growing up in Iowa*, ed. Clarence A. Andrews (Ames: Iowa State University Press, 1978), 116.
51 Sawtelle, 28–29.
52 Adams, *The Transformation of Rural Life*, xvii.

Bibliography

Adams, Jane. *The Transformation of Rural Life*. Chapel Hill: University of North Carolina Press, 1994.

Berry, Wendell. "Twelve Paragraphs on Biotechnology." In *Citizenship Papers*, edited by Wendell Berry, 53. Berkeley, CA: Counterpoint Press, 2003.

———. The Man Born to Farming." In *New Collected Poems*, edited by Wendell Berry, 115. Berkeley, CA: Counterpoint Press, 2012a.

———. "To Enrich the Earth." In *New Collected Poems*, edited by Wendell Berry, 125. Berkeley, CA: Counterpoint Press, 2012b.

Borgmann, Albert. *Technology and the Character of Contemporary Life*. Chicago, IL: University of Chicago Press, 1984.

Breimyer, Harold F. *Over-Fulfilled Expectations: A Life and an Era in Rural America*. Ames: Iowa State University Press, 1991.

Coates, Peter. *Nature: Western Attitudes since Ancient Times*. Berkeley: University of California Press, 1998.

Dobson, Andrew. *Green Political Thought*. New York: Routledge, 1990.

Featherlin, Erin (Chatterton) and Brett Chatterton. Interview by Holly Spangler. "Sudden Succession" in "Chatterton Family: A Corn Field Conversation." *Wallaces Farmer and Prairie Farmer-Confessions of a Farm Wife* Blogs, My Generation, January 18, 2016. http://farmprogress.com/blogs-chatterton-family-corn-field-conversation-10567 (accessed 9/12/16).

———. *American Agriculturalist,* January 17, 2016. http://www.americanagriculturist.com/blogs-chatterton-family-corn-field-conversation-10567 (accessed 1/18/17).

Food and Water Watch. "Food System." https://www.foodandwaterwatch.org/issues/food-system/ (accessed 9/28/23).

Friedel, Robert. "Why You Need to Understand Y2K." *Invention and Technology*, 15, no. 3 (Winter 2000): 24–31.

"Glen Brinkman with Case International 3388." Author's personal records. Charles City, Iowa, 1980–1981.

Hamilton, David. *Deep River: A Memoir of a Missouri Farm*. Columbia: University of Missouri Press, 2001.

Hansen, Ann Larkin. *The Organic Farming Manual*. North Adams, MA: Storey Publishing, 2010.

Haraway, Donna. "A Cyborg Manifesto: Science, Technology, and Social-Feminism in the Late Twentieth Century." In *Simians, Cyborgs and Women: The Reinvention of Nature*, edited by Donna Haraway, 149–181. New York: Routledge, 1991.

Hassebrock, Kenneth. *Rural Reminiscences: The Agony of Survival*. Ames: Iowa State University Press, 1990.

Hearst, James. "Young Poet on the Land." In *Growing up in Iowa,* edited by Clarence A. Andrews, 58–59. Ames: Iowa State University Press, 1978.

Jørgensen, Finn Arne. "The Backbone of Everyday Environmentalism: Cultural Scripting and Technological Systems." In *New Natures: Joining Environmental History with Science and Technology Studies*, edited by Dolly Jørgensen et al., 72–73. Pittsburg: University of Pittsburg Press, 2013.

Kalish, Mildred Armstrong. *Little Heathens: Hard Times and High Spirits on an Iowa Farm During the Great Depression*. New York: Random House, 2007.

Kingsolver, Barbara. *Animal, Vegetable, Miracle*. New York: HarperCollins Publishers, 2007.

Kirk, E.J. "Barley for Hogs Next Summer." *Wallaces' Farmer* 50, no. 7 (February 13, 1925): 222 (18).

Kirschenmann, Frederick. "Food as Relationship." *Journal of Hunger & Environmental Nutrition* 3, no. 2 (2008): 106–121.

Lockeretz, William, ed. *Organic Farming: An International History*. Cambridge, MA: CABI Publishing, 2007.

Mann, Linda. "The Enslaved Household of President Thomas Jefferson." *The White House Historical Association*, November 20, 2019. https://www.whitehousehistory. org/slavery-in-the-thomas-jefferson-white-house#:~:text=Despite%20working%20 tirelessly%20to%20establish,most%20of%20any%20U.S.%20president. (accessed 10/28/23).

McDonald, Julie. "Growing Up in Western Iowa." In *Growing Up in Iowa*, edited by Clarence A. Andrews, 114–123. Ames: Iowa State University Press, 1978.

Merchant, Carolyn. *The Death of Nature*. New York: HarperCollins Publishers, 1980.

Oldenziel, Ruth. *Making Technology Masculine*. Amsterdam: Amsterdam University Press, 1999.

Phillips, Ronald L. *Norman E. Borlaug 1914–2009, A Biographical Memoir*. National Academy of Sciences, 2013. http://www.nasonline.org/publications/biographical-memoirs/ memoir-pdfs/borlaug-norman.pdf (accessed 8/8/16).

Pingali, Prabhu L. "Green Revolution: Impacts, Limits, and the Path Ahead." *Proceedings of the National Academy of Sciences of the United States of America* 109, no. 31 (July 2012): 12302–12308.

Pollan, Michael. *The Omnivore's Dilemma*. New York: Penguin Press, 2006.

———. *In Defense of Fo*od. New York, Penguin Press, 2008.

Pursell, Carroll. "The Construction of Masculinity and Technology." *Polhem* 11 (1993a): 206–219.

———. "The Rise and Fall of the Appropriate Technology Movement in the United States, 1965–1985." *Technology and Culture* 34, no. 3 (1993b): 629–637.

Sawtelle, Emily Hoag. "The Advantages of Farm Life: A Study by Correspondence and Interviews with Eight Thousand Farm Women." Unpublished manuscript, U.S. Department of Agriculture, March 1924, 1. https://archive.org/stream/CAT31046460#page/n4/ mode/1up (accessed 9/29/16).

Schiebinger, Londa. "Why Mammals Are Called Mammals." *American Historical Review* 98, no. 2 (1993): 382–411.

Scot, Barbara J. *Prairie Reunion.* New York: Farrar, Straus and Giroux, 1995.

Seymour, John. *The Complete Book of Self-Sufficiency.* London: Corgi Books, 1978.

———. *The New Complete Book of Self-Sufficiency.* Illustration. London: Dorling Kindersley Publishers, Ltd, 2003.

Seymour, John and Will Sutherland. *The Self-Sufficient Life and How to Live It.* New York: DK Publishing, Inc., 2009.

Smith-Howard, Kendra. *Pure and Modern Milk: An Environmental History Since 1900.* New York: Oxford University Press, 2014.

Toxic Free NC. "Just and Sustainable Agriculture." https://toxicfreenc.org/programs/just-sustainable-agriculture/ (accessed 9/28/23).

———. "Women of Color Farmer's Network." https://toxicfreenc.org/fundraiser-2023-women-of-color-farmers-network/ (accessed 10/19/23).

Van Etten, Winifred M. "Three Worlds." In *Growing up in Iowa*, edited by Clarence A. Andrews, 141. Ames: Iowa State University Press, 1978.

Vos, Timothy. "Visions of the Middle Landscape: Organic Farming and the Politics of Nature." *Agriculture and Human Values* 17, no. 3 (2000): 246–247.

Wilson, E.O. *The Diversity of Life.* New York: W.W. Norton & Co., 1999.

10

"THERE THEY GO AGAIN"

Understanding Clashes Between Ultramodern Farmers and Organic Advocates Over Food Policy and Reform

> Suspicious of the expert, he [the farmer] is at the same time more receptive to new ideas, new machinery, new products, and new methods than either industrial management or labor.
>
> *Joseph Frazier Wall*, Iowa: A Bicentennial History, *1978*[1]

Throughout this book, I have presented the historical development of rural capitalistic modernity and globalized ultramodernity in the Midwest and have argued that farmers help to form and reinforce their identities through technological use. Here I argue that unarticulated disagreements over the relationship between gender, nature, and technology among those adhering to rural ultramodern and organic identities contributes to the debate over organic and sustainable foods.

One can gain greater insight into the clash in rural identities driving disagreements over food policy not at the USDA or the Congressional hearing, but at the Midwest tractor show. Farmers with rural modern identities embed within technology family histories or memories of past labor activities which defined family relationships. Photographing and using farm machinery both preserves memory and gives future generations a way to see how far they have come since the "good old days." Rather than associating a lack of technology with morality as with the organic discourse, the rural capitalistic identity tends to associate a less-technological past with poverty.[2] Carl Thomson, for example, in his account of attaining his first Ford Model T in rural Indiana in the 1920s recalled, "We was so stinking poor." Thompson's account, and the account of other rural actors recalling childhood, seems geared toward reminding the reader of an impoverished past to such an extent that the new technology almost fades into the background.[3] In another compilation of stories of residents of Buffalo Center, Iowa recalling life in the late

DOI: 10.4324/9781032637952-11

19th and early 20th centuries, farmers like John L. Howe focus on the hardships of draining wetlands, the dirty state of town streets, or the constant threats posed by nature such as snakes and pests.[4]

Such a view of history serves an important identity-maintaining function for rural Americans in the Corn Belt. Namely, the display of old technology, as well as stories about the hardships of past farming eras, forms part of a broader rural social practice of framing history extoling, in writer Barbara J. Scot's words, "the self-sufficiency of the earliest settlers.... These 'steady, hard-handed folk,' with 'pluck, spirit... and a firm grip upon the purse strings' replaced their early log cabins with white green-shuttered houses and wisely planted windbreaks northwest of them."[5] The comparison between old and new farm machinery, and the association of old machinery with poverty, serves as material evidence of rural denizen's self-image as having inherited these settler characteristics and as the agents of progress.[6] Old farm machinery stand as monuments to perceived inborn innovativeness.

Further, both men and women participate in this cultural practice at antique farm machinery shows in the Midwest (Figure 10.1). In Figure 10.1, a husband

Jim and Deb Kufahl of Cambridge, Wisconsin, proudly display their 1933 Farmall F-12 at the Rock River Thresheree in Janesville, Wisconsin.

FIGURE 10.1   Images of farmers, both men and women, embedding technology with cultural memory at a farm machinery reunion in the Corn Belt. Dave Mowitz, "Ageless Iron: Reunions that Revive the Past," *Successful Farming* (February 1992): 23–27, 30–33, 36–38.

and wife stand together at a 1992 jamboree in the middle of a row of Farmall tractors from the 1930s. The caption reads, "Jim and Deb Kufahl of Cambridge, Wisconsin, proudly display their 1933 Farmall F-12 at the Rock River Thresheree in Janesville, Wisconsin." The husband wears a Farmall baseball cap with his left arm around his wife and his right around the exhaust pipe sticking out of the top of the old tractor as if he regards the machine as another family member.[7]

The image comes from a 1992 article in *Successful Farming* reporting on a large farm machinery reunion featured men and women with a strong association between technology and the family (Figure 10.2). The author describes the event as, "family fun and sharing" stating

> Their show, like all machinery shows, will center around the family... The Jamboree also provides Fred Federler's family a chance to show off their 'relics.' The Chester, South Dakota, family all pitched in to sand and paint a one-row John Deere corn binder this fall that they hope to demonstrate at next year's show.

The article further explains that women played a crucial role in organizing these machine reunions. Lynette Briden, president of the "Western Minnesota Region" in an interview commented, "I have three daughters who are also very much involved in

FIGURE 10.2   Images of farmers embedding technology with cultural memory from the *Successful Farming* article "Ageless Iron." Dave Mowitz, "Ageless Iron: Reunions that Revive the Past," *Successful Farming* (February 1992), 23–27, 30–33, 36–38.

the show."[8] While the article reminds the reader that many of the reunions also gave "pioneer homemaking demonstrations," they also presented both men and women as embedding machinery with family legacy and a production ethos. For farmers, the family serves as a signal to outsiders that they should grant her way of life a high moral standing. Family-centered production also represents a sense of independence, which the farmer must heroically defend through conspicuous production.

One can observe what the dominant image of a moral relationship with nature and technology "looks like" for both men and women holding rural capitalistic modern and ultramodern identities in a 2008 print by Dave Barnhouse entitled "New Tractor for Show" (Figure 10.3). In the print of Barnhouse's own family in the past, one notices that the man drives the John Deere tractor, but he does so toward his family bathed in the center of the scene in warm light emanating from the barn. The woman, while she holds a child, stands outside next to another tractor rather than in a farmhouse. Other family members, three children, the family dog, and two grandparents, wait with the woman to greet the man mounted on the tractor as if he has heroically returned from the fields, a subtle representation of the Jeffersonian hero myth.[9] Barnhouse maintains an aspect of traditional German agrarianism by strategically placing his family in a prominent position in the scene to moralize rural identity. By making his own family the central focus of the painting, Barnhouse moralizes the farmer's technology use and embeds antique machinery with positive nostalgia about family legacy.

FIGURE 10.3 Dave Barnhouse, "New Tractor for Show," painting on canvas (2008) – See more at: http://www.galleryone.com/fineart/BARNE5.html (accessed 1/28/16).

I have also argued that women have used technology to perform their identities as independent and modern technical users within a moral framework as "family farmers" to dispute perceived urban-based gender identities. Thus, technologies for many women on farms represent moral values rather than immoral domination of women and nature. The discourse of organic and sustainable farm advocates alienates those in rural America ascribing to a rural modern identity because it frames their "moral" ways of making a living in "immoral" terms. The tractor representing progress, independence, family legacy, and the family itself headed by a benign "father figure" in rural America becomes, in the discourse of the organic movement and feminist/social justice interpretations, an unnatural industrial tool for the domination and destruction of women and nature.

As Barbara J. Scot, a feminist author, recalls in speaking with her mother in 1983 about her life as a self-described "Iowa "farmwife," ""I don't understand the lot of you.' Why is it so important to stay on the farm? The farm. That hallowed ground. The father, the son, and the holy farm.'" Scot went onto ask her mother, "'Were you trapped or were you there by choice?'" Scot's mother replied emphatically, "'*You'll never know how much.*'" Scot's autobiographical story proceeds to recall how she slowly abandoned a view of the farm as a male-dominated system of oppression of nature and women to adopt her mother's emotional tie to the land based on family legacy and Jeffersonian beliefs in the mythology of farming (although Scot retained ambivalence over "becoming my mother"). Scot noted such feelings about the farm even though her family "was always behind financially."[10]

The struggle within Scot's mind about how to think about herself and her family's farm mirrors unarticulated tensions in debates over the American farm more broadly. Namely, the incongruence of organic reformist and globalized ultramodern discourses and conceptions about technology underlies current debates over organic foods, sustainable agriculture, and genetically modified organisms preventing consensus over agricultural reform. One can view the debate over the reform of mainstream agriculture and global food systems as a clash of competing agrarian identities in which both sides use discourses of the family and independence to moralize historically and culturally embedded relationships with technology and nature. Ohio farmer Joe Everett, for example, returned to his family farm after working as a Navy electrician to keep his family's farm legacy going. Joe and his wife Casey embed the latest technologies with the morality of the family and family history. In an interview, Everett stated that "technology enhanced his own family farm and has made a big difference and has changed the way people farm across the nation and the world." The Ohio farmer proudly noted that "All our equipment now runs off satellite and GPS. You don't even need to steer, and it makes things a lot easier." In contrast, Everett explained, "When my grandpa started the farm, it took him a long time to do anything. Now we don't need as many people, and we can cover a lot more acres over a shorter period of time." The Ohioan also attributed his success to a financial officer and selling corn and soybeans to Cargill giving him access to Australian markets. Everett linked this ultramodern identity with morality by linking it to family legacy stating "I love the work ethic farming teaches kids. I feel like it's a lot of hard work and you are happy when you make money, but it's not everything." The real motive for Joe and Casey was that their kids would continue farming using the latest technology. Everett also made sure his audience knew he was proud to return to farming after obtaining an MBA and advanced electronics degree, a subtle way to combat urban yokel stereotypes. Hence, the Everetts posed with the whole family in front of a modern John Deere tractor, a common means of performing rural ultramodern identity and advocating for its morality (Figure 10.4).[11]

In the context described above, conversations about what modern farming is take on moral weight. As such, seemingly technical or scientific claims about the safety of genetically modified organisms or synthetic fertilizers become judgments about the righteousness of different agrarian ways of life and values. Rural identities in turn shape scientific claims and technological use. As a result, political positions over the reform of agriculture become solidified and immovable within a matrix of science, technology, and morality. For example, as one woman living on a grain farm in Illinois, Emily Webel, stated on her blog entitled *Confessions of a Farm Wife*,

> There's a craze going on. A food craze. An 'eat nuts and berries and twigs' and 'processed foods are the devil' craze, and while I agree, I would rather my kids eat fruit than fruit snacks, does that make corn bred to withstand drought that we planted evil? Does that genetic modification make us as farmers evil?

FIGURE 10.4   A photo of the Everett family in the Corn Belt displaying a family-centered ultramodern discourse. Beth Anspach, "Personal Journey: Military Skills Become Key Asset on the Farm, A Navy Veteran Returns to the Farm," *Dayton Daily News* (October 8, 2022).

Webel answered this question with a decided "No" because

> Scientists are in the lab, researching, and they're not evil scientists. They are just
> regular dudes who are wearing white coats and looking at CELLS. They're not
> figuring out a way to make the American public fatter. They have extensively
> studied this particular crop and have found a way for farmers like us to continue
> to survive during the driest of years and now the wettest of springs, and still
> harvest a corn crop so you folks can fuel up your SUVs with gas to get to Trader
> Joe's to purchase organic, non-GMO (supposedly) food and then make a stand
> on not eating conventionally grown food (sorry for the sarcasm, I'm grouchy
> today).

Webel concluded her blog post with some advice for those proffering organic
and sustainable discourses that articulates a common alienation of rural modern
identities,

> So before you post another shared "eat this not that" article on Facebook, check
> your sources, and think of my face, my husband's face, and know that we're
> not in cohoots [*sic*] with some big agricultural company, or trying to give you
> cancer or get you fatter or whatever. We're just trying to make a living in this
> crazy occupation that doesn't get a regular pay check [*sic*], is dependent upon
> the weather, and has the responsibility to fuel and feed a growing global need.
> Lucky us.[12]

Webel performs her morality under a rural ultramodern discourse on her webpage
with photos almost identical to that of the Everett family in Figure 10.4. She stands
with her husband and four children in a harvested cornfield in front of their newest
John Deere tractor and corn bin. Her husband, wearing jeans and a plaid shirt, holds
one of their children in his arms and her family members smile to pose as a happy
family with the tractor. While the Webel and Everett family photos are relatively
recent, I have shown that this kind of performance tying blissful family relationships to the latest technology has existed in the rural Midwest over multiple generations for most of the early 20th century (Figure 10.5). Interestingly, the photo
on Barbara Kingsolver's book jacket of *Animal, Vegetable, Miracle* also features
her smiling family but this time they all hold baskets of home-grown foods such as
eggs and onions. Kingsolver, her husband, and two daughters, stand on a staircase
surrounded by verdant vines and other plants but with no material objects other
than the wire basket held by one of her daughters. Importantly, the photo of Webel
and Kingsolver presents almost *identical scenes* of family bliss except Webel's
features technology prominently and Kingsolver's displays a conspicuous lack of
high-tech artifacts. Both Webel and Kingsolver engage in the Jeffersonian and German practice of employing the family to moralize their production processes but
embed technology with opposite meanings.[13]

FIGURE 10.5   Photos showing the multi-generational practice of performing and moralizing rural modern identities by associating family relationships with
the latest technologies. "Virginia, Dennis, and Doug Brinkman with Barn
and Corn Planter." Author's personal records. Charles City, Iowa, 1954;
"Dennis and Ann Brinkman with International Harvester." Author's personal records. Charles City, Iowa, 1975.

Other farmers, both male and female, exhibit the same rural cultural practice
displayed by the Webels and Everetts of performing for outsiders to demonstrate
the morality of ultramodernity. For example, at the 2016 Women in Ag Conference
in the Quad Cities (the border of northwestern Illinois and southeastern Iowa),
Webel and another self-described farm wife, Holly Spangler, held a panel featuring
Natasha Nichols, a blogger from Chicago. All three women introduced themselves
to their audience first as mothers of several children and then as urban or rural
dwellers. The two rural women saw Nichols as a worthy panelist because she valued her identity as a mother, rendering her moral, and because she represented an
unusual city dweller desiring to learn about "real" farm life. Thus, the panel itself
offered an opportunity for the rural women to perform for an outsider sympathetic
to ultramodernity and who could potentially relay this positive impression to other
urbanites through a blog.

The interview of Nichols revealed many instances of rural performance and
subtle resentment of organic reformism. For example, when one of the farmwives
asked how she first "got it," Nichols replied that when touring a large hog farm as
part of a "Farm Families tour" for Chicagoans interested in "learning where their
food came from," the owner of the farm, Steve Ward, stated

> I don't think lots of people when they're going on and on about the bad things
> that farmers do and how they don't trust farmers – we feed our children and our
> families with the food we raise, ya know, with the animals we raise.

Nichols characterized this statement as the "ah ha moment for me, where it really
solidified my trust in the food system in general and the fact that this is serious

work, ya know, this is their well-being and that they are really passionate about it." When asked by the farm wives if Nichols had any pre-conceived notions before going on the farm tour, she related that, "we had lots of people in our tour really really against GMOs, that was the big buzz word in our class" and that she found herself alone among her classmates in trusting genetically modified organism (GMO) food due to her biology degree. Nichols saw GMOs as a buzz word such as "gluten free" rather than an actual threat to her health or the environment.

One of the farmwives praised Nichols' GMO stance exclaiming "And you understand science, it sounds like, which would separate you from a lot [of people]." Nichols responded, "Yah, I'm a bit of a nerd that way" followed by a story about a fellow Chicagoan who saw one of her large tomatoes from her garden saying "Yah, you probably used those GMO seeds" causing laughter from the rural crowd. Nichols also elicited laughter from the audience of female farmers when she claimed that most anti-GMO urbanites confused GMOs with hybridization. Nichols also opined that making companies spend money to label items in the grocery store GMO-free was "stupid."

Nichols admitted that few of the urban women on the Farm Families tour who initially opposed GMOs "came around." One of the farmwives made an analogy between Nichols and British environmentalist and former anti-GMO advocate Mark Lynas who "was part of torching research plots and stuff, and he's come around, the last, I think two to three years, and said 'Oh wait, I've stopped and studied the science and this all makes sense.'" The three moms agreed that if GMO labeling legislation passed, urban dwellers would blame farmers for raising food prices and that "You're never ever gonna' make everyone happy." The farm wives then asked Nichols what grocery stores were like in Chicago as if Nichols lived in a foreign country.[14]

Other rural Americans exhibit Webel's resentment of urban views of farmers. Such rural antipathy is reminiscent of the rural-urban conflict of the 1920s. Wyoming farmer and rancher Linda M. Hasselstrom exhibited such rural resentment, "Now agricultural labor pollutes our water and soil; newspapers tell us so daily. Country people who love the land are suddenly its worst enemies." Hasselstrom ends her essay with a curious statement that almost sounds like a threat to those labeling farmers as the "enemies" of the land. She reminds environmentalists that farmers and ranchers have a deep appreciation for their past and that "folks like that can wait out a lot of social upheaval before they change their behavior."[15] In English professor David Hamilton's memoir about growing up on a farm in Missouri, which involved extensive drainage of bottomlands, he too expressed a defensive stance

...Farming means drainage, cutting ditches and setting culverts so the fields will dry out enough to work. Among my university friends and others, that has become a transgression of the worst order; those being the same people, often, who find transgression thrilling when it is their own adventure in science or art.[16]

Farmers voiced similar views of alienation with use of a family-centered discourse in an article about a debate over GMO foods held in Des Moines, Iowa in 2013 and attended by over 1500 anti-GMO advocates and large grain farmers. The *Wallaces Farmer* writer Rod Swoboda described how the only other Corn Belt farmer on the debate panel, Bill Horan, countered anti-GMO arguments about unknown effects of GMO crops on human health and the environment. Swoboda wrote

> The other farmer on the panel, Bill Horan, said yields and grain quality on his farm have improved thanks to biotech crops. And so has family life. A past president of the Iowa Corn Growers Association, Horan says he spent a large part of his summers as a kid with a hoe in hand, walking bean fields, hoeing weeds out by hand. "My kids got to play Little League baseball, swim team, dance lessons and, because of biotechnology, I got to go watch them. That was a luxury my parents never even dreamed about. Biotechnology offers me the opportunity to be a better husband and father, something that is hard to quantify."

Thus, Horan viewed the use of the latest technology as virtuous due to the improvement in family life. In contrast to his own family-centered moral lifestyle supported by technological modernization, Horan dismissed the anti-GMO advocates as "groups and individuals who make their living scaring people about food."[17] Horan, therefore, characterizes anti-GMO advocates as performing a less moral non-productive work that damages families.

Hence, while Swoboda interpreted the debate over GMOs as a technical disagreement over the correct or incorrect interpretation of scientific data, the argument is not about that at all. Rather, underlying these articulated technical arguments over the safety, efficiency, or even profitability of GMOs and organic or sustainable foods lay a debate over the morality of competing rural identities. Scientific and economic claims serve as "rational" proxy arguments standing in for clusters of clashing beliefs about morality, views of nature, and meanings of technology.[18]

These conflicting underlying value systems also explain why many organic farm advocates find themselves getting nowhere in rural America even after pointing out that organic and sustainable farming potentially brings greater efficiency than the methods used by mainstream farmers in the Corn Belt.[19] These claims by some organic advocates offends the identity of mainstream farmers who see themselves as moral and modern producers who "feed the world" through an efficient and globalized food system and "nature" as an ordered, productive, and controlled unit of production.[20] Farmers in the Corn Belt have regarded the exact straightness of corn rows, for example, as a crucial aesthetic virtue at least from the 1920s to the present.[21] As a result, more data or evidence produced by organic and sustainable advocates will not dislodge the views expressed by one farm journal editor that "biotechnology could mean the difference between an underdeveloped country feeding its population or millions starving because of grain disease. Anything that reduces the time it takes to develop new food crops is an incredible scientific advancement."[22]

Indeed, if the debate over organic and sustainable foods simply revolved around a technical problem over the efficiency or safety of farming methods, the articulated technical arguments proposed by Frederick L. Kirschenmann and other sustainable advocates since the 1970s would have led to more Corn Belt farmers drastically revising their farming methods by now. While most Midwest farmers have incorporated sustainability as a goal in using new technologies and methods, organizations such as Food and Water Watch and Toxic Free NC have largely failed to achieve significant system change or convince farmers to abandon their rural ultramodern identities.[23] Even when farmers in the Midwest adopt free-range livestock methods to access new consumers and resist vertical market integration, as with the farmers in the Fresh Air Pork Circle in Iowa, they often resist organic certification. In addition, the farmers raising free range pork in the Fresh Air Pork Circle only resisted corporate pork contracts that offended their notions of ultramodernity, not the use of technology in other areas of farming such as planting corn. The scholar studying this co-op noted that many of these pork growers rejected raising organic products even though they could have sold organically labeled meat for more money because "some farmers were suspicious of the organic movement itself." When asked about the organic movement, one farmer said, "I have always thought of organic as kind of like a 'cult.'" At meetings of the Fresh Air Pork Circle, farmers "stereotyped" their buyers in San Francisco, California "in hard, almost resentful terms, which poked fun at their urban lifestyles." Thus, farmers even resented outside urban actors who allowed them to sell free range pork and maintain their identities as independent hog raisers.[24] The negative reaction of farmers centers around the way they perceive identity through the lens of the pattern of audience, not around economics.

Some more recent urban advocates of reforming mainstream agriculture have acknowledged the alienation felt by many farmers. One *New York Times* article by Kim Severson in 2017, for example, recognized a

> Growing divide between those raised on the modern American food movement – which gained traction in the 1970s and drove a revival in cooking, local products and food justice-and a new generation excited about cellular proteins, Soylent, and app-based delivery services that are driven more by innovation than by pleasure.

But Severson reports that rather than bridging this divide by engaging farmers, some have looked to Kimbal Musk, Silicon Valley billionaire and brother of famous tycoon and engineer Elon Musk. After making his fortune by age 45 through technology companies such as Zip2, PayPal, and Tesla, the South African Musk turned to promoting what he called "real food" in big U.S. cities through a "network of businesses, education, and agricultural ventures." Clad in a white cowboy hat that many mock behind his back, Musk has sought to reform the food system by such methods as opening an organic restaurant in Boulder, Colorado with famous chef Hugo Matheson and designing modular curved plastic planters installed in schools to teach gardening.

As with traditional organic advocates that have eschewed technology, many farmers view Musk as a patronizing outsider who knows nothing about "real farming." Musk not only frames mainstream farming as morally wrong as many traditional organic advocates do, but he fails to recognize the existence of an ultramodern rural identity. Indeed, Musk has never lived on a farm in the Midwest or anywhere else and there is little evidence that he regards mainstream farmers as having any knowledge or experiences valuable for implementing technology-driven agricultural change. He seems to have become inspired to take up the cause of food-system reform while lying in bed recovering from a tubing injury when vacationing in Jackson Hole, Wyoming. At one conference in New York in 2017, for example, Musk declared, "food is one of the final frontiers that technology hasn't tackled yet. If we do it well, it will mean good food for all." On Twitter, one farmer spoke for many fellow Midwesterners when he responded, "You might want to visit a Farm Progress show. Or even a farm. I think you might have missed 70 years of Ag history. It's high-tech stuff bud."

Rather than seeking to learn from mainstream farmers, Musk often declares, "you couldn't design a worse food system than what we have."[25] Perhaps Musk reminds the reader of my Chapter 5 discussion of Eddie Albert in 1965 who discussed "getting back to the soil" from his organic garden on the grounds of his exclusive California mansion. Indeed, from the perspective of the ultramodern Corn Belt farmer, Musk simply presents an updated version of the outsider viewing the farmer as an ignorant rube who can only become truly modern by taking control of work processes away from him. To the Midwest agrarian, Musk and advocates like him, are nothing more than urban industrialists with computers telling farm families what to do and how to live their lives.

This underlying clash of identities explains why farmers failed to convince anti-GMO advocates and consumers at a recent Cass County Farm Bureau GMO forum in North Dakota that "GMO foods are safe to eat and good for the planet." At the meeting, the president and CEO of Peterson Farm Seed held up a two-to-three-inch stack of documents containing over 1700 scientific papers characterizing GMO foods as safe to eat. The *Wallaces Farmer* editor and part-time farmer Lon Tonneson observed that this display by the Peterson Farm Seed president failed to "hit a home run" but then claimed, "Hauling in a pallet load of the actual studies would have been the homerun."

In addition, farmers attending the meeting failed to persuade the audience of activists with other "rational" arguments. For example, Val Wagner, described as a North Dakota farm wife, argued that "farmers aren't pouring tanker loads of glyphosate on GMO corn and soybeans like some of the critics seem to think. It's more like a can of pop spread over a football field," and North Dakota farmer Mark Belter "pointed out that the Bt in some GMO corn hybrids is the same stuff that organic farmers spray on their crops." Tonneson seemed exasperated at the fact that none of the anti-GMO advocates or consumers left the forum persuaded by these scientific and "rational" arguments. Indeed, Tonneson regarded the crowd as "silly" as evidenced by asking questions such as

Why do you grow GMOs when GMOs are bad for the environment and bad for human health?

If you think glyphosate is so safe, will you drink a glass of it?

Why should we believe the government when it says GMOs are safe?

Aren't all university GMO studies funded by Monsanto?

I read on the Internet that GMOs are bad for you. Isn't it true?

If you grow GMOs, don't you contaminate your neighbors' crops?

Does Monsanto force you to buy its GMO seed?

Tonneson hoped that when farmers reading his article encountered these questions, they could "hit a home run," by presenting even more rational and scientific evidence of the safety of GMOs. Tonneson himself, however, recognized that many Corn Belt communities have held similar forums where farmers and seed representative present nearly identical arguments to anti-GMO audiences with little effect.[26] Indeed, anti-GMO organizations like Toxic Free NC have continued to characterize such evidence as "biased" due to funding or influence by biotech companies and "cherry-picked."[27] In spite of this acknowledgment, Tonneson continues to present the debate over the safety and environmental impact of GMO foods as a scientific, rather than a moral/cultural, one.

Another *Wallaces Farmer* writer and editor of *Indiana Prairie Farmer,* Tom Bechman, urged farmers to "stand up for agriculture in the GMO foods debate" as his wife did when judging food with an anti-GMO judge at the Indiana State Fair. Bechman's wife had grown up on a farm and "still has strong ties to agriculture." After making the point that the anti-GMO judge first incorrectly pronounced GMOs "HMOs" and told his wife "Whatever they are — they're in our food today and food just isn't the same," Bechman wrote, "This was her [his wife's] chance to defend agriculture to someone who obviously had misinformation, not facts. It was her chance to take a stand for agriculture, and she was up to the challenge." Bechman's wife "defended agriculture" with the following fact-based reply:

Let me clear up some things... First, GMO stands for genetically modified organism. They aren't just "something'" in your food. It's a trait, part of the genetic makeup of a plant that produces food.

Those traits include ones that help farmers better control insects and weeds that would otherwise require spraying more chemicals. If a corn plant has a trait inside itself that produces a substance toxic to a certain insect, then the insect will eat a small bite and die from the toxin. Without these traits in corn, the farmer might have to spray insecticides to control the same insects.

We're going to have to feed more people on less land as time goes along… Second, companies have invested years of testing and millions of dollars before any of these products with GMO traits ever come to market. Government agencies make them jump through all kinds of hoops. If a new GMO trait is labeled and comes onto the market, odds are that it's very safe or it never would have been approved in the first place.

Bechman related pride in his wife for this response because she "took a clear stand for agriculture." Bechman, however, did not relate to his farming audience whether the anti-GMO judge changed her views on GMOs after his wife's response. In fact, the anti-GMO judge remained silent throughout Bechman's wife's tirade, so one can only assume that the judge clung to her "misinformation."[28]

As with Tonneson and Emily Webel, Bechman's wife showed a level of alienation rarely elicited from mere "incorrect" presentation of scientific data. Thus, even though Bachman and his wife offer "the facts" to counter criticisms of GMOs, in actuality they aim to "defend" their identities as moral ultramodern producers. In addition, this unarticulated identity of rural globalized ultramodernity has become so embedded, such a part of the rural life, that it prevents Bechman and his wife from seeing how any rational person faced with these facts could possibly believe anything else. Thus, these rural ultramodern commentators can conceive of irrationality or misinformation as the only possible explanations for an anti-GMO stance even though the way they themselves approach technology, such as GMOs, is highly culturally and historically contingent. As Bechman asks his readers in another article, "Why do some people in this world ignore science and believe only bad things? Why do they think all big companies and industries like agriculture are up to no good? Where do they get these ideas?"[29]

For Tonneson and many other farmers, this ignorance of "the facts" occurs not because of a different identity held by organic advocates but because these critics have never experienced "real farming." As Tonneson explains in another article, "I usually sigh and roll my eyes when I hear the phrase 'sustainable agriculture.' There they go again, I think." From Tonneson's point of view,

Organic proponents, who probably have never set a foot in a plowed field, ignore the impact tillage has on soil health. The nation's farms were all virtually organic in the 1930s, and look what happened. We almost lost the ability to feed ourselves during the Dust Bowl.[30]

This same process of presenting "the facts" to defend an unarticulated identity also occurs when some organic advocates cannot understand why Corn Belt farmers overlook "clear" scientific evidence of the environmental impacts of monoculture or the high levels of petroleum use of contemporary farming. In this case, advocates such as Pollan and Berry can only conceive of these farmers as selfish, ignorant about environmental science, and/or fooled by greedy agribusiness corporations. Food and Water Watch, for example, gives little credit to the ability of farmers to evaluate evidence and make their own choices about GMOs based on their own lived experiences as producers claiming, "Nor is it easy for farmers to avoid planting GMOs. In our increasingly consolidated food industry, farmers have fewer and fewer options, and the advice they hear at every turn is "go GMO."[31] Such condescension implies that farmers are not modern independent producers "feeding the world" and preserving family legacies but fooled or controlled by greedy corporations to take part in an immoral act. Neither side appreciates their own or their adversary's underlying identities or the possibility that complex packages of discourse and identities impact the way they view or use technologies. As a result, the sides present "facts" with little effect and characterize the other side as silly or corrupt thus precluding any real solution to the "debate." In the end, such discussions devolve into attacking or defending the morality of opposing rural identities. The sides talk past one another.

My analysis explains why Tonneson, Bechman, Webel, and other Corn Belt farmers support the use of GMO crops so vehemently and view anti-GMO advocates with so much derision. Without a focus on rural identity, such views seem confusing. After all, selling 300 bushels of non-GMO corn promises a profit as well and many farmers grow non-GMO crops for European markets. But the "threat" posed by anti-GMO advocates goes beyond economics. Mainstream Corn Belt farmers view attacks on GMOs as a critique of their rural globalized ultramodern identities and the morality of their way of making a living. In short, anti-GMO voices from outside the farm attack rural performative use. Thus, Tonneson and Bechman's continued belief that the debate over GMOs involves a technical scientific question despite the failure of scientific and rational arguments to persuade critics. This continued presentation of facts makes more sense if one views Tonneson's faith in science as a proxy argument for the morality of ultramodernity.

I have shown that these clashes over the identities of rural people and an outside "other" and debates over the morality of rural ways of life have occurred in America since at least the 1920s when Progressive reformers sought to modernize agriculture according to an urban model. As such, scientific and rational arguments stand in for a more fundamental pattern of rural performance and resentment in relation to an outside threat. Throughout this book, I have presented this perceived dangerous "other" in several forms from the industrialist of the 1920s to the Soviet menace and, now, the organic and anti-GMO advocate. In each case, the farmer uses technology to perform his or her identity as a moral producer. When Tenneson, Webel, Bechman's wife, and other farmers hear a critique of GMOs, they therefore see it as an attack on themselves and their families as independent, technologically savvy producers. *Wallaces Farmer* editor and family farmer John Vogel expressed the view of most Corn Belt farmers, in response to speaking with a GMO food labeling advocate at a booth at the Iowa State Fair in 2013, much more succinctly by titling his article "Anti-GMOers Aim to Kill Food Abundance With GMO-Labeling." The advocate told Vogel "As an organic gardener, a mother, a grandmother, and someone with a food-sensitive chronic disease, I want and need to know what's in my food." Vogel tellingly responded: "Spare me the tears, please."[32]

To conclude, the notion of performative use not only can provide insight into the meaning farmers attach to material objects such as wind turbines; it can also reveal the use of technology in forming distinct identities that influence the way social actors respond to technoscientific controversies. The idea that people employ technology to form and reinforce their identities as an embodied practice allows the observer to view a policy debate about technology that appears narrow, such as the safety of GMOs, as a normative disagreement implicating a complex array of historically formed ideas about technology and the self. Technology plays a central role, therefore, not only in debates about the morality of high-tech farming, such as whether to use pesticides, but in reinforcing conflicting identities. The use of artifacts helps people answer both the questions "who am I" and "how should I and others behave?" The mundane everyday use of technology is ultimately a political act.[33]

Notes

1 Wall, *Iowa*, 128.
2 Breimyer, *Over-Fulfilled Expectations*, 50–51; Scot, *Prairie Reunion*, 141; see also Kalish, *Little Heathens*, 104.
3 Carl Thompson, "The Model T," in *Plain Talk*, ed. Carol Burke (West Lafayette, IN: Purdue University Press, 1983), 35; see also Adams, *The Transformation of Rural Life*, 1–2.
4 Wilson, *Buffalo Chips*, 5–16; it should be noted that John L. Howe is the author's great grandfather.
5 Scot, *Prairie Reunion*, 55.
6 For an example of the rural capitalistic view of technology as only leading to progress see Breimyer, *Over-Fulfilled Expectations*, 49.
7 Dave Mowitz, "Ageless Iron: Reunions that Revive the Past," *Successful Farming* (February 1992): 23–27, 30–33, 36–38.
8 Ibid.
9 Dave Barnhouse, "New Tractor for Show," painting on canvas (2008) – See more at: http://www.galleryone.com/fineart/BARNE5.html (accessed 1/28/16).
10 Scot, *Prairie Reunion*, 42, 53; for similar emotional attachments to the family farm among men see Hamilton, *Deep River*, 16.
11 Beth Anspach, "Personal Journey: Military Skills Become Key Asset on the Farm, A Navy Veteran Returns to the Farm," *Dayton Daily News*, October 8, 2022. See also Emily Webel, "Sticks and Stones May Break My Bones, But Will GMOs Really Hurt Me?" *Confessions of a Farm Wife*, April 25, 2013, http://webelfamilyfarm.blogspot.com/2013/04/sticks-and-stones-may-break-my-bones.html (accessed 4/17/15).
12 Emily Webel, "Sticks and Stones May Break My Bones, But Will GMOs Really Hurt Me?"
13 Webel; Kingsolver, *Animal, Vegetable, Miracle*, Back Jacket Cover.
14 Spangler, "Confessions of a Farm Wife: Vol. 18," *Wallaces Farmer and Prairie Farmer-Confessions of a Farm Wife* Blogs, My Generation, April 12, 2016, http://farmprogress.com/blogs-confessions-farm-wife-vol-18-10824#authorBio (accessed 9/20/16).

15 Linda M. Hasselstrom, "Addicted to Work," in *Rooted in the Land: Essays on Community and Place*, ed. William Vitek and Wes Jackson (New Haven, CT: Yale University Press, 1996), 66–75.

16 Hamilton, *Deep River*, 158.

17 Rod Swoboda, "The Great Debate on GMO Crops," *Wallaces Farmer,* 2013, http://farmprogress.com/blogs-great-debate-gmo-crops-7797-bpx_3 (accessed 4/17/15).

18 For a similar discussion of the use of proxy arguments in technoscientific policy debates, see Hirsh and Sovacool, "Wind Turbines and Invisible Technology," 705–734.

19 For a similar perspective of scientific controversies see Daniel Sarewitz, "How Science Makes Environmental Controversies Worse," *Environmental Science & Policy* 7 (2004): 385–403.

20 For a view of the ideal of an ordered and productive nature held by mainstream farmers, see "Farm Beautiful Contest," *agriculture.com*, 2013, http://community.agriculture.com/t5/contests/v2/contestpage/blog-id/farmbeautiful/tab/entries%3Amost-kudoed (accessed 4/19/2015). See also Frederick Kirschenmann's critique of mainstream farmer's view of nature in terms of fragments rather than in terms of relationships, Kirschenmann, "Food as Relationship," 106–121.

21 "Planting a Straight Corn Row," *Wallaces' Farmer* 50, no. 17 (April 24, 1925): 605 (3), 612 (10); "What Is the Secret to Planting Straight?" *Successful Farming* 95, no. 12 (December, 1997): 36L.

22 Harry Cline. "Anti-Biotech Crowd Convinced GMO Food Is Road to Extinction," *Western Farm Press* (September 27, 2011): 1–3.

23 *Food and Water Watch*, "Food System;" *Toxic Free NC*, "Just and Sustainable Agriculture."

24 Grey, "'Those Bastards Can Go to Hell!'" 172–174.

25 Kim Severson, "Kimball Musk Wants to Feed America, Silocon Valley-Style," *The New York Times* (October 16, 2017).

26 Lon Tonneson, "GMO Forum in Fargo Draws a Crowd," *Wallaces Farmer,* October 28, 2015, http://farmprogress.com/blogs-gmo-forum-fargo-draws-crowd-10331 (accessed 9/16/16).

27 Food and Water Watch, "Food System."

28 Tom Bechman, "Stand Up for Agriculture in the GMO Foods Debate, *Wallaces Farmer* and *Prairie Farmer,* August 15, 2016, Hoosier Perspectives, http://farmprogress.com/blogs-stand-agriculture-gmo-foods-debate-11225#authorBio (accessed 9/19/16).

29 Tom Bechman, "Have Marketing Techniques Fueled the Fire for Those Who Oppose Agriculture?" *Wallaces Farmer* and *Prairie Farmer,* April 7, 2016, Hoosier Perspectives, http://farmprogress.com/blogs-marketing-techniques-fueled-fire-those-oppose-agriculture-10793#authorBio (accessed 9/19/16).

30 Lon Tonneson, "Sustainable Agriculture: It's About Your Kids' Kids' Kids," *Wallaces Farmer,* September 12, 2016, Inside Dakota Ag, http://farmprogress.com/blogs-sustainable-agriculture-kids-kids-kids-11321#eAuthor (accessed 9/19/16).

31 *Food and Water Watch*, "GMOs Plant Seeds for Corporate Control," https://www.foodandwaterwatch.org/2021/03/02/gmos-plant-seeds-corporate-control/ (accessed 10/21/23).

32 John Vogel, "Anti-GMOers Aim To Kill Food Abundance with GMO Labeling." *Wallaces Farmer,* September 20, 2013, http://farmprogress.com/blogs-anti-gmoers-aim-kill-food-abundance-gmo-labeling-7637. (accessed 4/17/15); Richard White has also argued that the environmental movement more broadly has not reconciled the relationship between nature and work in Richard White, "Are You an Environmentalist or Do You Work for a Living?" in *Uncommon Ground*, ed. William Cronon (New York: W.W. Norton & Co., 1995), 171–185.

33 By "political," I mean a struggle over power and authority to speak for something, in this case the morality and form of food production.

Bibliography

Anspach, Beth. "Personal Journey: Military Skills Become Key Asset on the Farm, A Navy Veteran Returns to the Farm." *Dayton Daily News*, October 8, 2022.

Barnhouse, Dave. "New Tractor for Show," painting on canvas (2008). http://www.galleryone.com/fineart/BARNE5.html (accessed 1/28/16).

Bechman, Tom J. "Have Marketing Techniques Fueled the Fire for Those Who Oppose Agriculture?" *Wallaces Farmer* and *Prairie Farmer,* April 7, 2016, Hoosier Perspectives. http://farmprogress.com/blogs-marketing-techniques-fueled-fire-those-oppose-agriculture-10793#authorBio (accessed 9/19/16).

———. "Stand Up for Agriculture in the GMO Foods Debate." *Wallaces Farmer* and *Prairie Farmer,* August 15, 2016, Hoosier Perspectives. http://farmprogress.com/blogs-stand-agriculture-gmo-foods-debate-11225#authorBio (accessed 9/19/16).

Breimyer, Harold F. *Over-Fulfilled Expectations: A Life and an Era in Rural America.* Ames: Iowa State University Press, 1991.

Cline, Harry. "Anti-Biotech Crowd Convinced GMO Food Is Road to Extinction." *Western Farm Press,* September 27, 2011, 1–3.

"Farm Beautiful Contest." *Agriculture.com*, 2013. http://community.agriculture.com/t5/contests/v2/contestpage/blog-id/farmbeautiful/tab/entries%3Amost-kudoed. (accessed 4/19/2015).

Food and Water Watch. "Food System." https://www.foodandwaterwatch.org/issues/food-system/ (accessed 9/28/23).

———. "GMOs Plant Seeds For Corporate Control." https://www.foodandwaterwatch.org/2021/03/02/gmos-plant-seeds-corporate-control/ (accessed 10/21/23).

Grey, Mark A. "'Those Bastards Can Go to Hell!' Small-Farmer Resistance to Vertical Integration and Concentration in the Pork Industry." *Human Organization* 59, no. 2 (2000): 169–176.

Hamilton, David. *Deep River: A Memoir of a Missouri Farm.* Columbia: University of Missouri Press, 2001.

Hasselstrom Linda M. "Addicted to Work." In *Rooted in the Land: Essays on Community and Place*, edited by William Vitek and Wes Jackson, 66–75. New Haven, CT: Yale University Press, 1996.

Hirsh, Richard F. and Benjamin K. Sovacool. "Wind Turbines and Invisible Technology: Unarticulated Reasons for Local Opposition to Wind Energy." *Technology and Culture* 54, no. 4 (2013): 705–734.

Kingsolver, Barbara. *Animal, Vegetable, Miracle.* New York: HarperCollins Publishers, 2007.

Kirschenmann, Frederick. "Food as Relationship." *Journal of Hunger & Environmental Nutrition* 3, no. 2 (2008): 106–121.

Mowitz, Dave. "Ageless Iron: Reunions that Revive the Past." *Successful Farming* (Mid-February 1992): 23–27, 30–33, 36–38.

"Planting a Straight Corn Row." *Wallaces' Farmer* 50, no. 17 (April 24, 1925): 605 (3), 612 (10).

Sarewitz, Daniel. "How Science Makes Environmental Controversies Worse." *Environmental Science & Policy* 7 (2004): 385–403.

Scot, Barbara J. *Prairie Reunion.* New York: Farrar, Straus and Giroux, 1995.

Severson, Kim. "Kimball Musk Wants to Feed America, Silocon Valley-Style." *The New York Times,* October 16, 2017.

Spangler, Holly. "Confessions of a Farm Wife: Vol. 18." *Wallaces Farmer and Prairie Farmer-Confessions of a Farm Wife Blogs*, My Generation, April 12, 2016. http://farmprogress.com/blogs-confessions-farm-wife-vol-18-10824#authorBio (accessed 9/20/16).

Swoboda, Rod. "The Great Debate on GMO Crops." *Wallaces Farmer,* 2013. http://farmprogress.com/blogs-great-debate-gmo-crops-7797-bpx_3. (accessed 4/17/15).

Thompson, Carl. "The Model T." In *Plain Talk,* edited by Carol Burke, 35. West Lafayette, IN: Purdue University Press, 1983.

Tonneson, Lon. "GMO Forum in Fargo Draws a Crowd." *Wallaces Farmer,* October 28, 2015. http://farmprogress.com/blogs-gmo-forum-fargo-draws-crowd-10331 (accessed 9/16/16).

———. "Sustainable Agriculture: It's About Your Kids' Kids' Kids." *Wallaces Farmer,* September 12, 2016, Inside Dakota Ag. http://farmprogress.com/blogs-sustainable-agriculture-kids-kids-kids-11321#eAuthor (accessed 9/19/16).

Toxic Free NC. "Just and Sustainable Agriculture." https://toxicfreenc.org/programs/just-sustainable-agriculture/ (accessed 9/28/23).

Virginia, Dennis, and Doug Brinkman with Barn and Corn Planter. Author's personal records. Charles City, Iowa, 1954.

Vogel, John. "Anti-GMOers Aim to Kill Food Abundance with GMO Labeling." *Wallaces Farmer,* September 20, 2013. http://farmprogress.com/blogs-anti-gmoers-aim-kill-food-abundance-gmo-labeling-7637 (accessed 4/17/15).

Wall, Joseph Frazier. *Iowa: A Bicentennial History.* New York: W.W. Norton & Company, 1978.

Webel, Emily. "Sticks and Stones May Break My Bones, But Will GMOs Really Hurt Me?" *Confessions of a Farm Wife,* April 25, 2013. http://webelfamilyfarm.blogspot.com/2013/04/sticks-and-stones-may-break-my-bones.html. (accessed 4/17/15).

"What Is the Secret to Planting Straight?" *Successful Farming* 95, no. 12 (December 1997): 36L.

White, Richard. "Are You an Environmentalist or Do You Work for a Living?" In *Uncommon Ground,* edited by William Cronon, 171–185. New York: W.W. Norton & Co., 1995.

Wilson, Ruth. *Buffalo Chips: The History of a Town.* unpublished book, Buffalo Center, IA, year unknown.

CONCLUSION

Does it Still Run?

> The most worthy calling was to till the soil, and it was the natural order ordained by God that the farm would pass from father to son. My brother and I grew up in the fierce shadow of those beliefs, knowing that our father had failed, that our Uncle Jim had failed, and that maintaining the family farm was that duty-at-which-we-were-all-failing. And we were very, very sorry for it all.
>
> *Barbara J. Scot*, Prairie Reunion, *1995*[1]

The barn on the Brinkman farm had fallen out of use in the late 1970s, but in the years before his death in 2013, my grandfather invested thousands of dollars renovating it by covering it with aluminum siding. When asked in his last year about the future of farming, the old right-leaning farmer surprisingly expressed environmental concerns about soil depletion. He spent much of that year driving the old Farmall M performing tasks that did not really need to be done such as moving piles of gravel back and forth in the brutal December weather. Although he had a variety of ailments, he needed to drive the tractor and feel like a farmer in his last days. He reminded me of my great grandfather who, in his late nineties, told me over and over about the horses he used to farm with and all the skills you needed to handle large draft horses. When asked why he was out in the cold, my grandfather's response was always "to see if it still runs."

The notion that people do not use technologies for only articulated rational reasons but for the unspoken purpose of performing identity has formed my primary argument in this book. The identity-forming relationship we all have with material objects plays itself out as a process of embodiment on a daily or even hourly basis. For example, when I recently talked to my uncle on the phone, I had prepared to talk about my daughter or wife or my music or academic careers but he opened the

DOI: 10.4324/9781032637952-12

conversation with "I hear you got a new mower?!" I had indeed recently purchased a new Husqvarna riding lawn mower, but he wanted to discuss it more than I did. In fact, I did not want to talk about it at all, which prompted me to turn the notion of performative use on myself. I felt great pride when I first purchased the machine because I had earned the funds playing music, but this feeling had subsided in the ensuing months when I moved on to other thoughts and concerns (including writing this book). I realized that I had embedded my identity in my saxophone and manuscripts I had written, not in the lawn mower. Since my childhood working on the farm, my sense of self had derived from bundles of discourses and identities found among musicians and work colleagues, not ultramodern farmers. Thus, while I felt pride in my ability to buy the lawn mower, I did not perform my identity by using it. Rather, I used other material objects such as my saxophone or laptop to reinforce my sense of self.

In the end, I thoroughly dislike mowing my lawn. I cannot enjoy it more by rationally deciding to perform my identity while sitting on the new mower because, as with the farmers in this book, my process of performative use occurs on an unspoken level of forming embodied tastes. I like the mower I bought, but I regard using it as a boring distraction from other tasks I would rather do, such as playing my saxophone. In addition, I cannot associate the mower with my identity because I cannot change my social context on a daily basis. On some level, the structural influences on my identity such as economics or social practices have operated outside of my control. I also have no choice to determine my historical situatedness. My everyday experience with material objects involves performing on a stage with R&B and jazz musicians or writing and teaching, not farming among other Iowa agrarians. In other words, I do not use technology for purely "rational" or economic reasons either. As a result, I quickly changed the subject by asking my uncle about the weather in Iowa (a favorite topic of conversation among farmers). My uncle likely could not understand why I did not show greater interest in discussing my new mower.

To summarize more generally, I have argued that users of technology perform their identities every day through interacting with material objects in an unspoken embodied process. One can gain an understanding of these identities by examining discourses surrounding technology while accounting for the social context in which people speak and act. Such an analysis may reveal both people's explicit and unarticulated conceptions of self, motivating technology use. Further, identities change over time and, as such, have genealogies that depend on the historical experiences of the users of artifacts. Users of technology fail to articulate the performance of identities not only because they want to appear rational, but also because the process is deeply embodied and taken for granted.

My treatment of "performative use" does not adhere to a particular causal direction; rather, it views identity, technology, and structural factors (social, economic, material) as co-constructing one another. The world is simply too complex to claim one causal factor (such as economics) as fully determining how people use technology. Nor have I assumed monolithic identities among groups of users of technology but multiple and, often, contested discourses and identities in their social space. My analysis has also avoided conceiving of the process of identity formation in terms of "eras" as such an approach would employ an overly simplistic notion of periodization. Rather, I argued that prior discourses and identities may exist alongside newly formed conceptions and technology users may adopt these multiple conceptions of self.

One must recognize, however, that people are not free to choose among an infinite number of identities and discourses when using technology. A limited number of distinct and recognizable identities exist among a group of technological users, and use of material objects and other social factors limit the range of identities and discourses available. In turn, adopting a certain discourse-identity bundle limits how people may act in the world including how they use technology.

Further, people normally engage in performative use strategically to construct a moral image of themselves. While exceptions certainly exist, people generally desire to think that the way they live in the world, the way they make a living or interact with nature, is "good" or "right." In the case of the American Midwest farmer, I identified six distinct bundles of discourses and identities that have formed around the use of agricultural technology since the 18th century: *traditional German agrarianism, Jeffersonian agrarianism, urban industrialism, rural capitalistic modernity, rural globalized ultramodernity, and organic reformist discourse*. I have argued that technological change occurs not just from above by large institutions or famous inventors but from below by users of technology. Farmers or any other user of technology have agency according to my method of understanding technological change through the lens of performative use.

I intend to utilize this conclusion to offer some parting thoughts on how the social landscape currently may impact the future use of agricultural technology. I want to leave the reader with the argument that a new bundle of rural discourses and identities must emerge in the Midwest. In doing so, I do not intend to present an extensive or infallible argument; I only hope to propose a tentative hypothesis for the purposes of starting debate and discussion about what performative use may mean practically. In short, I seek to render the notion of performative use more relevant by starting a broader policy debate. Namely, I contend that the reformers of "industrial" agriculture and large grain farmers in the rural Midwest have established identities that make room for compromise, which in turn can lead to a new rural identity that incorporates elements of both. I suggest that by viewing each other as having common ground rather than as diametrically opposed, holders of these two identities can develop an updated rural identity.

The history of how urban industrialists and rural German/Jeffersonian agrarians worked together, in part through farm journals, to develop a rural capitalistic modernity in the 1920s and 1930s can serve as a model for forming a new rural identity. Performative use itself may provide a means for moving the policy debate away from "organic vs. industrial." Rural denizens do not have to choose between an organic and an industrial existence. Historically, new unforeseen identities have emerged. However, a new rural identity can only form if organic advocates accept the way farmers use technology performatively (rather than seeing farmers as "ignorant" or framing technology as immoral).

Farmers also need to understand that it may be time for a rural identity beyond ultramodernity. This rural sense of self has led to environmental problems, such as an over-reliance on petroleum-based agriculture, and ethical dilemmas (e.eg. the abuse of animals and people in large feedlots) without keeping young people on the farm. Both sides must cooperate to save what is important to them – much like Jeffersonian agrarians had to incorporate elements of urban industrialism to preserve their identities during the rural-urban conflict of 1920s. The remainder of this Conclusion will further develop my argument that a new rural sense of self, still based on historically formed identities, will, and should, emerge in the coming years. The most effective solution, I contend, to the policy debate between organic reformists and mainstream ultramodern Corn Belt farmers must occur at the site of performative use of technologies, not in the halls of a distant legislature or agency, through a reconciled rural identity. Farmers themselves must lead this process of forming a new rural sense of self.

Rationale for Altering Rural Globalized Ultramodernity

As I gathered my thoughts for a concluding chapter, I could not help but remember my grandfather in the last years of his life sitting on an old Farmall tractor in the biting December wind next to his remodeled but nonfunctional dairy barn, doing odd jobs in an attempt to hold onto something. The question remains, "Does it still run?" Or, more philosophically, is the American farmer trying to keep an empty barn going? Many memoirs of men and women who grew up in the Corn Belt exhibit a strange mixture of celebration of progress and a sense of melancholy.[2] These experienced agrarians understand that an unarticulated sense of self and way of life even more valuable than grain profit has been lost. They show acute awareness of the negative environmental consequences of ultramodernity that threaten to undermine their faith in progress and ability to maintain their ultramodern identities. Nor do farmers ignore the fact that the constantly shrinking number of farmers has led to dying rural communities across the Corn Belt.

After all, the ultramodern farmer is a smart and educated observer of the world around him. Many farmers operating large farms now have obtained degrees from some of the top agricultural colleges in the world. The USDA reported that "30 percent of the operators of million-dollar farms had college degrees in 2007, the same share as for all U.S. householders," as well as an increase in college education rates among all farmers.[3] Even my grandfather, who did not experience a day of college, worried that modern methods had depleted the soil. Similarly, author Barbara J. Scot in her return to her family's farm in Scotch Grove, Iowa in 1983 recalled her aunt's favorite tirade about how farmers used too many chemicals in raising chickens and corn, recalling that a few chickens she recently raised "grew so fast the bones in their legs didn't even hold up their bodies. They just rolled around on the floor."[4] Rural denizens, therefore, realize that ultramodern farming methods place a strain on nature in ways that may endanger the future of the "sacred" family farm.

In addition, the Midwest farmer often saw modernity in the 1920s as a way of preventing young people of both sexes from leaving rural America for urban centers. Farmwife Emma Gary Wallace, for example, told a story of a family able to keep their children from joining the rural-to-urban migration by modernizing their home,

> Eventually the home was entirely remodeled [*sic*] with hardwood floors, electric lights, bath-rooms with plumbing, furnace, and all the modern fittings of any home of refinment [*sic*]. The young people have no desire to go elsewhere.... There is a great deal to be said in favor of living in an atmosphere of prosperity....[5]

Farmers know that plans like Wallace's to keep young people on the farm through modernization have had little long-term success. Driving through much of rural Iowa, for example, one quickly realizes that much of the population is simply old. The agricultural census in 2012 showed that the largest age group of "principal operators" of farms in the U.S., with "farm" defined as "any place from which $1,000 of agricultural products were produced and sold," during the census year was the 55 to 64-year old age group. In addition, 701,255 of U.S. farmers reported an age of 65 or over while only 334,000 reported an age of 44 or younger.[6] The USDA webpage *Start2Farm.gov* recently reported that,

> The average age of a farmer today in America is 57 years of age. Five years ago it was 55. We have had an increase of 30% of the farmers over the age of 75 and a decrease in the number of farmers under the age of 25 by 20%.[7]

Although presenting slightly different numbers, the National Sustainable Agricultural Coalition (the Coalition) reported in 2014

> The new Census data continues to show the aging of the American farm population, with the average age of the American farmer increasing from 57.1 in 2007 to 58.3 in 2012. What's more concerning however, is the slow rate at which new farmers are entering agriculture, and the much faster rate at which older farmers are retiring from farming. On the whole, the U.S. farm population shrunk by roughly 4 percent in the last five years.

The Coalition also reported "20 percent fewer beginning farmers (those farmers who have been farming for ten years or less) in 2012 than there were five years earlier." While the 2012 Agricultural Census revealed that the Corn Belt had younger farmers than the South or West, the "youngest" state, Nebraska, still reported an average age of 55.7 years. In addition, the Corn Belt states of Iowa, Minnesota, Wisconsin, Illinois, Missouri, Indiana, and Ohio all reported a loss of new farmers (farmers with less than ten years of experience) from 2007 to 2012 with all states reporting a loss of more than 14%. Missouri and Indiana reported losses of over 25% of recently beginning farmers from 2007 to 2012.[8]

Ultramodernity has, in fact, worsened the exodus of young people form the rural Midwest as the growth of farms through use of the latest technologies has created barriers to entry by younger farmers. In 2017, only 4% of farms, those with over $1 million in annual sales, produced two-thirds of U.S. agricultural output. By 2022, the average age of the American farmer had risen to close to 60 at 57.5 years.[9]

Even Corn Belt farmers realize modernity has failed to keep young people on the farm. One farmer in Rexford, Kansas, for example, lamented, "This mega farmer deal, it's killing all the communities," and another farmer in Colby, Kansas regretted "If you don't have family in it, you're probably not going to get into it, because of the expense."[10] The USDA and the AFBF, acutely aware of this aging of the American farmer, have launched a concerted effort through *Start2Farm.gov* to provide resources that make it easier for younger people to start farming. In addition, the USDA administers the Beginning Farmer and Rancher Development Program offering grants "to provide U.S. beginning farmer and rancher producers and their families, as appropriate, with the knowledge, skills, and tools needed to make informed decisions for their operations, with the goal of enhancing success of beginning farmers and ranchers."[11]

This aging of farmers also accompanies the continued decrease in the number of farms. The 2012 Census counted 2.1 million farms in the U.S. in 2012, which was down 4.3% from the last agricultural census in 2007 continuing "a long-term trend of fewer farms." However, the land in farms only fell by 0.8% between 2007 and 2012 due to a 3.8% increase in farm size."[12] In keeping with this trend, the number of farms in the U.S. fell by 6,950 farms from 2020 to 2021.[13]

Thus, the American Corn Belt farmer and the USDA realize that modernity has not kept young people on the farm nor has it attracted young people back to farming. While the ultramodern farmer may see himself or herself as passing on an inborn capacity for innovation to his or her children, he or she also sees many farm children taking this ability to employers in the cities. One must ask if there are limits on how effective the use of new technologies can establish an ultramodern identity when only older users engage in the performance.

Rationale for Altering Organic Reformist Discourses and Identities

Farmers are not oblivious to the negative economic and environmental impacts of monoculture, dependence on gas powered machinery, pesticides, or inorganic fertilizers. I have argued organic reformist discourse often fails to acknowledge this rural environmental awareness and alienates rural denizens.[14] This book contends farmers use technologies to perform an ultramodern identity through a process that is deeply embedded historically.[15] Rural ultramodern identity regards science as leading to progress, not immoral agribusiness. Certainly, the ultramodern farmer knows much more about plant genetics or soil mechanics than most urbanites. The idea that farmers have ignorance of environmental conditions on their farms is problematic due to the tacit knowledge and their daily experiences with soil and climate on which their business depends.

Additionally, where does arguing that farmers ignore evidence from environmental science get us from a policy perspective? The use of "ignorance" to describe how farmers have not listened to more "enlightened" environmentalists strikes me as a tired and uncreative form of analysis that has been made with no effect ad nauseam by academics since at least the 1970s. The potentially more productive question is not how to make "ignorant" farmers pay attention to scientific evidence, but how to incorporate environmental science into a production ethos. In other words, how can legitimate concerns for sustainability fit into a cultural practice of performance that sees the farmer, technology, and the family as preserving morality? Regardless of one's beliefs about the proper relationship between humans and nature, one must concede that organic discourses have failed to establish themselves as dominant discourses in rural America.[16]

The Next Rural Identity Must Blend Organic Identity with Rural Ultramodernity through Performative Use

Thus, any agricultural policy that seeks to reform the relationship between food and nature can only succeed within the context of a new bundle rural identities and discourses. This new agrarian sense of self must incorporate past rural discourses and identities and it must come, in part, from Midwest farmers themselves. The farmers, male and female, and their technologies, their families, and their control over work processes must play a significant role in developing the next wave of dominant discourses and identities in the Corn Belt. Even when farmers, such as religion and philosophy professor Frederick L. Kirschenmann with his organic Iowa farm, discuss reform, they often do so using alienating language of the organic movement and incorporating condescending philosophy, which cause many farmers to "other" even an "inside" messenger.[17]

Farm journal editors and advertisers, even in the 1920s, realized they needed to appeal better to their farmer audiences to create mutual interests for the accomplishment of their goals. However, few more "sophisticated" scholars and scientists over a century later have been able to grasp this realization. Namely, the organic and sustainable foods movement has failed to make progress among farmers in the Midwest not because organic advocates are informed and farmers ignorant about science, but because the organic movement has not framed its discourse in a way that treats aspects of rural identities as moral and legitimate. The result is a rehashed rural-urban conflict that would have looked very familiar to the farmers writing into *Wallaces' Farmer* in the 1920s. As a result, the organic and sustainable foods movement will fail for the same reason farmers rejected urban industrialism advocated by Progressive reformers and urban businessmen in the past.

Perhaps much of the writing on reforming agriculture by authors such as Barbara Kingsolver and Michael Pollan or by organizations like Food and Water Watch critiquing contemporary practices, aims to form and reinforce organic discourses and identities rather than lead to real and meaningful change. A more thoughtful approach to agriculture and the globalized food system that takes rural relationships with technology more seriously, however, may reveal shared interests between organic food advocates and ultramodern farmers. Urban equipment manufacturers and Jeffersonian agrarians in the beginning of the 20th century worked together through farm journals to develop a modern rural identity. Yet, many could only conceive of a world in which urban industrialism opposed rural yeomanism and could not imagine the development of rural capitalistic modernity. Similarly, today's sustainable advocates and Corn Belt agrarians share many values. First, neither desire to eliminate the farmer and both grant the farm family a privileged moral position within society.

Second, neither side truly desires to render agrarian environments unsustainable. Most organic advocates and ultramodern farmers imbue land with deep, albeit different, cultural meanings. As Wyoming farmer and rancher Linda M. Hasselstrom explains,

> Country people generally realize that our families are not divisible from our land; our beliefs about both family and land grow out of everyday practice, rather than theory... The crustiest, most conservative, most antienvironment [*sic*] ranchers will say, with an oddly gentle note, 'I love this country.' Others say, 'If you take care of the land, it will take care of you.' Or 'We don't really own the land; we're just taking care of it for the next generation.'

While Hasselstrom lives in the plains rather than the Corn Belt, farmers in the Midwest often express similar sentiments while associating family legacy with the land's productivity. As liberal studies professor and environmentalist William Vitek has noted, for farm families across rural regions, "land is not merely scenery and hiking trails, or resources in need of extraction." Rather, land "becomes part of people's lives, intermingled with buying and selling, working and playing, living and dying." Advocates such as Vitek, Berry, and Kingsolver adopting organic identities approve of this rural relationship to land found within Jeffersonian and German agrarianism, modern capitalistic modernity, and ultramodernity.[18]

Third, both organic advocates and ultramodern farmers see work as a virtue. In other words, both groups practice conspicuous production, although they do so through different performances that attach dissimilar meanings to nature and technology.[19] Therefore, *both* farmers and organic advocates must devise a way of incorporating a sustainable or environmental ethos into rural ultramodernity. The farmer's need for an agrarian identity that maintains rural values while attracting young people back to the farm must incorporate sustainability. Farmers need more than Emma Gary Wallace's new furniture to avoid a landscape where aging agrarians preserve empty barns. Rural Americans must realize that updated farmers require new identities that go beyond modern tendencies to blindly associate technology with progress. Organic and sustainable food advocates must acknowledge that change will only occur through updating identities and discourses in a way that respects deeply embedded cultural ideas about morality, the family, and technology in rural America. Reformers must acknowledge and embrace the rural cultural practice of performative technology use.

While I cannot predict the outlines of this new rural identity, the welcoming acceptance of wind turbines by farmers (see Chapter 8) provides one case study of how rural performative use may align with the goals of sustainability. Importantly, the case of PIMBY attitudes among farmers suggests that rural modern identity and discourse does not necessarily exclude the type of environmental reform, in the form of renewable energy production, central to most organic identities. Farmers utilizing wind turbines often incorporate the environmental benefits of the machines into ultramodern discourses extoling progress, technology, and innovation. However, ignoring the historical process of performative use of technologies to form rural identities has already led to a system of what Michael Pollan calls "big organic."[20] Such large-scale farms technically meet organic regulations while resembling "industrialized farms" in too many ways to satisfy most organic advocates.[21] Simply passing organic regulations without reconciling organic and ultramodern identities affecting the way people use technology on a daily basis has not led to meaningful change in systems.[22] Only with an updated bundle of discourses and identities can "it still run." As with the final scene of the DeKalb Farm Bureau's 1923 pageant "Forward! Farm Bureau," "the future of an idea" must build on previous acts, and its success depends on the actors giving a competent, if not stellar, performance.[23]

My last name, Brinkman (or its original spelling Brinkmann), is a German name meaning a person who lives on the "brink" of a field, i.e. a farmer. This knowledge makes me think again about my family and their farm photos, which I generally do not enjoy looking at for reasons I never fully understood. Maybe, I thought, there is something about the passage of time between then and now that makes me uncomfortable. Or perhaps they just make me feel old. Nevertheless, I promised myself that when I finished this book, I would take time to go back to look at some of my family's farm photos, almost all of which show me or my relatives posing with technology. I wanted to find out if I would look at these pictures differently than I had before embarking on this study. Surely, writing this book would give me fresh perspective on my own life as a first-generation city dweller. I thought that perhaps my thoughts when viewing pictures of old tractors and combines would change all these years later in a way that eschews emotional ways of thinking such as nostalgia and idealism.

Much to my surprise, my reaction to my old family farm photos has changed very little. I cannot help but romanticize a photo of my late grandfather proudly posing in front of his giant grain elevators or perched on a new combine with my uncle about to harvest corn. Although I wrote this book, I cannot bring myself to view these objects as cold or impersonal units of production. Instead, the photos make me remember looking out over a field with my grandfather with a cool Midwest breeze gently waving the corn in front of a beautiful sunset. The corn bin makes me think of how the grain smelled as I shoveled it into an auger and how my grandfather would crack a joke when we finished, displaying a wonderfully dry Midwest sense of humor. I remember feelings of pride and optimism watching my family "feed the world." The material objects make me think about the little human details of life that one can only really know by using those artifacts.

One photo that I took made me particularly nostalgic. It shows my dad in the 1990s sitting on the Farmall tractor he drove as a child (Figure 11.1). In the photo, he is about as old as I am now, but he sits on the machine as he did when he was a teenager before leaving the farm, going to college, and working as a successful engineer in Washington, D.C. I think what strikes me is that he must have sat on

FIGURE 11.1 "Doug Brinkman sitting on a 1930s Farmall M tractor." Author's personal records. Charles City, Iowa, 1995.

the tractor to reconnect with memories and events that I will never know about. I cannot recapture the smells, sounds, and feelings of those days, but maybe he can. After writing this book, I now understand that my dad once used the machine to perform an identity of his rural self. By grasping the wheel later in his life, he reconnected with that identity. The tractor was part of what it meant to be him.

While looking at the photo, I was reminded of David Hamilton's quote early in the book, "To farm is to hold onto something, and a farm is land to grasp." When I first read this passage, I thought it overly sentimental. After all, I often found the farm in Iowa isolated and boring. Where I grew up on the East Coast (also my current home) offers much more in terms of culture and excitement than rural Iowa. In my adulthood in the city, I have done interesting things with a more diverse group of people than I ever would have experienced had I grown up in "the middle of nowhere" in northeastern Iowa.

Yet, the longer I looked at this picture of my dad on the Farmall tractor, I started to understand why family photos have made me so uncomfortable all these years. It is the sense that even though my dad physically grips the wheel in the photo, we have not "held onto something."[24] Indeed, maybe we have lost grasp of "something," an identity that can never be replaced.

My dad and I never did start the old tractor that day. After taking the photo, I naturally sat in the seat, but I eventually had to let go of the wheel and leave the Farmall sitting unused in the shed. It is difficult to walk away from technology containing so much meaning without noting a sense of loss. Or perhaps my father and I have simply replaced a rural identity with other senses of self and other technologies. In one of the last conversations I had with my grandfather, I asked about his life. He had started out with a team of oxen, farming just a few acres and ended up with a vast expanse of land tilled with the latest tractors and harvested with the newest combines. To my surprise, he simply gazed into the distance, sighed, and replied, "It was a lot of work." His response calls one to ask what causes people to romanticize farming and the use of agricultural technologies. In fact, my grandfather more often spoke sentimentally of his brother's experience of flying fighter planes in World War II rather than his own years of plowing with a tractor.

It is difficult not to romanticize farming. But why does farming evoke such sentimentalism? In other words, the question remains whether one form of performative use is really more moral or worthy of preserving than others. If some uses of material objects for the purpose of forming identities are more valuable, what criteria does one use to establish such a hierarchy of use? Perhaps those of us from farm families have too strongly embraced Jeffersonian agrarianism and, as a result, we are quite full of ourselves. Alternatively, maybe there is something inherently valuable about performative use on an American farm that is worth preserving. I do not pretend to offer an answer to these questions. I only hope to pose them for the reader to decide.

Notes

1 Scot, *Prairie Reunion*, 57–58.
2 High rates of depression and suicide among farmers have recently prompted Midwest states to train frontline mental health experts. Kendall Crawford, "How Some Midwestern States Are Building a New Frontline to Help Farmers with Stress," *NPR*, October 5, 2022, https://www.npr.org/2022/10/05/1126806917/how-some-midwestern-states-are-building-a-new-frontline-to-help-farmers-with-str (accessed 7/21/23).
3 U.S. Department of Agriculture, Economic Research Service, *Share of Principal Farm Operators with College Degrees has Increased,* October 18, 2012, http://www.ers.usda.gov/data-products/chart-gallery/detail.aspx?chartId=32868 (accessed 4/25/16).
4 Scot, *Prairie Reunion*, 48–49.
5 Emma Gary Wallace, "Making the Young People Contented."
6 U.S. Department of Agriculture, *2012 Census of Agriculture Preliminary Report Highlights: U.S. Farms and Farmers,* February, 2014, 1–4, https://www.agcensus.usda.gov/Publications/2012/Online_Resources/Highlights/Farm_Demographics (accessed 3/2/16).
7 U.S. Department of Agriculture for Beginning Farmers and Ranchers, *Farm Demographics: Introduction to Farm Demographics,* http://www.start2farm.gov/usda/knowledge (accessed 3/2/16).
8 National Sustainable Agriculture Coalition, *2012 Census Drilldown: Beginning Farmers and Ranchers,* May 28, 2014, http://sustainableagriculture.net/blog/2012census-bfr-drilldown/ (accessed 3/2/16); see also U.S. Department of Agriculture, *2012 Census of Agriculture: State Level* 1, Chapter 1, February, 2014, http://agcensus.usda.gov/Publications/2012/Full_Report/Volume_1,_Chapter_1_State_Level/ (accessed 3/2/16).
9 Ximena Bustillo, "There Aren't Enough Young Farmers. Congress Is Looking to Change That," *NPR*, September 1, 2022, https://www.npr.org/2022/09/01/1120100449/farm-bill-not-enough-young-farmers-congress (accessed 7/31/23); U.S. Department of Agriculture, *2022 Census of Agriculture Impacts the Next Generations of Farmers,* February 22, 2023, https://www.usda.gov/media/blog/2023/02/22/2022-census-agriculture-impacts-next-generations-farmers (accessed 10/22/23).
10 Bunge, "Plowed Under."
11 U.S. Department of Agriculture for Beginning Farmers and Ranchers.
12 U.S. Department of Agriculture, *2012 Census of Agriculture Preliminary Report Highlights*.
13 U.S. Department of Agriculture, *Farms and Land in Farms 2021 Summary,* February 2022, https://www.nass.usda.gov/Publications/Todays_Reports/reports/fnlo0222.pdf (accessed 10/28/23).
14 As a further example, a recent piece by historian Frank Uekotter, attributed the adoption of monocultured corn by German farmers to "ignorance." This notion is not only insulting to the capacities of the ultramodern farmer, but it is too "easy" because it implies that the reform of Western agriculture simply requires more science. Frank Uekotter, "Farming and Not Knowing: Agnotology Meets Environmental History," in *New Natures: Joining Environmental History with Science and Technology Studies*, ed. By Dolly Jørgensen et al. (Pittsburg, PA: University of Pittsburg Press, 2013), 37–50.
15 My approach agrees with the STS scholar Sheila Jasanoff's approach in her work on science and the law, as taking place within complex historical and cultural contexts. See for example, Sheila Jasanoff, *Science at the Bar, Law, Science and Technology in America* (Cambridge, MA: Harvard University Press, 1995).
16 Uekotter, 37–50.
17 Kirschenmann, 106–121; *Leopold Center for Sustainable Agriculture*.

18 William Vitek, "Rediscovering the Landscape," in *Rooted in the Land: Essays on Community and Place*, ed. William Vitek and Wes Jackson (New Haven, CT: Yale University Press, 1996), 3.
19 Hasselstrom, "Addicted to Work," 68.
20 By "big organic," Pollan Means Large Farms that Technically Meet Organic Regulations but Resemble a "Factory Farm." Pollan, *The Omnivore's Dilemma*, Chapter 9.
21 Ibid.
22 For organic regulations, see U.S. Department of Agriculture, *National Organic Program*, https://www.ams.usda.gov/about-ams/programs-offices/national-organic-program.
23 Wood, "A Pioneer Farm Bureau Celebrates," 3.
24 Hamilton, *Deep River*, 155–156.

Bibliography

Bunge, Jacob. "Plowed Under: Supersized Family Farms are Gobbling Up American Agriculture." *The Wall Street Journal*, October 23, 2017.

Bustillo, Ximena. "There Aren't Enough Young Farmers. Congress Is Looking to Change That." *NPR*, September 1, 2022. https://www.npr.org/2022/09/01/1120100449/farm-bill-not-enough-young-farmers-congress (accessed 7/31/23).

Crawford, Kendall. "How Some Midwestern States Are Building a New Frontline to Help Farmers with Stress." *NPR*, October 5, 2022. https://www.npr.org/2022/10/05/1126806917/how-some-midwestern-states-are-building-a-new-frontline-to-help-farmers-with-str (accessed 7/21/23).

"Doug Brinkman sitting on a 1930s Farmall M tractor." Author's personal records. Charles City, Iowa, 1995.

Hamilton, David. *Deep River: A Memoir of a Missouri Farm*. Columbia: University of Missouri Press, 2001.

Hasselstrom Linda M. "Addicted to Work." In *Rooted in the Land: Essays on Community and Place*, edited by William Vitek and Wes Jackson, 66–75. New Haven, CT: Yale University Press, 1996.

Jasanoff, Sheila. *Science at the Bar, Law, Science and Technology in America*. Cambridge, MA: Harvard University Press, 1995.

Kirschenmann, Frederick. "Food as Relationship." *Journal of Hunger & Environmental Nutrition* 3, no. 2 (2008): 106–121.

Leopold Center for Sustainable Agriculture. Iowa State University. http://www.leopold.iastate.edu/about/staff (accessed 1/27/16).

National Sustainable Agriculture Coalition. *2012 Census Drilldown: Beginning Farmers and Ranchers*, May 28, 2014. http://sustainableagriculture.net/blog/2012census-bfr-drilldown/ (accessed 3/2/16).

Pollan, Michael. *The Omnivore's Dilemma*. New York: Penguin Press, 2006.

Scot, Barbara J. *Prairie Reunion*. New York: Farrar, Straus and Giroux, 1995.

Uekotter, Frank. "Farming and Not Knowing: Agnotology Meets Environmental History." In *New Natures: Joining Environmental History with Science and Technology Studies*, edited by Dolly JØrgensen et al., 37–50. Pittsburg, PA: University of Pittsburg Press, 2013.

U.S. Department of Agriculture. *2012 Census of Agriculture Preliminary Report Highlights: U.S. Farms and Farmers*, February 2014, 1–4. https://www.agcensus.usda.gov/Publications/2012/Online_Resources/Highlights/Farm_Demographics (accessed 3/2/16).

————. *2012 Census of Agriculture: State Level* 1, Chapter 1, February 2014. http://ag-census.usda.gov/Publications/2012/Full_Report/Volume_1,_Chapter_1_State_Level/ (accessed 3/2/16).

————. Economic Research Service, *Share of Principal Farm Operators with College Degrees Has Increased,* October 18, 2012. http://www.ers.usda.gov/data-products/chart-gallery/detail.aspx?chartId=32868 (accessed 4/25/16).

————. *Farms and Land in Farms 2021 Summary,* February 2022. https://www.nass.usda.gov/Publications/Todays_Reports/reports/fnlo0222.pdf (accessed 10/28/23).

————. *2022 Census of Agriculture Impacts the Next Generations of Farmers,* February 22, 2023. https://www.usda.gov/media/blog/2023/02/22/2022-census-agriculture-impacts-next-generations-farmers (accessed 10/22/23).

U.S. Department of Agriculture. *National Organic Program.* https://www.ams.usda.gov/about-ams/programs-offices/national-organic-program.

U.S. Department of Agriculture for Beginning Farmers and Ranchers. *Farm Demographics: Introduction to Farm Demographics.* http://www.start2farm.gov/usda/knowledge (accessed 3/2/16).

Vitek, William. "Rediscovering the Landscape," In *Rooted in the Land: Essays on Community and Place,* edited by William Vitek and Wes Jackson, 1–14. New Haven, CT: Yale University Press, 1996.

Wallace, Emma Gary. "Making the Young People Contented." *Better Farming* 46, no. 1 (January 1923): 4.

Wood, Mabel Travis. "A Pioneer Farm Bureau Celebrates." *Better Farming* 46, no. 4 (August 1923): 3.

INDEX

Note: *Italic* page numbers refer to figures and page numbers followed by "n" denote endnotes

Printed in the United States
by Baker & Taylor Publisher Services

Printed in the United States
by Baker & Taylor Publisher Services